Engineering Optimization: Methods and Applications

Optimization carries great significance in both human affairs and the laws of nature. It refers to a positive and intrinsically human concept of minimization or maximization to achieve the best or most favorable outcome from a given situation. Besides, as the resources are becoming scarce there is a need to develop methods and techniques which will make the systems extract maximum from minimum use of these resources, i.e. maximum utilization of available resources with minimum investment or cost of any kind. The resources could be any, such as land, materials, machines, personnel, skills, time, etc. The disciplines such as mechanical, civil, electrical, chemical, computer engineering as well as the interdisciplinary streams such as automobile, structural, biomedical, industrial, environmental engineering, etc. involve in applying scientific approaches and techniques in designing and developing efficient systems to get the optimum and desired output. The multifaceted processes involved are designing, manufacturing, operations, inspection and testing, forecasting, scheduling, costing, networking, reliability enhancement, etc. There are several deterministic and approximation-based optimization methods that have been developed by the researchers, such as branch-and-bound techniques, simplex methods, approximation and Artificial Intelligence-based methods such as evolutionary methods, Swarm-based methods, physics-based methods, socio-inspired methods, etc. The associated examples are Genetic Algorithms, Differential Evolution, Ant Colony Optimization, Particle Swarm Optimization, Artificial Bee Colony, Grey Wolf Optimizer, Political Optimizer, Cohort Intelligence, League Championship Algorithm, etc. These techniques have certain advantages and limitations and their performance significantly varies when dealing with a certain class of problems including continuous, discrete, and combinatorial domains, hard and soft constrained problems, problems with static and dynamic in nature, optimal control, and different types of linear and nonlinear problems, etc. There are several problem-specific heuristic methods are also existing in the literature.

This series aims to provide a platform for a broad discussion on the development of novel optimization methods, modifications over the existing methods including hybridization of the existing methods as well as applying existing optimization methods for solving a variety of problems from engineering streams. This series publishes authored and edited books, monographs, and textbooks. The series will serve as an authoritative source for a broad audience of individuals involved in research and product development and will be of value to researchers and advanced undergraduate and graduate students in engineering optimization methods and associated applications.

Shahin Jalili

Cultural Algorithms

Recent Advances

Springer

Shahin Jalili
School of Engineering
University of Aberdeen
Aberdeen, UK

ISSN 2731-4049 ISSN 2731-4057 (electronic)
Engineering Optimization: Methods and Applications
ISBN 978-981-19-4632-5 ISBN 978-981-19-4633-2 (eBook)
https://doi.org/10.1007/978-981-19-4633-2

This Springer imprint is published by the registered company Springer Nature Singapore Pte Ltd.
The registered company address is: 152 Beach Road, #21-01/04 Gateway East, Singapore 189721, Singapore

Preface

According to the biocultural evolutionary theory, genes and culture are two interacting forms of inheritance and overall human evolution can be viewed as the product of the changes in biological and cultural traits. Both genetic and cultural evolutionary processes include a form of information transmission. In genetic evolution, the genetic information embedded within the genes is being vertically transmitted from parents to offspring. In cultural evolution, the cultural variants such as beliefs, habits, skills, traditions, and preferences, pass from one generation to the next and they can be changed or even replaced by new ones over time due to changes in the cultural environment. In comparison to genetic evolution, the information transmission in cultural evolution is a much more complicated process. The information flow in cultural evolution includes both vertical and horizontal transmissions in which the cultural information is not only vertically inherited from parents to offspring, but also offspring have the opportunity to socially learn and acquire information from other members of the society.

Cultural algorithms (CAs) are meta-heuristic numerical optimisation algorithms inspired by the abovementioned biocultural evolutionary theory. CAs have some characteristic features that make them unique in comparison to other evolutionary algorithms (EAs). They model the biocultural evolutionary theory to perform the search process for global optima, in which both types of vertical and horizontal learning behaviours of individuals are modelled. Since their emergence, CAs have been extended and successfully employed to solve a wide variety of problems in different branches of science and technology.

The main aim of this book is to explore the recent advances in the algorithmic framework of CAs and their applications to the different problems in the literature. While the main approach of the book is to briefly discuss and explain the application studies and algorithmic details of CAs, the detailed mathematical formulations and algorithmic pseudo-codes are also discussed in each chapter to provide a clear explanations for the different concepts. The book is mainly aimed at postgraduate students and researchers in computer science and engineering subjects with research interests in optimisation and meta-heuristic algorithms. Throughout the book, it is assumed that the readers are familiar with the basic concepts of optimisation theory and meta-heuristic algorithms. However, the author tried to provide relevant references in each chapter to assist the readers in understanding the contents of the book.

The book comprises nine different chapters divided into three parts. Part I contains the basic concepts of standard CAs and their theoretical background. The first part explains how the basic concepts in biocultural evolutionary theory have been employed to develop standard CAs. Part II discusses the applications of CAs to a wide range of real-world problems and presents their detailed mathematical formulations, including decision variables, objective functions, and constraints. Part III investigates the different variants of CAs developed in literature and their algorithmic details. The third part includes a comprehensive survey and detailed pseudo-codes of different extended, hybrid, and multi-population versions of CAs. The last chapter of the book presents the application study of CAs to the real-world structural optimisation problems.

Although a significant effort has been made to minimise the errors and typos in the book, the author warmly welcomes receiving feedback, suggestions, and comments from readers on the contents of the book.

The author would like to thank the series editors, Prof. Amir H. Gandomi and Dr. Anand J. Kulkarni, as well as the publishing editor, Ms. Kamiya Khatter, and her colleagues in Springer Nature for their efforts and supports during the production process of this book.

Aberdeen, Scotland, UK Shahin Jalili
April 2022

Contents

About the Author

Shahin Jalili is a Research Fellow in the School of Engineering at the University of Aberdeen working on renewable energies. His research interests focus on problems that link engineering, computer and mathematical sciences, with particular emphasis on optimization methods and numerical algorithms. He has developed a series of numerical approaches for various optimization problems in different branches of engineering science, ranging from the optimum design of skeletal and composite structures to optimum scheduling of offshore wind operations and performance optimization of transportation networks. He has also worked on the mathematical aspects of finite element analysis and developed a set of new structural and sensitivity reanalysis formulations based on the polynomial vector extrapolation methods to reduce the computational complexity of structural optimisation.

Part I
Foundations

Chapter 1
Introduction to Stochastic Optimisation

For since the fabric of the universe is most perfect and the work of a most wise Creator, nothing at all takes place in the universe in which some rule of maximum or minimum does not appear.
—Leonhard Euler

Abstract Modern stochastic heuristic and meta-heuristic optimisation methods are efficient tools to deal with the "black-box" problems in which the objective and constraint functions cannot be expressed as explicit functions of decision variables. This chapter presents a brief introduction to available conventional and stochastic optimisation approaches in the literature and discusses their applicability in dealing with real-world problems in science and technology. The chapter also provides a taxonomy of meta-heuristics based on their sources of inspiration.

Keywords Optimisation · Heuristic · Meta-heuristic · Hyper-heuristic · Evolutionary algorithms

1.1 Introduction

The optimisation is everywhere. The universe minimises the efforts during its evolution. The principle of minimum energy states that the internal energy of a closed physical system, with constant external parameters and entropy, always tends to approach a minimum value to satisfy equilibrium conditions. Fermat's principle in optical physics, which is also referred to as the principle of least time, states that a ray always chooses the path with minimum travel time between two points. The human evolutionary process can be viewed as an optimisation process in which human biology tries to enhance human adaptability to harsh environmental conditions over different generations. Ants deposit pheromone as they travel and indirectly communicate with each other to find the shortest path between their nest and food resources.

In our daily life, we find ourselves in a position in which we have to make decisions to achieve certain goals. We consciously or unconsciously try to make the best possible judgement in all circumstances and optimise our decisions to achieve

S. Jalili, *Cultural Algorithms*, Engineering Optimization: Methods and Applications,
https://doi.org/10.1007/978-981-19-4633-2_1

the most favourable outcomes. This tendency for optimisation takes different forms of behaviours in literature, art, business, etc. For example, poets make an effort to adopt the best combination of words with different rhythms and styles to deliver their thoughts, ideas, emotions, and messages to the readers. In the music improvisation process, the musicians try different combinations of music pitches to achieve the best possible harmony. In stock markets, the shareholders compete to take efficient investment strategies in their trades, minimise their losses, and maximise their profits. The manufacturing companies always plan to minimise their production costs and maximise their profit margin and market share. Numerous additional examples can be found in which the human plays the role of the optimiser. Despite our innate tendency for optimising our decisions, our capabilities in taking the right decisions with the most promising outcomes are limited. In some cases, the available time for making the judgements is quite short, and we are not able to react on time in response to dynamic environmental conditions. On the other hand, there is a huge number of complex situations in which human is unable to make the best possible judgements, no matter how long it would take. Nowadays, the complexity of problems in science and technology further highlights the fact that we desperately need powerful decision-making technologies to achieve different goals with minimum effort and energy.

With recent rapid technological development, numerous complex optimisation problems have emerged in different branches of science and technology. Early real-world optimisation problems were relatively easy to solve and handle. Hence, the classical techniques in applied mathematics were almost capable of dealing with these problems in a relatively efficient way. However, the complexity of human artefacts has been rapidly and consistently increased over time. Today's optimisation problems in different branches of science and technology are categorised as highly nonlinear problems with discrete and continuous variables under numerous equality and inequality constraints.

The main goal of this chapter is to present a brief introduction to available optimisation approaches in the literature. In Sect. 1.2, the conventional optimisation techniques are briefly reviewed, and their limitations in dealing with complex problems are discussed. Section 1.3 provides a brief review of stochastic heuristic and meta-heuristic algorithms as well as their taxonomy. This section categorises the meta-heuristics based on their sources of inspiration. As the main focus of this book is on CAs, the chapter also discusses the position of CAs in the taxonomy of meta-heuristic algorithms. In this chapter, it is assumed that the readers are familiar with the basic optimisation terminology. However, the readers are referred to the relevant references for more details.

1.2 Conventional Optimisation Methods

Conventional mathematical approaches are efficient tools to solve different types of optimisation problems. Linear programming deals with the problems in which the objective and constraints are linear functions of decision variables. Dantzig's simplex algorithm is one of the popular linear programming approaches (Dantzig 2016). The linear integer programming techniques have been developed for problems in which some or all of the decisions variables can only take integer values. There are a wide variety of problems in which mixed types of variables, including discrete and continuous, should be optimised, such as scheduling and production planning problems.

The dynamic programming method developed by Bellman in the 1950s is another popular optimisation approach in literature (Bellman 1966). The main idea of the approach is to break down the original difficult problem into several sub-problems which are easy to handle and deal with. The dynamic programming recursively calculates the optimal solutions for these sub-problems and then uses acquired information to obtain the global optimal solution for the original problem.

Many problems in science and technology are highly nonlinear. Nonlinear programming is a branch of mathematical science that focuses on optimisation problems with nonlinear objective and constraint functions. Although these techniques are capable of solving a set of practical problems, most real-world optimisation problems are much more difficult to be formulated and addressed by the nonlinear programming approaches (Foulds 2012). There are a series of outstanding monographs in the literature on the conventional optimisation methods, such as those by Luenberger and Ye (2021), Denardo (2012), and Foulds (2012).

The conventional optimisation approaches use gradient information of objective and constraint functions to search and locate the optimum solution for the problem at hand. Despite their remarkable performance in simple problems, their applicability to complicated problems is challenging from different perspectives. The conventional approaches perform the search process for optimal solutions based on the information acquired from derivations of objective and constraint functions. However, the calculation of gradient information is not always an easy task. Most of the modern optimisation problems belong to the category of "black-box" problems. In these problems, the objective and constraint functions cannot be expressed as explicit functions of decision variables, which make it difficult or sometimes even impossible to calculate their sensitivities for a given variable. The type of decision variables is also another source of difficulties in conventional techniques. Calculation of gradient information for discrete variables may be possible; however, it could be difficult or time-consuming in some cases. Another challenge in the application of conventional techniques is that they have been developed for the problems with single optimum solutions. While most of the modern optimisation problems are highly non-convex and nonlinear with multiple local optimum points. In such problems, the performance of classical techniques is highly sensitive to the initial solution, and they can easily get stuck into local optimums without efficient exploration of search space.

The abovementioned challenges intensified the efforts to develop alternative tools to deal with complex nonlinear problems. In the past decades, stochastic heuristic and meta-heuristic algorithms have emerged which are capable of providing optimal or near-optimal solutions for real-world problems within a reasonable time. Compared to the conventional approaches, heuristics and meta-heuristics are simple and easy to implement, and more importantly, they are independent of gradient information. These characteristic features make them attractive tools for "black-box" problems. By taking the advantage of their stochastic nature, "hopefully", they can escape from the local optimum points in the search space and enhance their chance of converging to the global optimum solutions. The main goal of the next section is to provide a brief literature survey on the historical development of heuristic and meta-heuristic algorithms as well as their taxonomy.

1.3 Modern Stochastic Methods

One of the characteristic features of conventional optimisation approaches is their deterministic nature. This means that the same outputs are obtained for a given initial solution or starting point over different independent runs. Due to the limited capabilities of conventional approaches and the difficulties in their applications to real-world problems, stochastic heuristic and meta-heuristic optimisation methods have been developed to solve difficult problems in science and technology. They are suitable tools to find near-optimal, or hopefully optimal, solutions for complex problems within a reasonable time. These techniques do not necessarily guarantee the achievement of global optimum for different problems. Rather, they are expected to provide near-optimal solutions for a certain range of problems, for which there is no available algorithm to find the optimal solution in polynomial time. The literature survey reveals the absence of general agreement on the exact definitions for heuristic and meta-heuristic algorithms. However, this chapter distinguishes between heuristics and meta-heuristics. This section provides a brief review of heuristics and meta-heuristics.

1.3.1 Heuristics

The heuristic[1] is a Greek word that means "find" or "discover". The main aim of heuristic optimisation methods is to discover optimum or near-optimum solutions for complex problems with relatively less computation effort. They are capable of finding near-optimum solutions for NP-hard problems in which the conventional approaches fail to provide meaningful results. Nowadays, they are the only available options to deal with a range of complex real-world problems. The heuristics work in a stochastic trial-and-error manner. They include a set of rules that are employed in a stochastic manner to discover useful clues about the optimum solution in the search space. The heuristics have a simple framework that makes them easy to understand and implement. In comparison to conventional approaches, they do not need gradient information of objective and constraint functions, and they are less sensitive to the

[1] εὑρίσκω.

initial solutions. By taking advantage of their stochastic behaviour, they can provide an efficient exploration of search space in highly nonlinear problems.

To decide whether a heuristic-based approach should be used for a given problem, a set of trade-off criteria is usually considered, including optimality, completeness, accuracy and precision, and execution time (Pearl 1984). The optimality criterion raises the question that whether the global optimum solution is needed for the problem at hand. In some problems, the global optimum exists, and it is necessary to discover it. While in others, finding an exact global optimum solution is not vital and only an approximation of the optimum solution is needed. Due to the stochastic nature of heuristic methods, they do not always guarantee the achievement of the global optimum solution. The completeness criterion discusses whether multiple global optimums exist in the problem. Although heuristics could potentially converge to different final solutions in different runs, they are usually designed in a way to provide a single output solution that can limit their applications to the problems with multiple global optimums. Accuracy and precision is another important criterion that highlights the question that whether the heuristics will be able to provide the final solution with an acceptable level of accuracy. This is particularly important in engineering optimisation problems. In most cases, a certain level of error can be tolerated. However, the errors in some cases could be irrationally large. The last criterion, i.e., execution time, considers whether the selected heuristic exhibits a faster convergence rate in comparison to the conventional methods. Some of the heuristics may not provide a clear advantage over conventional approaches in terms of convergence speed and required computational effort. Hence, the selection of the most promising heuristic for the problem at hand is vital.

As Rothlauf (2011) argues, heuristics can be categorised into construction and improvement heuristics. The construction heuristics start with an uncomplete solution and gradually construct the full solution for the problem over multiple steps. In these approaches, the search process is terminated as soon as a complete solution is constructed for the problem. On the other hand, the improvement heuristics start with a complete solution and gradually modify the value of each variable to achieve better results based on a given performance measure. As was mentioned earlier, heuristics are problem-specific tools. A wide variety of heuristics can be found in the literature that has been developed for different types of problems. One of the popular problems in computing science is the Travelling Salesman Problem (TSP) for which numerous heuristic techniques have been developed. The nearest neighbour, nearest insertion, cheapest insertion, and furthest insertion are examples of construction heuristics designed for TSP (Rosenkrantz et al. 1977; Rothlauf 2011). The two-opt, *k*-opt, and Lin–Kernighan heuristics belong to the category of improvement heuristics developed for the TSP problem (Rothlauf 2011).

1.3.2 Meta-Heuristics

Meta-heuristics are a new generation of stochastic optimisation techniques. A meta-heuristic can be described as an upper-level strategy that simultaneously takes the advantage of different heuristics to efficiently explore the search space and find optimum solutions. Several definitions for meta-heuristics are available in the literature. For example, Voß et al. (2012) state the following definition for meta-heuristics:

> "A metaheuristic is an iterative master process that guides and modifies the operations of subordinate heuristics to efficiently produce high quality solutions. It may manipulate a complete (or incomplete) single solution or a collection of solutions at each iteration. The subordinate heuristics may be high (or low) level procedures, or a simple local search, or just a construction method."

There are basic differences between heuristics and meta-heuristics. Contrary to heuristic approaches, which have been developed for certain problems, meta-heuristics are not problem-specific tools, and they can be applied to a large set of problems in science and technology. Meta-heuristics try to keep the balance between the exploitation (or intensification) and exploration (or diversification) mechanisms. The former refers to the capability of a search method in further improving the quality of the best solution, while the latter is related to the ability of the algorithm to search different regions of search space and escape from local optimum points.

1.3.3 No Free Lunch Theorems

As was discussed in previous sections, heuristics and meta-heuristics do not necessarily guarantee the consistent achievement of the global optimum solution for all kinds of problems. The reality is that their performance is very sensitive to the nature of the problem. They may provide highly satisfiable results for a given class of problems, while they may fail to exhibit an efficient performance in others. In the 1990s, Wolpert and Macready (1997) provided a set of theorems, called no free lunch theorems, to explain that there is no unique heuristic or meta-heuristic approach capable of exhibiting equally better performance than all others for all problem types. According to these theorems, although a specific heuristic or meta-heuristic algorithm can perform better than others for a given problem, there are other classes of problems in which the same algorithm might show weaker performance than other algorithms. Generally speaking, these theorems state that the average performance of all heuristic and meta-heuristic algorithms for different static and dynamic optimisation problems is almost the same. Hence, the main aim of algorithm designers should be the design of a heuristic or meta-heuristic approach that is capable of solving most types of problems in an efficient way, rather than all types of problems. This was the motivation for the researchers to develop a significant number of meta-heuristics to solve different types of problems in recent decades.

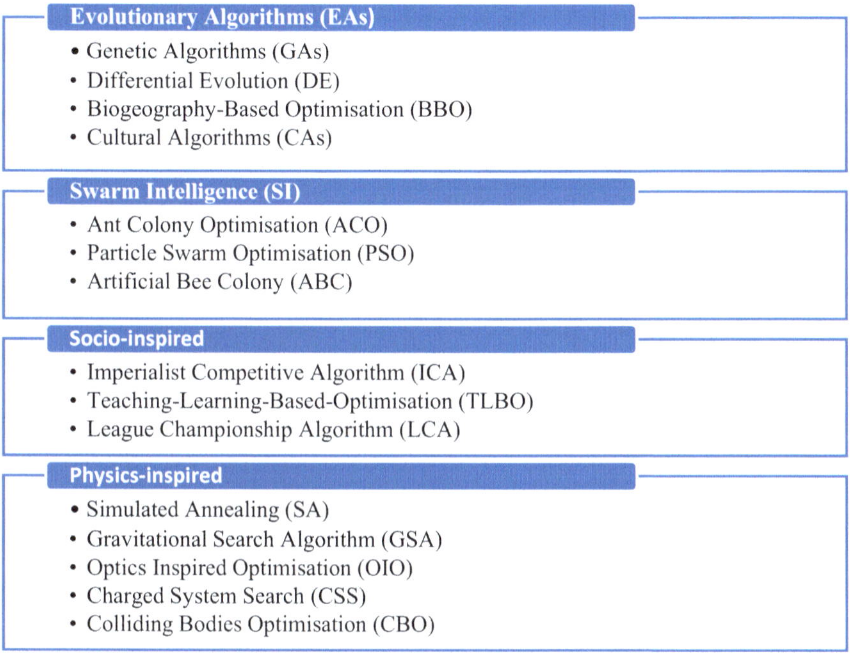

Fig. 1.1 Taxonomy of meta-heuristics based on their sources of inspiration

The emergence of meta-heuristic algorithms goes back to the 1960s, when Holland (1975) and his fellow researchers at the University of Michigan developed Genetic Algorithms (GAs) (Goldberg 1986) based on the principle of natural selection in Darwin's evolutionary theory of biological species. Following the successful application of GAs, a significant number of new meta-heuristics inspired by different phenomena in nature have been developed, such as Simulated Annealing (SA), Tabu Search (TS), Ant Colony Optimisation (ACO), Differential Evolution (DE), Particle Swarm Optimisation (PSO), Harmony Search (HS), Biogeography-based Optimisation (BBO), and Teaching–Learning-Based Optimisation (TLBO). Meta-heuristics can be categorised based on their sources of inspiration. In the subsequent sections, a taxonomy of meta-heuristics are presented based on their sources of inspiration as shown in Fig. 1.1.

1.3.4 Evolutionary Algorithms (EAs)

EAs have been developed mainly based on Darwin's evolutionary theory of biological species. GAs developed by Holland (1975) is the pioneering and most prominent EA. The GAs adapt the biological evolutionary concepts, such as mutation, cross-over, and selection, to stochastically perform the search process for optimal solutions. In

GAs, the solution for the problem is encoded as a chromosome consisting of a set of genes that are evolved over different generations. According to the principle of natural selection in the theory of evolution, the genes of individuals with better adaptation ability are more likely to be transmitted to the next generations. Mutation refers to the errors and damages that can happen in genes during the transmission process, which can potentially lead to random changes in genes and genetic diversity in the population (Newson et al. 2007). GAs model these processes to gradually evolve a population of solutions. These algorithms have found a vast amount of applications in real-world problems (Katoch et al. 2021).

The DE algorithm proposed by Storn and Price (1997a) is another EA. Storn and Price developed DE when they were trying to modify the Genetic Annealing algorithm (Price et al. 2006a). The modifications of the Genetic Annealing algorithm resulted in a new mutation equation on which DE has been developed. The DE uses three main operators, including mutation, cross-over, and selection, in which the evolutionary process is modelled based on the weighted differences between the individual vectors in the search space.

BBO introduced by Simon (2008) is another EA that was developed based on the probabilistic mathematical models in biogeography science (Wilson and MacArthur 1967). The algorithm assumes the solution candidates as a set of habitats in which the species perform the emigration and immigration processes. In BBO terminology, the position of each habitat is represented by Suitability Index Variables (SIVs), and the corresponding fitness is indicated by Habitat Suitability Index (HSI). BBO simulates the emigration and immigration processes based on the migration and mutation operators. The migration and mutation operators use a set of emigration and immigration rates which are obtained based on migration models from biogeography science. A recent literature survey performed by Ma et al. (2017) reveals that BBO has attracted relatively remarkable attention from different research communities over the past decade.

CAs developed by Reynolds and his colleagues can be categorised as an EA, which have been developed based on the biocultural evolution theory (Reynolds and Rolnick 1995a, b; Ostrowski and Reynolds 1999). According to this theory, genes and culture can be viewed as two interacting forms of inheritance that form the overall evolution of the human species. This means that human behaviour is a product of two different and interacting evolutionary processes, genetic and cultural evolutions. CAs have some features that make them unique in comparison to other EAs. The conventional EAs work on the population level in which vertical genetic inheritance mechanism is modelled. Contrary to conventional EAs, CAs employ a dual inheritance system that is consisted of two parallel spaces, population and belief spaces. The population space models the genetic evolution and represents the micro-evolutionary level. While the belief space simulates the cultural evolution process of individuals in the population space and can be viewed as a macro-evolutionary level. The belief space includes a network of cultural knowledge that can be used by individuals during the decision-making process to find better results for the problem. During the solution-finding process, the population and belief spaces exchange information based on communication protocols. Chapter 2 will present the theoretical background on cultural evolution, and Chap. 3 will explain the full details of CAs.

1.3.5 Swarm Intelligence (SI)

SI refers to the family of mate-heuristic algorithms inspired by the collective behaviour of a group or swarm of species in biological systems, such as ants and birds. The basic characteristic features of these cooperation models observed in biological systems are their decentralised and self-organised nature. In these systems, a set of agents interact with each other and their environment. Their behaviours are not controlled by any centralised forces, and they are free to locally and globally interact within the swarm in a relatively random manner. Despite the simple individual behaviour of agents, their communication and collective interactions lead to the emergence of a global complex behaviour that is much more complicated. SI meta-heuristic techniques stimulate this intelligent behaviour to model the search process for the global optima in optimisation problems.

ACO developed by Dorigo et al. (2006) is a SI technique inspired by the foraging behaviour of ants in finding the shortest path between their nest and food sources. The ants indirectly communicate with each other through a chemical process, called pheromone deposition. They can deposit pheromone on the ground to inform other ants about potential danger or trail the paths between their nest and food sources. In ACO, the solution candidates are modelled as ants which deposit pheromone on parts of the search that could potentially result in better fitness values.

PSO originally introduced by Kennedy and Eberhart (1995) is another popular SI algorithm. The algorithm imitates the collective behaviour of a bird flock or fish school in finding food sources. In PSO, each solution candidate is represented as a particle in the search space that moves with a variable velocity. PSO considers the personal experience gained by every single particle as well as global experience acquired by the whole swarm to update the positions and velocities of particles in each iteration. The algorithm has been a quite popular approach for solving a wide variety of problems and, occasionally, its different variants have been developed by researchers (Parsopoulos and Vrahatis 2010; Mirjalili et al. 2020).

1.3.6 Socio-inspired Algorithms

The socio-inspired meta-heuristic algorithms adopt the social learning behaviours observed in real human societies to perform the search process for global optima. Imperialist Competitive Algorithm (ICA), TLBO, and League Championship Algorithm (LCA) are examples of this category of algorithms. ICA introduced by

Atashpaz-Gargari et al. (2007) imitates the imperialist competition process in political science, TLBO developed by Rao et al. (2011) simulates the interactions between teacher and students in a class, and LCA presented by Kashan (2009) models the competition of teams in a sports league to achieve most favourable outcomes. The socio-inspired algorithms have been employed to solve various problems in literature (Hosseini and al Khaled 2014; Rao 2016; Jalili et al. 2017; Husseinzadeh Kashan et al. 2018).

1.3.7 Physics Inspired Algorithms

Several meta-heuristics model a given physical phenomenon observed in nature. The algorithms in this category adopt the governing equations in physics to build the search operators. Examples include, but are not limited to, SA, Gravitational Search Algorithm (GSA), Optics Inspired Optimisation (OIO), Charged System Search (CSS), and Colliding Bodies Optimisation (CBO).

SA (van Laarhoven and Aarts 1987) is a single point meta-heuristic algorithm inspired by the annealing process in metallurgy in which a material is being heated and gradually cooled down to improve its physical and chemical properties. GSA introduced by Rashedi et al. (2009) metaphorically models Newtonian gravity and the laws of motion into the searching process, in which a set of particles interact with each other based on their masses. In GSA, the particles with given masses calculated based on the fitness values represent a set of solutions for the problem. In a similar approach, Kaveh and Talatahari (2010) adopted the Coulomb law from electrostatics and the Newtonian laws of mechanics to develop a new meta-heuristic, called CSS. In CSS, a set of charged particles are solution candidates for the problem that impose electrical forces on each other according to Coulomb's law. CBO is another physics-inspired algorithm presented by Kaveh and Mahdavi (2014), in which the governing equations in the one-dimensional collision process of moving bodies are adopted to design a new meta-heuristic.

OIO is a population-based algorithm that models the optical phenomena observed in spherical mirrors (Kashan 2015; Jalili and Husseinzadeh Kashan 2018, 2019). The algorithm uses the governing equations in optical physics which explain how the images are formed in concave and convex mirrors. OIO treats the surface of the objective function as a wavey reflecting surface consisting of concave and convex parts on which the artificial images are formed. The algorithm models each solution candidate as an artificial light point that forms an artificial image on the function surface which represents a new solution for the problem.

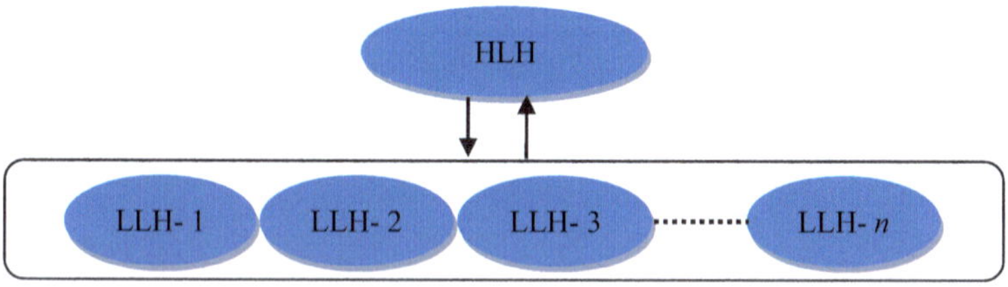

Fig. 1.2 General framework of hyper-heuristics

1.3.8 Hyper-Heuristics

According to the no free lunch theorems explained in Sect. 1.3.3, there is no single meta-heuristic algorithm capable of exhibiting equally efficient performance for different problems with different features in the search space. During the past decade, this has been intensified the efforts to develop more general and efficient algorithms that can be applied to most types of problems with a relatively stable performance. The emergence of hyper-heuristics is an example of such efforts. The logic behind the hyper-heuristics is to perform the search process in the heuristic space, rather than in the solution space (Burke et al. 2010). In the hyper-heuristic framework, a set of Low-Level Heuristics (LLHs) perform the search process in the solution space which are controlled by a High-Level Hyper-heuristic (HLH) strategy as shown in Fig. 1.2. In the high-level strategy, the algorithm decides which heuristic in the lower level should be applied to perform the search process at a given time in the solution space. The high-level strategy provides the learning capability for the algorithm to apply the most efficient heuristic strategy depending on the feedback received from the application of different LLHs in the past. The learning process in hyper-heuristics can be online or offline (Burke et al. 2013a). It is also possible to construct a hyper-heuristic without learning capability, in which the LLH is selected in a purely random manner. The learning process in hyper-heuristics can potentially make them capable of dealing with different types of problems. The LLH can be any constructive or improvement heuristics. In literature, various HLHs have been developed, such as choice function, greedy selection, Multi-Armed Bandit (MAB), Hidden Markov Model (HMM), and Monte Carlo Tree Search (MCTS) methods (Choong et al. 2018). Although most of the hyper-heuristics in literature employ heuristics in the lower level, population-based meta-heuristics can also be used in the lower level to perform the search within the solution space.

1.4 Summary

This chapter provided a brief review of deterministic and stochastic optimisation approaches. The chapter started with a short overview of conventional optimisation approaches available from applied mathematics. The conventional approaches use gradient information of objective and constraint functions to search and locate the optimum solutions. Although the conventional approaches are efficient and primary options to solve relatively simple problems, their application to highly nonlinear problems is challenging from different perspectives.

Modern stochastic heuristic and meta-heuristics optimisation methods are efficient tools to deal with the "black-box" problems in which the objective and constraint functions cannot be expressed as explicit functions of decision variables. These approaches do not necessarily guarantee the achievement of global optimum for different problems. Rather, they are expected to provide near-optimal solutions for a certain range of problems, for which there is no available algorithm to find the optimal solution in polynomial time. The chapter discussed the no free lunch theorems in stochastic optimisation which state that although a specific heuristic or meta-heuristic algorithm can perform better than others for a given problem, there are other classes of problems in which the same algorithm might show weaker performance than other algorithms. According to these theorems, the main aim of algorithm designers should be the design of a heuristic or meta-heuristic approach that is capable of solving most types of problems in an equally efficient manner, rather than all types of problems.

During the past decades, a significant number of meta-heuristic algorithms have been developed to solve optimisation problems in different branches of science and technology. Based on their sources of inspiration, the chapter categorised meta-heuristics into EAs, SI, socio-inspired, and physics-inspired algorithms. It was discussed that CAs belong to the category of EAs that model the biocultural evolution theory into the searching process for global optima. In the eyes of biocultural evolution theory, genes and culture are two interacting forms of inheritance that form the overall evolution of the human species. The CAs employ a dual inheritance system that makes them unique in comparison to other EAs. As the main focus of this book is on CAs, the theoretical background on biocultural evolutionary theory will be discussed in Chap. 2, and the full algorithmic details of CAs will be presented in Chap. 3.

References

Atashpaz-Gargari E, Lucas C (2007) Imperialist competitive algorithm: an algorithm for optimization inspired by imperialistic competition. In: 2007 IEEE congress on evolutionary computation. IEEE, pp 4661–4667

Bellman R (1966) Dynamic programming. Science (1979) 153:34–37

Burke EK, Gendreau M, Hyde M et al (2013) Hyper-heuristics: a survey of the state of the art. J Oper Res Soc 64:1695–1724. https://doi.org/10.1057/jors.2013.71

Burke EK, Hyde M, Kendall G, et al (2010) A classification of hyper-heuristic approaches. pp 449–468

Choong SS, Wong L-P, Lim CP (2018) Automatic design of hyper-heuristic based on reinforcement learning. Inf Sci 436–437:89–107. https://doi.org/10.1016/j.ins.2018.01.005

Dantzig G (2016) Linear programming and extensions. Princeton University Press

Denardo E v (2012) Dynamic programming: models and applications. Courier Corporation

Dorigo M, Birattari M, Stutzle T (2006) Ant colony optimization. IEEE Comput Intell Mag 1:28–39. https://doi.org/10.1109/MCI.2006.329691
Foulds LR (2012) Optimization techniques: an introduction. Springer Science & Business Media
Goldberg DE (1986) The genetic algorithm approach: why, how, and what next? Adaptive and learning systems. Springer, US, pp 247–253
Holland J (1975) Adaptation in natural and artificial systems. University of Michigan Press, Ann Arbor
Hosseini S, al Khaled A (2014) A survey on the imperialist competitive algorithm metaheuristic: Implementation in engineering domain and directions for future research. Appl Soft Comput 24:1078–1094. https://doi.org/10.1016/j.asoc.2014.08.024
Husseinzadeh Kashan A, Jalili S, Karimiyan S (2018) Optimum structural design with discrete variables using league championship algorithm. Civil Eng Infrastruct J 51:253–275
Jalili S, Husseinzadeh Kashan A (2018) Optimum discrete design of steel tower structures using optics inspired optimization method. Struct Des Tall Spec Build 27:e1466. https://doi.org/10.1002/tal.1466
Jalili S, Husseinzadeh Kashan A (2019) An optics inspired optimization method for optimal design of truss structures. Struct Des Tall Spec Build 28:e1598. https://doi.org/10.1002/tal.1598
Jalili S, Kashan AH, Hosseinzadeh Y (2017) League championship algorithms for optimum design of pin-jointed structures. J Comput Civ Eng 31:04016048. https://doi.org/10.1061/(ASCE)CP.1943-5487.0000617
Kashan AH (2015) An effective algorithm for constrained optimization based on optics inspired optimization (OIO). Comput Aided Des 63:52–71. https://doi.org/10.1016/j.cad.2014.12.007
Kashan AH (2009) League championship algorithm: a new algorithm for numerical function optimization. In: 2009 International conference of soft computing and pattern recognition. IEEE, pp 43–48
Katoch S, Chauhan SS, Kumar V (2021) A review on genetic algorithm: past, present, and future. Multimedia Tools Appl 80:8091–8126. https://doi.org/10.1007/s11042-020-10139-6
Kaveh A, Mahdavi VR (2014) Colliding bodies optimization: a novel meta-heuristic method. Comput Struct 139:18–27. https://doi.org/10.1016/j.compstruc.2014.04.005
Kaveh A, Talatahari S (2010) A novel heuristic optimization method: charged system search. Acta Mech 213:267–289. https://doi.org/10.1007/s00707-009-0270-4
Kennedy J, Eberhart R (1995) Particle swarm optimization. In: Proceedings of ICNN'95—international conference on neural networks. IEEE, pp 1942–1948
Luenberger DG, Ye Y (2021) Linear and nonlinear programming. Springer, Cham
Ma H, Simon D, Siarry P et al (2017) Biogeography-based optimization: a 10-year review. IEEE Trans Emerg Top Comput Intell 1:391–407. https://doi.org/10.1109/TETCI.2017.2739124
Mirjalili S, Song Dong J, Lewis A, Sadiq AS (2020) Particle swarm optimization: theory, literature review, and application in airfoil design. pp 167–184
Newson L, Richerson PJ, Boyd R (2007) Cultural evolution and the shaping of cultural diversity. Handbook of cultural psychology 454–476
Ostrowski DA, Reynolds RG (1999) Knowledge-based software testing agent using evolutionary learning with cultural algorithms. In: Proceedings of the 1999 congress on evolutionary computation-CEC99 (Cat. No. 99TH8406). IEEE, pp 1657–1663
Parsopoulos KE, Vrahatis MN (2010) Particle swarm optimization and intelligence: advances and applications: advances and applications. IGI global
Pearl J (1984) Heuristics: intelligent search strategies for computer problem solving. Addison-Wesley Longman Publishing Co., Inc.
Price K, Storn RM, Lampinen JA (2006a) Differential evolution: a practical approach to global optimization. Springer Science & Business Media
Rao RV (2016) Teaching-learning-based optimization algorithm. Teaching learning based optimization algorithm. Springer, Cham, pp 9–39

Rao RV, Savsani VJ, Vakharia DP (2011) Teaching–learning-based optimization: a novel method for constrained mechanical design optimization problems. Comput Aided Des 43:303–315. https://doi.org/10.1016/j.cad.2010.12.015
Rashedi E, Nezamabadi-pour H, Saryazdi S (2009) GSA: a gravitational search algorithm. Inf Sci 179:2232–2248. https://doi.org/10.1016/j.ins.2009.03.004
Reynolds RG, Rolnick SR (1995a) Learning the parameters for a gradient-based approach to image segmentation using cultural algorithms. In: Proceedings 1st international symposium on intelligence in neural and biological systems. INBS'95. IEEE Computation Social Press, pp 240–247
Reynolds RG, Rolnick SR (1995b) Learning the parameters for a gradient-based approach to image segmentation from the results of a region growing approach using cultural algorithms. In: Proceedings of 1995b IEEE international conference on evolutionary computation. IEEE, pp 819–824
Rosenkrantz DJ, Stearns RE, Lewis IPM (1977) An analysis of several heuristics for the traveling salesman problem. SIAM J Comput 6:563–581. https://doi.org/10.1137/0206041
Rothlauf F (2011) Design of modern heuristics. Springer, Berlin
Simon D (2008) Biogeography-based optimization. IEEE Trans Evol Comput 12:702–713. https://doi.org/10.1109/TEVC.2008.919004
Storn R, Price K (1997) Differential evolution—a simple and efficient heuristic for global optimization over continuous spaces. J Global Optim 11:341–359. https://doi.org/10.1023/A:1008202821328
van Laarhoven PJM, Aarts EHL (1987) Simulated annealing. Simulated annealing: theory and applications. Springer, Netherlands, pp 7–15
Voß S, Martello S, Osman IH, Roucairol C (2012) Meta-heuristics: advances and trends in local search paradigms for optimization. Springer Science & Business Media
Wilson EO, MacArthur RH (1967) The theory of island biogeography. JSTOR
Wolpert DH, Macready WG (1997) No free lunch theorems for optimization. IEEE Trans Evol Comput 1:67–82. https://doi.org/10.1109/4235.585893

Chapter 2
Introduction to Culture

Abstract Culture can be defined as a source of information with the potential capability of affecting the behavioural patterns of members of a population. This information can be any kind of mental state, such as beliefs, habits, skills, traditions, art, etc., which are transmissible to others using teaching, imitation, and other forms of social interactions. Based on the biocultural evolution theory, genes and culture can be viewed as two interacting forms of inheritance. This means that human behaviour is a product of two different and interacting evolutionary processes, genetic evolution and cultural evolution. In this chapter, the basic and general concepts of human cultural evolution in real human societies are discussed.

Keywords Culture · Cultural evolution · Genetic evolution · Social learning · Cultural variants

2.1 Introduction

The study of human evolution highlights the fact that enormous progress has been made by humankind in improving the quality of life over a very short time. After the emergence of modern humans in Africa about 60,000 years ago, they have rapidly spread across the globe with limited primary technology and communication skills. They have exhibited remarkable success in efficiently dealing with the harsh conditions in nature and rapidly adapting themselves to different environmental challenges. During human history, they have consistently acquired new knowledge about how to deal with environmental challenges, and most importantly, they have transferred this knowledge over different generations. This successively accumulated amount of knowledge and information has always significantly improved the life quality of humans and enhanced their ability to take the control of the environment.

One of the major turning points in modern human history is the invention of agricultural technology (Gowdy and Krall 2014). Within a relatively short period after widespread agricultural adoption, the lifestyle of the human species turned upside down. They started to form large groups and develop their communication skills instead of living in isolated small groups with limited communication. Agricultural knowledge, technology, skills, and practices have been successively modified,

S. Jalili, *Cultural Algorithms*, Engineering Optimization: Methods and Applications,
https://doi.org/10.1007/978-981-19-4633-2_2

enhanced, and transmitted to the next generations which resulted in a remarkable improvement in human life quality and significant population growth. The advances in agricultural technology formed small permanent settlements and rapidly expanded the human population from 6 to 200 million within 1000 years (Biraben 2003; Cox et al. 2009; Bocquet-Appel 2011). This was facilitated the formation of cities, kingdoms, political institutions, and social organisations (Altman and Mesoudi 2019). The current cultural diversity in the world can be viewed as a result of this population expansion, which created different languages, traditions, beliefs, and other cultural features (Newson et al. 2007). The rapid population expansion resulting from agricultural development also brought new challenges to human life. Emergence of new disease patterns in human societies was one example of these challenges, which continue to exist and threaten human life (Armelagos et al. 2005). However, the accumulated knowledge and ongoing advancements in medical science made humankind able to survive, take the control of disease, and continue to expand its population size further.

One of the important unique characteristics of the human species in survival and adaptation is their reliance on tools (Muehlenbein 2015). The researches on human evolution reveal that initial human artefacts were not more complex than those created by other species. However, with the emergence of complex human societies, the human technology has experienced rapid development during only the past 10,000 years (Toth and Schick 2015). Alongside technology, humans have been consistently improved their communication and information exchange skills. Languages have been played a pivotal role in this sense. Throughout history, humankind has spoken more than six and a half thousand languages (Nettle 1998). Some of the languages have been entirely lost to history, while some others have been enriched and evolved. However, most of them are threatened with extinction in a near future (Pinker 1995; Nettle 1998).

The above mentioned developments in human life can be studied by human evolutionary theories. Genetic evolution explains the changes in genes that can control the biological traits of a human. However, genetic evolution cannot fully explain human success in comparison to other species. The human is a social species with strong social interactions and learning mechanisms. Based on the biocultural evolution theory, genes and culture can be viewed as two interacting forms of inheritance. This means that human behaviour is a product of two different and interacting evolutionary processes, genetic evolution and cultural evolution.

The success of humans stems from the fact that they have the power to socially learn and pass their experiences to the next generations. In comparison to other species, the capability of humans in social learning is unique. The accumulated information acquired by humans has been transmitted over time and helped them to enhance their adaptation capability against environmental conditions and improve the quality of their life. This information transmission process is provided by culture. Humans can transmit their experiences, skills, beliefs, and other cultural variants to the next generations. The culture and its variants evolve during the time and support humans to make the most productive decisions in their life. The intention of this chapter is not to provide a comprehensive analysis of the cultural evolution, as the goal of the book is to study the numerical aspects of CAs. Rather, the main goal of this chapter is to describe the basic concepts used within CAs by providing a brief introduction to the cultural evolution of real human societies.

The chapter starts with different definitions and interpretations in literature for culture during the past two centuries. Most characteristics features of culture are provided in Sect. 2.3 based on the latest research in social science. The social learning process of humans differs from those observed in animals. A short discussion on the main differences between the learning abilities of humans and animals is presented in Sect. 2.4.

Despite to extent of disagreement on the definition of culture, there is a strong agreement between anthropologists and archaeologists that culture evolves. In Sect. 2.5, the cultural evolution is briefly explained from the eyes of "kinetic theorists", and some similarities between the genetic and cultural evolution processes are discussed. The section discusses the theory developed by Richerson & Boyd (2008) which tries to explain that cultural evolution includes both vertical and horizontal information transmission mechanisms.

In cultural evolution, the offspring receive and learn information not only from their parents but also from other individuals through their social interactions. In Sect. 2.6, the social learning biases that can be exhibited by individuals in their social interactions are briefly discussed. The cultural variants, such as skills, habits, preferences, beliefs, and traditions, can influence the behaviours and decisions of the individuals in society. Section 2.7 discusses which individuals in society can potentially change or replace these cultural variants. Finally, a summary of the contents provided in this chapter is presented in Sect. 2.8.

2.2 What is the Culture?

Providing a general definition of *culture* is difficult and controversial. Since the 1880s, there is ongoing debates and discussions between anthropologists and archaeologists on how culture can be defined, described, or interpreted. One of the early definitions was provided by Edward Burnett Tylor in the 1880s, a pioneer anthropologist, known as the founder of cultural anthropology. He states that *"Culture, or Civilisation, taken in its wide ethnographic sense, is that complex whole which includes knowledge, belief, art, morals, law, custom, and any other capabilities and habits acquired by man as a member of society"* (Tylor 1871). Following Tylor's definition, a range of definitions was developed in literature by considering different aspects of culture.

A critical review performed by Kroeber and Kluckhohn (1952) in the 1950s revealed 164 different definitions for culture. This review suggested that the available definitions for culture can be sorted into descriptive, historical, normative, psychological, structural, genetic, and incomplete categories (Lyman 2008) as shown in Fig. 2.1. The descriptive aspect of culture represents a set of ideas, beliefs, traditions, etc. As its name suggests, the historical aspect tries to define how the descriptive cultural contents are transmitted between the different generations during the

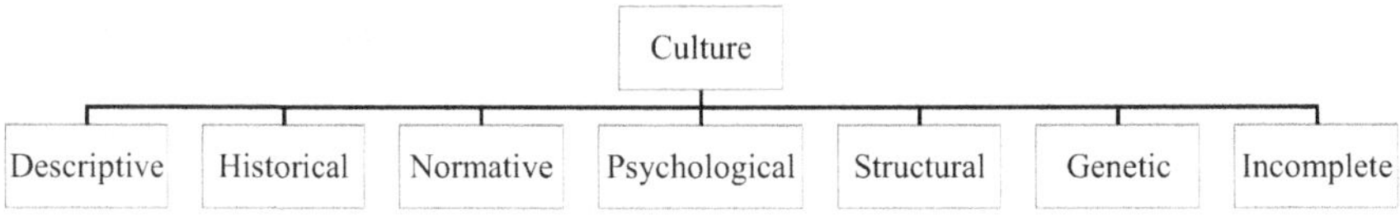

Fig. 2.1 Categorisation of definitions for the concept of culture investigated by Kroeber and Kluckhohn (1952)

time. The normative aspect is a set of values and prescriptions that determine the behaviours of the people and guide them in different cultural situations (Baldwin et al. 2006). Psychology-based definitions focus on the learning process and non-genetic transmission of knowledge and skills between individuals. The structural definitions of culture try to describe the connection patterns between distinct descriptive elements, such as ideas, habits, skills, preferences, beliefs, and traditions. The origin and existence of culture are addressed by genetic-based definitions (Baldwin et al. 2006). All other types of definitions of culture are assumed as incomplete definitions.

Kroeber and Kluckhohn (1952) provided a more comprehensive definition of culture by synthesising a variety of definitions in literature as follows (Baldwin et al. 2006):

> "*Culture consists of patterns, explicit and implicit, of and for behaviour acquired and transmitted by symbols, constituting the distinctive achievements of human groups, including their embodiments in artifacts; the essential core of culture consists of traditional (i.e., historically derived and selected) ideas and especially their attached values; culture systems may, on the one hand, be considered as product of action, on the other as conditioning elements of further action.*"

This overall definition provided by Kroeber and Kluckhohn (1952) covers all of the mentioned aspects observed in the preceding definitions, i.e., descriptive, historical, normative, psychological, structural, genetic, and incomplete. For a more detailed analysis of the definitions of culture, interested readers are referred to Baldwin et al. (2006).

2.3 Characteristic Features of Culture

Although the definition of culture is rather controversial, Holmes (2020a) argues in a book on cultural psychology that there is a striking agreement among socialists on some characteristic features of culture. In that book, some notable characteristics of culture are discussed, including its transmissibility, shared meanings, adaptivity, symbolic nature, and performability.

Transmissibility refers to the cultural knowledge transmission between different successive generations. While the shared meanings emphasise that the individuals within the same community have different understanding and knowledge about their culture. In addition, their capabilities to follow and behave according to the social norms are not the same (Holmes 2020a). The culture is dynamic with remarkable potential adaptivity that can constantly change in response to the changes in environmental conditions. The culture includes a system of shared meanings represented by a set of symbols that individuals use to communicate and translate into meaningful actions, feelings, and perceptions (Geertz et al. 1973; Holmes 2020a). As its name suggests, performability means that the individuals in a community have the freedom to implement their cultural models and develop their cultural meaning in their daily interactions. These characteristic features of culture presented by Holmes (2020a) can help us to have an overall understanding of the culture in this book.

2.4 Cultural Learning in Humans and Animals

As was mentioned earlier, the culture is transmissible. In human culture, the different skills, habits, preferences, traits, beliefs, and traditions pass from one generation to the next. The question is do only humans have this capability? The brief answer is no. Scientific research obtained from a variety of controlled experiments and field observations revealed that the animals are also capable of transferring their culture through social learning or imitation (Roper 1986; Whiten 2019). However, the human and non-human cultural learning processes are entirely different.

Holmes (2020b) identifies three main differences between human and non-human cultural learning. The first difference is the high capacity of human individuals in imitating others in their social world. In comparison to non-human primates, the human individual is remarkably more capable of acquiring knowledge in different forms, such as skills, habits, preferences, traits, and traditions, from other individuals. Recent comparative studies on the social learning capabilities of humans and chimpanzees showed that humans are undoubtedly superior in this sense (Whiten and Custance 1996; Newson et al. 2007; Tomasello 2009). The second difference is that the individuals in human societies can be educated through the teaching process, which is not the case for non-human primates. The third is related to the capabilities of humans in transmitting cultural variants, such as behavioural standards, to the next generations (Holmes 2020b). Humans can create/modify the cultural variants over time and pass them to the next generations. However, the creativity and capability of non-human primates to make changes in their cultural variants are limited.

2.5 Cultural Evolution

Culture gradually changes over time. Due to changes in the cultural environment, the cultural variants can be changed or even replaced by new ones over time. Cultural evolution research tries to identify different forces that can affect the existence or emergence of cultural variants (Newson et al. 2007). In comparison to genetic evolution, cultural evolution is a more complicated process. In cultural evolution, the information transmission process happens not only between parents and offspring but also between the offspring and other individuals. The individuals are free to choose, accept, or reject the information that they receive. There are several theories on cultural evolution. The common assumption of almost all theories is that the culture evolves. However, each theory describes the cultural evolution process in a different way.

Lewens (2015) provides a threefold taxonomy of cultural evolutionary theories, including historic, selection, and kinetic theories. As Lewens (2015) states, the historical theories express that the culture gradually evolves over generations. The selectionists consider that the different cultural variants, such as beliefs, habits, preferences, traditions, and skills, compete in a Darwinian struggle (Lewens 2015). The kinetic theorist investigates the interaction between the individuals within the population by considering the typical vertical information transmission from the parents to offspring as well as horizontal information transmission between offspring and other members of the population. In this chapter, the population-based theory developed by Richerson and Boyd (2008), which is recognised as a kinetic cultural evolution theory, is briefly discussed.

As cultural evolution is a rather complex process, cultural evolutionists have been tried to develop mathematical models and perform different empirical experiments to investigate how information is transmitted from one generation to the next. These models and empirical experiments do not necessarily provide accurate results, rather they try to derive some sensible results on how the information is transmitted (Newson et al. 2007). Newson et al. (2007) discuss the similarities between cultural evolutionary and meteorological models. They argue that both models provide approximate approaches to describe the cultural evolution and forecast future weather conditions, respectively.

Cultural evolution is guided by the cultural variants acquired by individuals. One of the key distinctions between the genetic and cultural evolutionary processes is the way information is transferred between different generations, which is significantly more complicated in cultural evolution. In genetic evolution, the genetic information embedded within the genes is transferred from parents to offspring, which is usually called "vertical transmission" in literature. However, the information transmission mechanism in the cultural evolutionary process is not that simple. In the latter case, the individuals acquire cultural or behavioural information not only from their parents but also from their friends, schools, social groups, kins, etc. This information transmission mechanism in human society is called "biased transmission" by Richerson and Boyd (2008). The biased transmission can appear in human behaviours in different forms. Interested readers are referred to (Boyd and Richerson 1988) for more details about the biased transmission.

In the human genetic evolutionary process described by Darwin, the biological variants are inherited from parents to offspring. The transmission process of biological variants is affected by different forces including natural selection, mutation, drift, and migration. Newson et al. (2007) argue that the forces affecting the cultural changes in human societies are similar to those mentioned in genetic evolution. However, they are more complicated.

Individuals in a human society transmit cultural variants, such as beliefs, traits, behaviours, traditions, habits, preferences, and so on, to their offspring. The parents with fewer offspring would have less chance to spread and transmit their information to the next generations. This represents the natural selection force in cultural evolution.

The quality of teaching and learning processes could cause remarkable alterations in cultural variants over time. The individuals do not have an equal capacity to properly or completely transmit their information to offspring. The capacities of offspring to learn cultural variants are different as well. This means that the amount and accuracy of the transmitted information would be affected by the capabilities of parents and offspring in transmitting and learning the cultural variants. This may cause the emergence of different forms of cultural variants in the population. Newson et al. (2007) categorised this type of change in cultural variants as the mutation force.

The extent of social interactions between the individuals also plays a pivotal role in cultural changes. Some individuals have more social interactions which could potentially boost their ability to learn and share their information with other members of society. Hence, the isolated individuals with limited social activities would have less chance to learn/transmit the information from/to others. Newson et al. (2007) categorised the changes in cultural variants due to the extent of social interactions of individuals as the drift force.

Migration is another factor that can influence cultural changes. Each society has its own characteristics cultural variants. The migrant individuals can transmit their cultural variants to the new society and potentially affect the current cultural variants in the host society.

Table 2.1 compares the driving forces in genetic and cultural evolutionary processes (Newson et al. 2007).

2.6 Social Learning Biases

As was discussed in previous sections, the information transmission mechanisms in cultural and genetic evolutionary processes are different. The information flow in genetic evolutionary processes are vertical, which means that the information is transmitted from parents to offspring through genes. However, this mechanism in cultural evolution is much more complicated, in which the information flow is biased and includes both vertical and horizontal transmissions. In cultural evolution, the offspring receive and learn information not only from their parents but also from other individuals through their social interactions. Providing a general model to describe the information transmission process between the individuals in the population is challenging, as the overall behaviours of individuals in different circumstances depend on a variety of uncertain personal and environmental parameters. In this section, the learning biases that can be exhibited by individuals in their social interactions are briefly discussed.

Table. 2.1 Driving forces in genetic and cultural evolutionary processes [based on Newson et al. (2007)]

Influencing factors	Genetic evolution	Cultural evolution
Natural selection	The biological traits of individuals with better adaptability capabilities to the environment have more chance to appear in the next generations	The parents with fewer offspring would have less chance to spread and transmit their information to the next generations
Mutation	The transmission process of genes between the parents and offspring is affected by errors and damages which can potentially lead to genetic diversity	The different capabilities of individuals in teaching and social learning can lead to changes in cultural variants of a society
Drift	As the population sizes are not infinite, the parental genes are not always transmitted to offspring generation based on a uniform probability pattern. This uncertainty in parental genes distribution is not related to their effects on the offspring. Drift can potentially lead to the extinction of some gene variants and the reduction of genetic variation	The transmission of cultural information is not a uniform deterministic process. The extent of social interactions of individuals probabilistically affects the frequency of appearance of cultural variants in the next generations
Migration	Genes are migrated from one population into another. Migration can affect the gene frequencies and reduce the effect of natural selection and drift	The migration process of individuals brings new cultural information to the population which could affect the prevalence of older cultural variants

The adoption criteria of a given cultural variant are entirely different for individuals with different social backgrounds. Here, some examples are provided to show the complexity of the cultural adoption behaviours. For example, most of the individuals in the population might like to follow the most popular cultural variants in society, while a small portion of the population may be willing to adopt distinct cultural variants. In some circumstances, the decisions that individuals make in their daily life may be affected by the decisions previously made by other members of society. As an example, some conservative individuals tend to minimise the risk of their decisions by evaluating the results already obtained by others in adopting any specific cultural variant. On the other hand, some individuals are capable enough to independently make their own decisions in adopting the cultural variants. These are only a few examples of complexities in the cultural adoption behaviours of individuals in human society.

To explain the cultural variant adoption behaviour of individuals, Boyd and Richerson (1985) propose three "learning biases" as follows:

- **Content-based bias**: In this case, the individuals adopt their preferred variants from the available set of cultural variants based on their own perceptions and decisions. They try to adopt the most appropriate variant based on their own experience to achieve the most promising results and improve their life quality.
- **Model-based bias:** In this case, the individual follows the decisions made by successful or most "prestigious" individuals in the population, without evaluating the outcomes and consequences of their decisions. This means that the overall characteristic features of "prestigious" individuals, such as their confidence, happiness, and success, convince the individual to simply imitate their decisions.
- **Frequency-based bias**: As its name suggests, the individuals adopt the most popular variants in the population. They evaluate the consequences of similar decisions already taken by other individuals. This evaluation is performed to see whether other individuals are happy with their decisions. If these decisions have already resulted in remarkable achievements for other members, the individuals simply start to imitate and make similar decisions in adopting the cultural variants. In some ways, the frequency-based bias can be viewed as a general conservative behaviour of individuals in society, which minimises the risk of potential troubles. However, it confines the individuals to the popular variants and does not allow them to adopt distinct cultural concepts.

In addition to the abovementioned learning biases, Newson et al. (2007) discuss that communicator biases may also happen among individuals. The interested readers are referred to Newson et al. (2007) for details about communicator biases.

2.7 Who Can Change Cultural Variants?

As was explained in previous sections, the cultural variants change over time. They may become inefficient, inappropriate, impracticable, and redundant over time due to different changes in the physical or cultural environment (Newson et al. 2007). In some circumstances, the individuals have the power to occasionally modify current cultural variants or even replace them with new ones which are more suitable and efficient. Of course, not all individuals in society are capable of changing these variants directly. When available cultural variants for a given generation become inefficient, (some) elite individuals accept the potential risk and consequences of modification or invention of cultural variants. The invention of a new cultural variant is a difficult task, as its legitimacy may be questioned or challenged by society. Newson et al. (2007) use the term "innovation" for the emergence of new cultural variants in society by individuals. They also argue that the changes in cultural variants may be affected by the decision-making processes of a group of individuals within a network.

2.8 Summary

CAs have been developed based on the biocultural evolution theory, in which genes and culture are two interacting forms of inheritance. The main goal of this chapter was to provide the basic concepts in cultural evolutionary theory. Although providing a general definition of culture is difficult and controversial, the chapter started with a very brief survey of the definitions and interpretations provided by different researchers. Kroeber and Kluckhohn (1952) provided a comprehensive definition for the culture, which states that *"culture consists of patterns, explicit and implicit, of and for behaviour acquired and transmitted by symbols, constituting the distinctive achievements of human groups, including their embodiments in artifacts; the essential core of culture consists of traditional (i.e., historically derived and selected) ideas and especially their attached values; culture systems may, on the one hand, be considered as product of action, on the other as conditioning elements of further action"*.

Despite difficulties in providing a general definition for culture, the chapter discussed characteristics features of culture from a social science perspective, including its transmissibility, shared meanings, adaptivity, symbolic nature, and performability. Social learning is not a unique behaviour to humans. The researches have been showed that the animals are also capable of transferring their culture through social learning or imitation. However, the superiority of humans in social learning is remarkable in comparison to other species. Moreover, the ability to teach, modify, and transmission of cultural variants to the next generations is unique to humans.

The cultural variants change over time. Over different generations, the cultural variants, such as skills, beliefs, habits, preferences, traditions, and technology, change in response to different environmental conditions. Some cultural variants are modified to improve the decisions of the individuals, while others may be entirely discarded and replaced by new ones. To investigate how culture evolves over generations, a kinetic cultural evolution theory developed by Richerson and Boyd (2008) was briefly reviewed. In genetic evolution, the genetic information embedded within the genes is vertically transferred from parents to offspring. However, the information flow in cultural evolution includes both vertical and horizontal transmissions. This means that the cultural information is not only vertically inherited from parents to offspring, but also offspring have the opportunity to socially learn and acquire information from other members of the society. Similar to genetic evolution, it was discussed that the forces affecting the cultural evolutionary process in human societies can be categorised into natural selection, mutation, drift, and migration forces.

The complexity of the information transmission process in cultural evolution makes it difficult to understand and model the behaviours and decisions of individuals in society. The overall behaviours of individuals in different circumstances depend on a variety of uncertain personal and environmental parameters. The adoption models of cultural variants, including content-based, model-based, and frequency-based biases, were presented and the learning behaviours of individuals were briefly discussed. The content-based bias explains how individuals adopt their preferred variants from the available set of cultural variants based on their own perceptions and decisions. On the contrary, the model-based and frequency-based biases describe how the decisions of individuals can be affected by the social status and experiences of other members of society. Moreover, it was discussed that the elite individuals in the society can occasionally modify the inefficient cultural variants or even invent new ones with better cultural performance. As was mentioned earlier, the main goal of the chapter was to provide brief information about basic concepts in cultural evolution. Hence, the contents of this chapter would not be enough to gain sound knowledge in this area. Hence, the interested readers are strongly encouraged to read the references cited in this chapter.

References

Altman A, Mesoudi A (2019) Understanding agriculture within the frameworks of cumulative cultural evolution, gene-culture co-evolution, and cultural niche construction. Hum Ecol 47:483–497. https://doi.org/10.1007/s10745-019-00090-y

Armelagos GJ, Brown PJ, Turner B (2005) Evolutionary, historical and political economic perspectives on health and disease. Soc Sci Med 61:755–765. https://doi.org/10.1016/j.socscimed.2004.08.066

Baldwin JR, Faulkner SL, Hecht ML, Lindsley SL (2006) Redefining culture: perspectives across the disciplines. Routledge

Biraben J-N (2003) The rising numbers of humankind. Population et societies:1–4

Bocquet-Appel J-P (2011) When the world's population took off: the springboard of the neolithic demographic transition. Science (1979) 333:560–561

Boyd R, Richerson PJ (1985) Culture and the evolutionary process. Press, Chicago

Boyd R, Richerson PJ (1988) Culture and the evolutionary process. University of Chicago press

Cox MP, Morales DA, Woerner AE et al (2009) Autosomal resequence data reveal late stone age signals of population expansion in sub-Saharan African foraging and farming populations. PLoS ONE 4:e6366

Geertz C et al (1973) The interpretation of cultures. Basic books

Gowdy J, Krall L (2014) Agriculture as a major evolutionary transition to human ultra sociality. J Bioecon 16:179–202. https://doi.org/10.1007/s10818-013-9156-6

Holmes RM (2020a) Cultural psychology: exploring culture and mind in diverse communities. Oxford University Press

Holmes RM (2020b) Cultural evolution and cultural ecology. In: Cultural psychology. Oxford University Press, pp 70–114

Kroeber AL, Kluckhohn C (1952) Culture: a critical review of concepts and definitions. In: Papers Peabody Museum of archaeology & ethnology. Harvard University

Lewens T (2015) What is cultural evolutionary theory? In: Cultural evolution. Oxford University Press, pp 7–24

Lyman RL (2008) Culture, concept and definitions. In: Encyclopedia of archaeology. Elsevier, pp 1070–1075

Muehlenbein MP (2015) Basics in human evolution. Academic Press

Nettle D (1998) Explaining global patterns of language diversity. J Anthropol Archaeol 17:354–374. https://doi.org/10.1006/jaar.1998.0328

Newson L, Richerson PJ, Boyd R (2007) Cultural evolution and the shaping of cultural diversity. Handbook of cultural psychology:454–476

Pinker S (1995) The language instinct: the new science of language and mind. Penguin UK
Richerson PJ, Boyd R (2008) Not by genes alone: how culture transformed human evolution. University of Chicago press
Roper TJ (1986) Cultural evolution of feeding behaviour in animals. Sci Prog (1933-), pp 571–583
Tomasello M (2009) The cultural origins of human cognition. Harvard University Press
Toth N, Schick K (2015) Evolution of tool use. In: Basics in human evolution. Elsevier, pp 193–208
Tylor EB (1871) Primitive culture-researches into the development of mythology, philosophy, religion, language, art and custom, London: Murray JJ
Whiten A (2019) Cultural evolution in animals. Annu Rev Ecol Evol Syst 50:27–48. https://doi.org/10.1146/annurev-ecolsys-110218-025040
Whiten A, Custance D (1996) Studies of imitation in chimpanzees and children

Chapter 3
Cultural Algorithms (CAs)

Abstract CAs are EAs modelled based on the biocultural evolutionary theory in real societies. The idea of using multiple sources of knowledge during the search process has emerged with the development of CAs. They model a dual inheritance system observed from the human cultural evolution, in which the belief space represents the macro-evolutionary level and the population space performs the micro-evolutionary level. The belief space includes a network of knowledge components obtained from the evolution process of individuals in the population space, in which each knowledge category represents a collection of problem-specific information. This chapter presents the basic definitions, framework, operators, and formulations of CAs.

Keywords Cultural algorithms · Belief space · Knowledge · Communication protocols

3.1 Introduction

As was discussed in Sect. 1.3.4 of Chap. 1, EAs are meta-heuristic algorithms developed based on the evolutionary process of biological species in nature. EAs model the candidate solutions for the problem as a set of individuals that are evolved based on Darwin's evolutionary theory over different generations. Each generation of individuals produces offspring through a set of operators inspired by the natural selection principle, such as mutation, cross-over, and selection. In biological evolutionary theory, natural selection describes the survival and reproduction of individuals due to the differences in phenotype. According to this theory, the individuals with biological traits better suited to the environment will have more chances to survive and produce more offspring for the next generations. These traits are heritable and passes from parents to offspring.

CAs belong to the category of EAs. However, due to their characteristic features, CAs differ from conventional EAs. The conventional EAs work on the population level in which the changes in individuals happen based on biological evolutionary operators. In Chap. 2, it was argued that the learning process in biological evolution is vertical, which means that the genetic information embedded within the genes is

S. Jalili, *Cultural Algorithms*, Engineering Optimization: Methods and Applications,
https://doi.org/10.1007/978-981-19-4633-2_3

transferred from parents to offspring. Conventional EAs model this vertical information transmission process in their frameworks. These algorithms usually construct new individual solutions for the problem by modelling the inheritance process of favourable traits between the parents and offspring.

Based on the biocultural evolution theory discussed in Chap. 2, human evolution is a product of two interacting evolutionary processes, genetic evolution and cultural evolution. In the eyes of biocultural evolution theory, genes and culture are two interacting forms of inheritance. In addition to the improvements in biological traits, humans culturally evolve through social learning and cultural information transmission over different generations. They can transmit their technology, skills, beliefs, traditions, and other cultural variants to the next generations. All these cultural variants play a pivotal role in human adaptation to environmental conditions and improve their life quality. Cultural variants change or even replaced by new ones over time. Cultural evolution theory describes how these changes happen and identifies forces that can affect the existence or emergence of cultural variants.

Contrary to the genetic evolution in which the genetic information embedded within the genes is being vertically transmitted from parents to offspring, the cultural variants in the cultural evolutionary process are transmitted based on both vertical and horizontal transmission processes. In a real human society, offspring not only learn the cultural variants from their parents but also from other individuals through social interactions. As an evolutionary meta-heuristic algorithm, CAs employ the basic concepts in the biocultural evolution theory to develop a numerical optimisation approach, in which both types of vertical and horizontal learning behaviours of individuals are modelled. The main aim of this chapter is to describe the basic framework, concepts, operators, and algorithmic details of CAs.

The contents of this chapter are organised as follows. Section 3.2 introduces the overall framework of the CAs. Sections 3.3 and 3.4 discuss the population and belief spaces in CAs. The communication protocols between the population and belief spaces are presented in Sect. 3.5. Section 3.6 explains how the different knowledge components of belief space are updated by the accepted individuals in each generation. The main algorithmic steps of CAs are presented in Sect. 3.7. In Sect. 3.8, an illustrative example is selected to show how the different components of CAs work. In the end, Sect. 3.9 provides a summary of the contents discussed in this chapter.

3.2 Overall Framework

The overall framework of conventional EAs consists of a single population space. The information transmission follows a vertical pattern from the parents to the offspring. The offspring solutions are constructed through the modification of some variables of parent individuals. This information flow mechanism may sometimes offer limited benefit for the whole population, as it ignores the interaction between the individuals. CAs are different from this perspective. They try to model both types of information transmission processes observed in human evolution. Based on biocultural evolution theory, the overall human evolution is a product of two evolutionary processes, i.e., genetic evolution and cultural evolution.

CAs try to take the advantage of biocultural evolution theory to construct a new meta-heuristic algorithm in which both vertical and horizontal information flow mechanisms are available for the individuals within the population space. To this end, the main framework of CAs consists of two spaces as shown in Fig. 3.1, the belief and population spaces. Like other EAs, population space contains a set of individual vectors representing potential solutions for the problem. The belief space includes a network of cultural variants or knowledge that are acquired by individuals over different generations. These knowledge components can be used by individuals within the population space to simulate the model-based and frequency-based social learning biases as discussed in Sect. 2.6 in Chap. 2.

These population and belief spaces exchange information about the problem at hand through a set of communication protocols, including the acceptance and influence operators. The acceptance operator determines which individuals should participate in the modification process of cultural variants within the belief space. The cultural variants/knowledge within the belief space are adjusted based on the experience of individuals. On the other hand, the influence operator simulates the cultural evolution in the population space. The variations happen within the population space using influence functions. The influence functions determine to what extent the decisions and behaviours of individuals in the population space can be affected

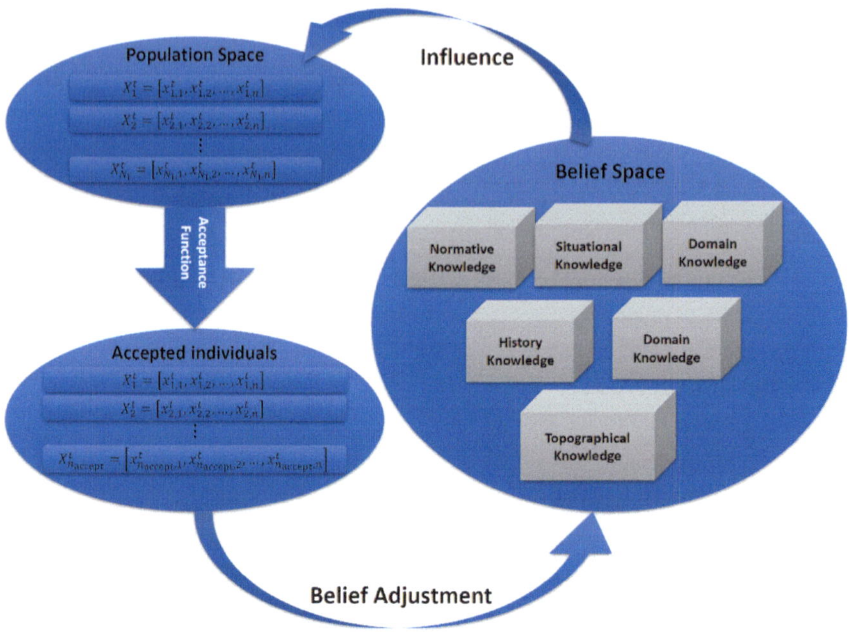

Fig. 3.1 Overall framework of CAs

by cultural variants. These functions produce new offspring solutions by modelling the vertical and horizontal information transmissions between the individuals within the population space.

Let us $\mathcal{P}^t$ be the population space and $\mathcal{B}^t$ be the belief space at a given time t. The overall algorithmic process of CAs can be explained as follows. Similar to other meta-heuristics, CAs start with a population space $\mathcal{P}^0$ consisting of random individuals initialised within the search space. The belief space $\mathcal{B}^0$ is also initialised by assuming a set of appropriate values. Then, the fitness values of the individuals are calculated by using the fitness function represented by Obj(). By using the acceptance operator indicated by Accept(), a set of individuals from population space are selected to adjust the cultural variants in the belief space using the update function denoted by Update(). The offspring solutions are generated by influence functions represented by Influence() based on a biocultural logic. According to this brief explanation, the general pseudo-code of CAs can be summarised as Algorithm 3.1. The detailed descriptions for different components of CAs are presented in subsequent sections.

```
Algorithm 3.1. General pseudo-code of CAs
Set t = 0;
Initialise the population space P^0 with a set of random individuals within the search space;
Initialise the belief space B^0 with suitable initial assumptions for cultural variants;
while stopping criteria are not met do
    Set t = t + 1;
    Calculate the fitness values of individuals in P^t using Obj();
    Select a set of individuals from P^t using Accept();
    Employ the selected individuals to adjust the cultural variants and obtain B^{t+1} using the
    Update();
    Generate the offspring solutions and form the new population space P^{t+1} based on
    Influence();
End
```

3.3 Population Space

The population space in CAs is similar to those in other EAs. The population space $\mathcal{P}^t$ of CAs is consisted of a set of N_I individuals indicated by vectors $X_1^t, X_2^t, \ldots, X_{N_I}^t$. Each individual is an n-dimensional vector represented by $X_i = \left[x_{i,1}^t, x_{i,2}^t, \ldots, x_{i,n}^t\right]$ where $x_{i,j}$, $j = 1, 2, .., n$ denotes the jth variable of the ith individual and n is the dimension of the problem at hand.

3.4 Belief Space

As was discussed in Chap. 2, a set of cultural variants, such as beliefs, habits, preferences, traditions, and skills, are transmitted during the cultural evolutionary process. CAs model this cultural information within the belief space. The belief space contains a network of cultural variants or knowledge. These cultural knowledge components include the experiences and collective behaviour acquired by individuals over different generations. The belief space facilitates the transmission of cultural information from one generation to another one. It can be viewed as an information repository that can potentially be used to affect and improve the behaviours and decisions of the individuals within the population space.

The belief space of CAs can be designed to store a variety of information about the search space as well as the behaviours of the individuals. The primary version of CAs was developed based on a single cultural knowledge, called *situational knowledge* (Reynolds 1979). Situational knowledge stores information about the behaviours of the best individuals over different generations. Further research on CAs led to the design of additional knowledge components, including normative, domain, history, and topographical knowledge components. These knowledge components collect different types of information about the behaviours of individuals and the characteristics of search space that can be potentially used by individuals to make the best possible decisions. The types of knowledge components employed within CAs depend on the type of the problem. The recent literature survey reveals that the different combinations of knowledge components have been employed within CAs to deal with different problems as summarised in Table 3.1 (Maheri et al. 2021). As it can be seen from Table 3.1, the normative and situational knowledge components are two mostly used knowledge types within CAs. However, in the following subsections, all types of knowledge components will be described.

The belief space in CAs can be formulated as a quintuple as follows:

$$\mathcal{B}^t = \left(\mathcal{S}^t, \mathcal{N}^t, \mathcal{C}^t, \mathcal{D}^t, \mathcal{H}^t\right) \tag{3.1}$$

where $\mathcal{B}^t$ indicates the belief space at time t; $\mathcal{S}^t$, $\mathcal{N}^t$, $\mathcal{C}^t$, $\mathcal{D}^t$, and $\mathcal{H}^t$ are the situational, normative, topographical, domain, and history knowledge components at time t. Subsequent sections will provide detailed explanations for different knowledge components.

3.4.1 *Situational Knowledge*

The *situational knowledge* represents the individual with the best fitness value within the population space at a given generation t. Let us represent the situational knowledge component at generation t by vector $\mathcal{S}^t = \left[\mathcal{s}_1, \mathcal{s}_2, \ldots, \mathcal{s}_n\right]$, in which $\mathcal{s}_j, j = 1, 2, \ldots, n$ represents the jth variable of situational vector at iteration t. In

Table 3.1 Types of knowledge components employed within CAs in literature

References	Situational	Normative	Domain	History	Topographical
Haldar and Chakraborty (2014)	✓			✓	
Haldar and Chakraborty (2015)	✓			✓	
Arpaia et al. (2007)	✓	✓			
Jalili and Hosseinzadeh (2015)	✓	✓			
Jalili et al. (2019)	✓	✓			
Jafari et al. (2019)	✓	✓	✓		
Ali et al. (2018)	✓	✓			
Yan et al. (2017a)	✓	✓			
Liu and Lin (2015)	✓	✓			
Guedria and Hassine (2019)	✓	✓			
Ali and Reynolds (2014)	✓	✓	✓	✓	✓
Chen et al. (2008)		✓			
Wang et al. (2015)	✓	✓			
Haikal and El-Hosseni (2011)	✓			✓	
Xu et al. (2010)	✓	✓			
Xu et al. (2012)	✓	✓			
Tremayne et al. (2009)	✓	✓		✓	✓
Gu and Wu (2010)	✓				
Gao et al. (2020)	✓	✓			
Chen et al. (2013)	✓				
Ya-Li and Li-qing (2010)	✓	✓			
Goudarzi et al. (2016, 2017)	✓	✓			
Bhattacharya et al. (2012a)		✓			
Ali et al. (2016b)	✓	✓		✓	✓
Wu et al. (2010)	✓	✓			
Stanley et al. (2019)	✓	✓	✓	✓	✓
Awad et al. (2013)	✓	✓	✓	✓	✓
Abdolrazzagh-Nezhad et al. (2020)	✓	✓			
Lin et al. (2009b)		✓			
Oloruntoba et al. (2019)	✓	✓			
Sun et al. (2009)	✓	✓			
Mohammadhosseini et al. (2019)	✓	✓			
Coello and Becerra (2004)	✓	✓			
Chung and Reynolds (1997)	✓	✓			
Coelho and Alotto (2009)	✓	✓			

(continued)

Table 3.1 (continued)

References	Situational	Normative	Domain	History	Topographical
Oliveira De Freitas et al. (2018)	✓	✓			✓
Khan et al. (2014)	✓	✓			
Khan et al. (2017)	✓	✓			
Guo et al. (2013b)	✓	✓			
Guo et al. (2013a)	✓	✓			
Gao and Diao (2011)	✓	✓			
Gao and Xu (2013)		✓			
Digalakis and Margaritis (2002)	✓	✓			
Gao et al. (2019)	✓	✓			
Khodabakhshian and Hemmati (2013)	✓	✓			
Bhattacharya et al. (2012b)	✓	✓			
Lu et al. (2011)	✓			✓	
Yuan et al. (2008, 2009)	✓	✓			
Cai et al. (2018)	✓	✓			✓
Coelho et al. (2006)		✓			
Xie et al. (2009)	✓	✓			
Ali et al. (2014a)	✓	✓	✓		✓
Zhou et al. (2019)	✓	✓			
Yan et al. (2017b)		✓			
Wang et al. (2017)	✓	✓	✓		
Pan et al. (2010)		✓			
Chung and Reynolds (1998)	✓	✓			
Becerra and Coello (2004)	✓	✓		✓	✓
Pierezan et al. (2019)		✓			
Shah and Kobti (2020)	✓	✓			
Mao et al. (2020)	✓	✓			✓
Goli et al. (2019)					
Mojab et al. (2019)	✓	✓	✓		
Murugadass and Sivakumar (2020)	✓	✓			
Xue (2020)	✓	✓			
Muhamediyeva (2020)	✓	✓			
Amalo et al. (2020)		✓			
Selvarajah et al. (2020)		✓			✓
Xue et al. (2011)	✓	✓			

(continued)

Table 3.1 (continued)

References	Situational	Normative	Domain	History	Topographical
Guo et al. (2011)		✓			✓
Gao et al. (2006)		✓			
Gao and Diao (2010)	✓	✓			
Gao et al. (2007)	✓				
Nguyen and Yao (2006)	✓	✓			
Ali et al. (2011a)	✓	✓	✓	✓	✓
Guo et al. (2018)	✓	✓			✓
Kim and Cho (2012)	✓	✓		✓	✓
Guo et al. (2012)	✓	✓			
Reynolds and Zhu (2001)	✓	✓			
He and Xu (2011)	✓	✓			
Sun et al. (2010)	✓	✓			
Awad et al. (2017)	✓	✓	✓		✓
Ali et al. (2016a)	✓	✓			✓
Ali and Awad (2014)	✓	✓	✓	✓	✓
Ali et al. (2012)	✓	✓	✓	✓	✓
Cortés Rivera et al. (2007)	✓				
Zhang and Zhu (2012)	✓	✓			
Yang and Gu (2014)	✓	✓			✓
Becerra and Coello (2005)	✓				
Alami et al. (2007)	✓				
Lagos et al. (2014)	✓	✓		✓	
Cabrera et al. (2011)	✓				
Srinivasan and Ramakrishnan (2012)	✓	✓	✓	✓	✓
Crawford et al. (2013)	✓			✓	
Crawford et al. (2007)	✓	✓			
Soza et al. (2011)	✓	✓	✓		
Soza et al. (2007)	✓	✓	✓		
Zadeh and Kobti (2015)	✓	✓			
Chen and Yang (2015)	✓	✓	✓	✓	✓
Terán et al. (2017)	✓	✓			
Coelho and Alotto (2009)		✓			
Xue and Guo (2007)	✓	✓		✓	✓
Ali et al. (2014b)	✓	✓	✓	✓	✓
Ali et al. (2011b)	✓	✓	✓	✓	✓

(continued)

Table 3.1 (continued)

References	Situational	Normative	Domain	History	Topographical
Best et al. (2010)	✓	✓	✓	✓	✓
Reynolds and Ali (2008)	✓	✓	✓	✓	✓
Reynolds et al. (2008)	✓	✓	✓	✓	✓
Reynolds and Peng (2005)	✓	✓	✓	✓	✓

some cases, more than one situational vector are stored by selecting a set of fittest individuals at each generation. In this case, the situational knowledge will include a set of vectors which can be represented by $\mathcal{S}_1^t, \mathcal{S}_2^t, \ldots, \mathcal{S}_{n_s}^t$ in which n_s is the number of stored situational vectors. Without loss of generality, it is assumed in this chapter that only one situational vector is stored within the situational knowledge component.

3.4.2 Normative Knowledge

The idea of the *normative knowledge* component was introduced by Chung and Reynolds (1998). This provides a set of behavioural standards for individuals in the population space to guide and improve their decisions. This provides the opportunity to socially learn and adjust the behaviours of individuals to achieve more favourable results. This is a form of horizontal information transmission mechanism in which the individuals can socially learn and acquire information about popular and more productive behaviours experienced by successful individuals in the population. Individuals need to follow these behavioural standards to improve their fitness values.

Mathematically speaking, the normative knowledge component provides a set of intervals for each dimension of the problem at hand. These intervals are obtained based on the experience acquired by elite individuals over previous generations. This knowledge component can be a source of information to guide the variations in the population and help the individuals to produce more fittest offspring.

The normative knowledge component includes a set of information for each dimension of the problem which can be expressed by the following list of elements:

$$\mathcal{N}^t = \left(\mathscr{n}_1^t, \mathscr{n}_2^t, \ldots, \mathscr{n}_n^t\right) \tag{3.2}$$

in which:

$$\mathscr{n}_j^t = \left(I_j^t, L_j^t, U_j^t\right), j = 1, 2, \ldots, n \tag{3.3}$$

In Eqs. (3.2) and (3.3), $\mathcal{N}^t$ indicates the normative knowledge component at time t, $\mathscr{n}_j^t$ represents the normative knowledge for the jth dimension at time t,

$I_j^t = \left[x_{\min,j}^t, x_{\max,j}^t\right]$ is the behavioural interval for the jth dimension at time t, $x_{\min,j}^t$ and $x_{\max,j}^t$ are the lower and upper behavioural bounds in the jth dimension at time t, respectively, L_j^t and U_j^t are the lower and upper fitness bounds in the jth dimension at time t, respectively, and n is the number of dimensions of the problem at hand.

The behavioural intervals provided by the normative knowledge component somehow represent the productive regions of the search space. They provide a guiding mechanism that determines the productive parts of search space. The individuals who are outside these behavioural intervals should adjust their behaviours to satisfy the behavioural norms of the population and improve the outcomes of their decisions. In the influence functions, it will be discussed how the behaviours of individuals are affected by these normative intervals.

3.4.3 History Knowledge

In some of the optimisation problems, the search landscapes change over time. The changes in the search space can lead to significant changes in global optimum value and its geometrical location within the search space. To keep the track of these changes, the *history (or temporal) knowledge* component was introduced by Saleem that stores the history of shifts in the search space. This knowledge component stores the best solution obtained before every shift, the distance change that happened in each dimension as well as the direction of changes in the search space. The history knowledge component can be written as a list as follows (Ali et al. 2016b):

$$\mathcal{H}^t = \hbar_1, \hbar_2, \ldots, \hbar_{K^t} \tag{3.4}$$

in which:

$$\hbar_k = (E_k, O_k, D_k), k = 1, 2, \ldots, K^t \tag{3.5}$$

In Eqs. (3.4) and (3.5), $\mathcal{H}^t$ represents the history knowledge component at time t, $\hbar_k$ indicates the history knowledge for the kth shift in the search space, K^t indicates the number of stored shifts happened in the search space until iteration t, $E_k = \left[e_{k,1}, e_{k,2}, \ldots, e_{k,n}\right]$ represents the latest best solution obtained before the kth shift happened, $O_k = \left[o_{k,1}, o_{k,2}, \ldots, o_{k,n}\right]$ indicates the directional change in the search region at the kth shift, and $D_k = \left[d_{k,1}, d_{k,2}, \ldots, d_{k,n}\right]$ is the distance change in each dimension as a result of the shift in the search space.

3.4.4 Topographical Knowledge

The *topographical knowledge* component introduced by Jin and Reynolds (1999a, b) is a region-based schema that stores a set of information about the different regions of the search space. This knowledge component provides a hierarchical architecture that includes a set of "belief cells" in each dimension of the problem. In each belief cell, a set of information is stored, such as the fitness, feasibility/infeasibility, and frequency of cell occupation. It can be viewed as a memory in which the experience and information about the search space acquired by individuals are stored and transmitted to the next generations. The size and the number of belief cells depend on the problem type and feature of the problem at hand. During the solution process, the algorithm may increase the number of cells in a given part of the search space due to changes in the function surface. On the other hand, the fewer number of cells may be enough for the regions of the search space in which the function surface has a uniform pattern.

The topographical knowledge component considers the search space as a multi-dimensional grid or array represented by $C^t = c_1^t, c_2^t, \ldots, c_k^t$, where each element c_i^t represents a belief cell that stores a set of information and k is the number of the belief cells in the topographical knowledge component. The implementation of topographical knowledge depends on the problem type, and it can be extended to store a variety of information about the search space. For example, Jin and Reynolds (1999a, b) designed the following topographical knowledge component:

$$c_i^t = (\text{Class}_i, \text{Cnt1}_i, \text{Cnt2}_i, W_i, \text{Pos}_i, \text{Csize}_i), i = 1, 2, \ldots, n_c \tag{3.6}$$

where c_i^t represents the i th belief cell in the search space, Class_i indicates the feasibility, infeasibility, or semi-feasibility of the i th cell which can be unknown at the beginning of search process due to the absence of information, Cnt1_i and Cnt2_i represent the number of feasible and unfeasible individuals within the i th belief cell, W_i indicates the weight of the ith belief cell that is proportional to the fitness values of individuals within the cell, Pos_i is a n-dimensional vector representing the position of the cell in the search space, and Csize_i is another n-dimensional vector that indicates the size of the ith cell in each dimeson of search space.

Topographical knowledge can play a pivotal role in guiding the individuals within the population. Based on the regional information, individuals can be guided towards less visited cells to gain new knowledge for the population. When a given cell provides new useful information, the cell can be further exploited by diving it into several sub-cells which may potentially lead to additional useful information. On the other hand, this provides the required information for individuals to abandon less productive cells and avoid unnecessary computational efforts. Hence, the proper application of topographical knowledge can reduce the computational efforts, enhance the exploration ability of the algorithm, and enhance the quality of final solutions.

Despite the abovementioned advantages of topographical knowledge, an unnecessary amount of belief cells can lead to high computational expenses for the algorithm.

The update process of the belief cells can be computationally expensive, particularly when a large number of information types are stored within each cell. The degree of nonlinearity of the objective and constraints functions may also challenge the application of topographical knowledge. The challenge stems from the fact that the sudden changes on objective function surface or feasibility conditions will push the algorithm to create excessively additional sub-belief cells which in turn will lead to an exponential increase in computational effort required for the update process. To deal with this computational challenge, some studies focused on the application of the *k-d* ternary tree-based approach for the topographical knowledge component (bin Peng and Reynolds 2004; Ali et al. 2016c, b).

3.4.5 Domain Knowledge

The *domain knowledge* component developed by Reynolds and Saleem (2005) is similar to the situational knowledge component. It stores the individuals with the best fitness function values over different generations to form an archive that can play a role of memory for the evolution process. The domain knowledge can be expressed as a list with situational vectors obtained over different generations as follows:

$$\mathcal{D}^t = \left(\mathcal{S}^1, \mathcal{S}^2, \ldots, \mathcal{S}^t\right) \tag{3.7}$$

where $\mathcal{D}^t$ represents the domain knowledge at time t, and $\mathcal{S}^t$ is the situational knowledge vector obtained at time t.

3.5 Communication Protocols

In CAs, the population and belief spaces constantly exchange information over the different generations to model the biocultural evolutionary process. As it can be seen from Fig. 3.1, the interaction between the belief and population spaces is provided by the communication protocols, including acceptance and influence operators. The acceptance operator determines which individuals from the population space should be selected to form or modify the cultural variants stored within the belief space. The influence operator transmits the cultural information to the population space and generates new offspring by modelling vertical and horizontal learning processes. In the following subsection, these operators are explained in detail.

3.5.1 *Acceptance Function*

As was discussed in Sect. 2.7 of Chap. 2, the cultural variants in a real human society may become inefficient, inappropriate, impracticable, and redundant over time due to different changes in the physical or cultural environment. Some individuals in the population can occasionally modify the cultural variants or even replace them with new suitable and efficient ones. It was also discussed that not all individuals are capable of modifying cultural variants. This process is usually performed by some elite individuals in the society whose characteristic features distinguish them from other individuals.

Similar to cultural evolution in real societies, CAs update the cultural knowledge components of the belief space by using a set of elite individuals with better fitness function values in the population space. To this end, CAs employ the acceptance function that determines which individuals should be selected to update the belief space. The Accept() in Algorithm 3.1 represents this acceptance function in CAs. There are a variety of acceptance functions in the literature that have been developed for CAs. Some of them simply choose a certain number of fittest individuals to modify the belief space at each generation. While other approaches define the number of accepted individuals as a function of time. The former approaches are called static acceptance functions, while the latter approaches are referred to as dynamic acceptance methods (Engelbrecht 2007). In this book, the number of accepted individuals to update the belief space is represented by n_{accept}. Literature survey reveals that fuzzy acceptance functions have been proposed by researchers that can be categorised into dynamic methods (Ali et al. 2018). The fuzzy acceptance functions will be discussed in Sect. 7.7.1 of Chap. 7.

3.5.2 *Influence Function*

As was discussed in Chap. 2, the information transmission process in cultural evolution is much more complicated than that in genetic evolution. The information flow in genetic evolution follows a vertical pattern in which the biological information is transmitted directly from parents to offspring through the genes. On the other hand, the information flow in cultural evolution is biased, and both types of vertical and horizontal transmissions can happen between the individuals in the population space. In a real human society, individuals not only learn and acquire information from their parents but also can socially learn from other non-kin individuals through different forms of social interactions.

Inspired by the social learning concept in cultural evolution, CAs model both the vertical and horizontal learning mechanisms during the evolution of individuals within the population space. In each iteration, the offspring are generated by considering both genetic and cultural information transmission processes. The individuals in CAs not only inherit the genetic information from their parent solutions but also take the advantage of accumulated cultural information within the belief space acquired by previous generations. In this section, it is shown how individuals in CAs learn both genetic and cultural information.

The role of influence functions is to produce a generation of offspring from the current individuals. These functions use the knowledge components stored within the belief space as well as the genetic information of individual parents. Depending on the problem's nature, the influence functions can be designed based on various knowledge components. Some influence functions employ a single knowledge component, while others may use multiple types of information in the belief space. The main challenge in designing influence functions with several knowledge components is the adjustment of the extent of their contributions to the learning process of individuals. Various influence functions developed in literature will be discussed in Chap. 7. However, the most prominent influence functions employed in standard CAs are presented in this section.

Reynolds et al. (1997) developed a set of influence functions based on the situational and normative knowledge components as follows:

- **Influence function based on normative knowledge component with random search direction**

This influence function modifies the step size in each dimension of the problem based on the size of the normative interval in the corresponding dimension. However, the direction of the search is purely random. This influence function is expressed as follows:

$$x_{i,j}^{t+1} = x_{i,j}^{t} + \text{size}\left(I_j^t\right) N_{i,j}(0, 1), \ \ j = 1, 2, \ldots, n \tag{3.8}$$

where $x_{i,j}^{t+1}$ represents the j th variable of the i th individual at iteration $t + 1$, $\text{size}\left(I_j^t\right) = x_{\max,j}^{t} - x_{\min,j}^{t}$ indicates the size of the normative interval for the j th variable, and $N_{i,j}(0, 1)$ denotes normal distribution function with a mean value of 0 and standard deviation of 1.

- **Influence function based on situational knowledge component with a random step size**

This influence function uses the situational knowledge vector to determine the direction of the search. In this case, the individuals try to imitate the decisions made by the most successful individual within the population. The situational knowledge component can be viewed as the experience acquired by the most successful individual so far. Hence, this influence function uses situational knowledge to choose the search direction. This is very similar to the model-based information bias explained in Sect. 2.6 of Chap. 2, if it is not the same. However, the step size is purely random in this case. This influence function can be written as:

$$x_{i,j}^{t+1} = \begin{cases} x_{i,j}^t + \left|\sigma_{i,j}^t N_{i,j}(0,1)\right|, \text{if} x_{i,j}^t < S_j \\ x_{i,j}^t - \left|\sigma_{i,j}^t N_{i,j}(0,1)\right|, \text{if} x_{i,j}^t > S_j \\ x_{i,j}^t + \sigma_{i,j}^t N_{i,j}(0,1), \text{otherwise} \end{cases}, j = 1, 2, \ldots, n \tag{3.9}$$

In Eq. (3.9), S_j represents the jth variable of the situational knowledge vector at time t, and $\sigma_{i,j}^t$ indicates the strategy parameter of the jth variable of ith individual at time t.

- **Influence function based on normative and situational knowledge components**

In this influence function, the direction of search is selected based on the situational knowledge component and the step size is determined based on the sizes of normative intervals. This influence function can be expressed as:

$$x_{i,j}^{t+1} = \begin{cases} x_{i,j}^t + \left|\text{size}\left(I_j^t\right) N_{i,j}(0,1)\right|, \text{ if} x_{i,j}^t < s_j^t \\ x_{i,j}^t - \left|\text{size}\left(I_j^t\right) N_{i,j}(0,1)\right|, \text{ if} x_{i,j}^t > s_j^t \\ x_{i,j}^t + \text{size}\left(I_j^t\right) N_{i,j}(0,1), \quad \text{otherwise} \end{cases} \tag{3.10}$$

- **Influence function based on the normative knowledge component**

The idea of this influence function is to use normative knowledge component to determine both the step size and search direction as follows:

$$x_{i,j}^{t+1} = \begin{cases} x_{i,j}^t + \left|\text{size}\left(I_j^t\right) N_{i,j}(0,1)\right|, \text{ if} x_{i,j}^t < x_{\text{min},j}^t \\ x_{i,j}^t - \left|\text{size}\left(I_j^t\right) N_{i,j}(0,1)\right|, \text{ if} x_{i,j}^t > x_{\text{max},j}^t \\ x_{i,j}^t + \beta \text{size}\left(I_j^t\right) N_{i,j}(0,1), \quad \text{otherwise} \end{cases}. \tag{3.11}$$

where β indicate a positive constant parameter.

From Eqs. (3.8), (3.9), (3.10), and (3.11), it can be seen that the new individuals are generated based on both genetic and cultural inheritance mechanisms. The first terms in the above equations represent the genetically inherited features, while the second terms indicate the culturally inherited features. The former can be explained as the vertical information transmission between the parents and offspring. On the other hand, the latter models the horizontal learning process between the offspring and other individuals in the population, in which the cultural information is indirectly transmitted through the cultural knowledge components stored in the belief space.

3.6 Cultural Adjustment

The discussions provided in Chap. 2 revealed that the cultural variants change during the cultural evolution process. To simulate the cultural changes, CAs employ a set of formulas to update knowledge components in the belief space. These formulas represent the Update() function in Algorithm 3.1. The Update() function updates the knowledge components based on the accepted individuals. The update process for some of the knowledge components is problem dependent and it is very difficult to explain them in a general way. Hence, the update functions for the two most popular knowledge components, including situational and normative components, are explained as follows:

- **Situational Knowledge**

To update the situational knowledge component, the Update() function performs a greedy process between the previous situational vector and accepted individuals as follows:

$$\mathcal{S}^{t+1} = \begin{cases} X_l^t & if\ f\left(X_l^t\right) <;\ f\left(\mathcal{S}^t\right) \\ \mathcal{S}^t & otherwise \end{cases},\ l = 1, 2, \ldots, n_{\text{accept}} \tag{3.12}$$

where $\mathcal{S}^t$ and $\mathcal{S}^{t+1}$ are thc situational knowledge vectors at time t and $t + 1$, respectively, X_l^t indicate the lth accepted individual at iteration t, and n_{accept} is the number of accepted individuals obtained by Accept() function explained in Sect. 3.5.1.

- **Normative Knowledge**

As explained in Sect. 3.4.2, the variables within the normative knowledge components include normative intervals and fitness bounds. As a result of cultural evolution, the normative knowledge component should be updated by the accepted individuals. The update process of normative variables performed by Update() function be expressed by following equations:

$$x_{\min,j}^{t+1} = \begin{cases} x_{l,j}^t & \text{if } x_{l,j}^t \leq x_{\min,j}^t \text{ or } f\left(X_l^t\right) < L_j^t \\ x_{\min,j}^t & \text{otherwise} \end{cases} \tag{3.13}$$

$$x_{\max,j}^{t+1} = \begin{cases} x_{l,j}^t & \text{if } x_{l,j}^t \geq x_{\max,j}^t \text{ or } f\left(X_l^t\right) < U_j^t \\ x_{\max,j}^t & \text{otherwise} \end{cases} \tag{3.14}$$

$$L_j^{t+1} = \begin{cases} f\left(X_l^t\right) & \text{if } x_{l,j}^t \leq x_{\min,j}^t \text{ or } f\left(X_l^t\right) < L_j^t \\ L_j^t & \text{otherwise} \end{cases} \tag{3.15}$$

$$U_j^{t+1} = \begin{cases} f\left(X_l^t\right) & \text{if } x_{l,j}^t \geq x_{\max,j}^t \text{ or } f\left(X_l^t\right) < U_j^t \\ U_j^t & \text{otherwise} \end{cases} \tag{3.16}$$

where $x_{l,j}^{t}$ represents the lth accepted individual at time t in which $l = 1, 2, \ldots, n_{\text{accept}}$, subscript j indicates the j th dimension, and $j = 1, 2, \ldots, n$. The definitions for the rest of the parameters were explained in Sect. 3.4.2.

3.7 Algorithmic Steps

In this section, the main algorithmic steps of CAs are presented based on the explanations provided in previous sections. Let us assume CAs are used for a minimisation problem with the objective function represented by $f(X = [x_1, x_2, \ldots, x_n])$. The problem consists of n decision variables. With this assumption, the main steps of CAs are described as follows.

Step 1: Initialisation

Like other meta-heuristic algorithms, CAs start the solution-finding process with N_{I} number of individuals randomly generated within the search space as follows:

$$x_{i,j}^{0} = \phi\left(l_j, u_j\right), i = 1, 2, \ldots, N_{\text{I}}, j = 1, 2, \ldots, n \tag{3.17}$$

where $x_{i,j}^{0}$ represents the j th variable of the i th individual in the population space at the initialisation phase, l_j and u_j are the lower and upper bounds for the j th dimension of the problem, respectively, and $\phi(a, b)$ is a uniform random function that generates a random number between a and b, and N_{I} is the population size. The generated random individuals will form the initial population space represented by $\mathcal{P}^0$.

The knowledge components of the belief space should also be initialised with appropriate values. The initial belief space can be written as:

$$\mathcal{B}^0 = \left(\mathcal{S}^0, \mathcal{N}^0, \mathcal{C}^0, \mathcal{D}^0, \mathcal{H}^0\right) \tag{3.18}$$

Some of the knowledge components can be initialised as empty variables. However, others may need appropriate values. As an illustrative example, the normative variables should be initialised as $x_{\min,j}^{0} = -\infty$ and $x_{\max,j}^{0} = \infty$ to consider the most possible widest interval and the normative bounds should be initialised as $L_j^0 = \infty$ and $U_j^0 = \infty$, where $j = 1, 2, \ldots, n$.

Step 2: Evaluation

The second step evaluates the fitness function of initialised individuals in the previous step.

Step 3: Adjustment of Beliefs

The main goal of this step is to update the knowledge components within the belief space based on the fitness function values calculated in the previous step. To update

Table 3.2 Internal parameters of standard CAs

Algorithm	Internal parameters
CA-I	$N_{\mathrm{I}}, n_{\mathrm{accept}}$
CA-II	$N_{\mathrm{I}}, n_{\mathrm{accept}}, \sigma_{i,j}^{t}$
CA-III	$N_{\mathrm{I}}, n_{\mathrm{accept}}$
CA-IV	$N_{\mathrm{I}}, n_{\mathrm{accept}}, \beta$

the belief space, the Eqs. (3.12), (3.13), (3.14), (3.15), and (3.16) explained in Sect. 3.6 should be employed.

Step 4: Reproduction

The fourth step is focused on the reproduction of individuals. The offspring individuals are generated by using one of the influence functions in Eqs. (3.8), (3.9), (3.10), and (3.11), as explained in Sect. 3.5.2. The offspring solutions will be used to form the new population space for the next generation indicated by $\mathcal{P}^{t+1}$.

Step 5: Termination

Steps 2–4 should be repeated until the termination criteria are satisfied.

Typically, meta-heuristic algorithms have a set of internal parameters that could potentially affect their performance. The numerical performance of CAs can also be affected by the values considered for the internal parameters. Depending on the type of the influence function, the number of internal parameters would be different. Table 3.2 lists the internal parameters for each version of CA. In this table, CA-I, CA-II, CA-III, and CA-IV refer to the CA with influence functions in Eqs. (3.8), (3.9), (3.10), and (3.11), respectively. The values of these parameters should be carefully selected to achieve the most promising results.

The MATLAB code for the CAs is presented in Appendix 1. The code assumes that the belief space consists of normative and situational knowledge components. The code also models all four influence functions available in the standard CAs.

3.8 Illustrative Example

In this section, the algorithmic steps of CAs will be illustrated using a simple minimisation problem of the Ackley function in two-dimensional search space as shown in Fig. 3.2, which can be written as follows:

Find x and y.

To minimise:

$$f(x, y) = -20 \exp\left(-0.2\sqrt{0.5\left(x^2 + y^2\right)}\right) - \exp(0.5(\cos 2\pi x + \cos 2\pi y)) + 20 + e \tag{3.19}$$

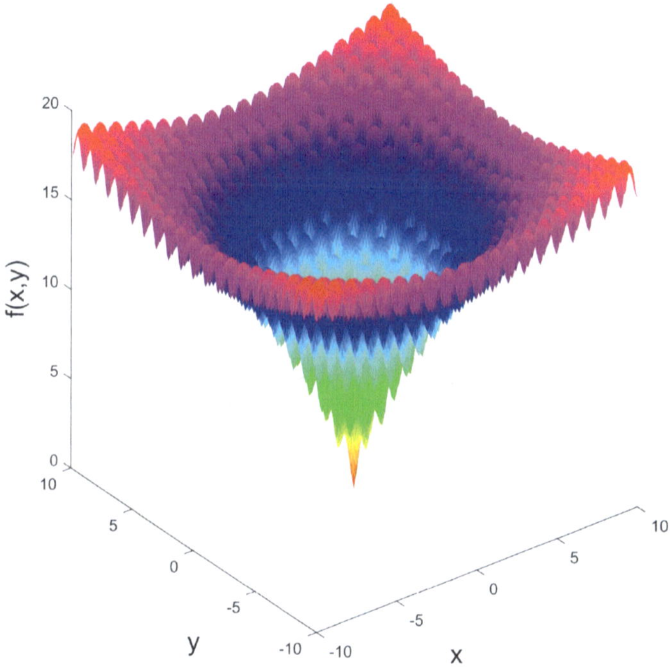

Fig. 3.2 Two-dimensional form of Ackley function

where x and y indicate the decision variables and $f(.)$ represents the objective function. The allowable bounds for each dimension of the problem are considered as $-6 \leq x \leq 6$ and $-6 \leq y \leq 6$.. The global optimum solution for this problem is located at point 0 with zero function value, i.e., $f(0, 0) = 0$.

In order the keep the illustration process as simple as possible, it is assumed that the population space consists of 5 individuals (i.e., $N_\text{I} = 5$) and the belief space includes the situational and normative knowledge components. The population space is influenced by the influence function presented in Eq. (3.10). It is also assumed that 20% of individuals within the population space are accepted to adjust the belief space at each generation (i.e., $n_\text{accept} = 0.20N_\text{I} = 1$).

The first step of the algorithm is the initialisation of population space with random individuals in the search space, where the normative bounds, as well as situational variables, are initialised as large real values. Initial belief space, represented by $\mathcal{B}^0 = \left(\mathcal{S}^0, \mathcal{N}^0\right)$, is also initialised by assuming appropriate values for the situational and normative knowledge components as follows:

$$\mathcal{S}^0 = \left[1.80 \times 10^{308} 1.80 \times 10^{308}\right] \tag{3.20}$$

$$\mathcal{N}^0 = n_1^0, n_2^0 \tag{3.21}$$

Table 3.3 Initial population $\mathcal{P}^0$ for Ackley function

No. of individual	x	y	$f(X_i)$
1	−3.9083	−0.3130	9.9767
2	2.5292	−1.5356	9.1813
3	−0.4380	0.5897	4.2790
4	2.0278	3.8892	9.5611
5	−0.7768	4.2849	10.9396

in which:

$$\mathcal{N}_1^0 = \left(\left[-1.80 \times 10^{308}\, 1.80 \times 10^{308}\right], \infty, \infty\right) \tag{3.22}$$

$$\mathcal{N}_2^0 = \left(\left[-1.80 \times 10^{308}\, 1.80 \times 10^{308}\right], \infty, \infty\right) \tag{3.23}$$

where the normative bounds, as well as situational variables, are initialised as large real values. The second step evaluates the objective functions of the individuals. Table 3.3 lists the individuals alongside their objective function values.

In the third step, the knowledge components of belief space are updated based on the Eqs. (3.12), (3.13), (3.14), (3.15), and (3.16) explained in Sect. 3.6. The updated knowledge components of belief space are calculated based on the information received from the population space as follows:

$$\mathcal{S}^1 = [-0.4380\, 0.5897],\ f\left(\mathcal{S}^1\right) = 4.2790 \tag{3.24}$$

$$\mathcal{N}_1^1 = \left(\left[-0.4380\, 1.80 \times 10^{308}\right], 4.2790, \infty\right) \tag{3.25}$$

$$\mathcal{N}_2^1 = \left(\left[0.5897\, 1.80 \times 10^{308}\right], 4.2790, \infty\right) \tag{3.26}$$

where $\mathcal{S}^1$ is the situational knowledge at the first iteration, and, $\mathcal{N}_1^1$ and $\mathcal{N}_2^1$ are the normative knowledge components in the x and y dimensions, respectively. The objective function value of situational knowledge is equal to $f\left(\mathcal{S}^1\right) = 4.2790$, which indicates the best solution found by the first iteration. From Eqs. (3.25) and (3.26), it can be seen that the upper normative bounds in both dimensions have not changed yet.

In the fourth step, the offspring solutions are generated based on the influence function in Eq. (3.10). Table 3.4 lists the new population space at the first iteration. As it can be observed from Table 3.4, the fourth offspring individual has achieved an objective function value better than the situational vector, which means that the situational vector will be updated by this individual.

As the termination criterion is not satisfied, the algorithm should get back to the second step and evaluate the objective function values. Then, the belief space should be re-updated based on the information obtained from the new population space.

Table 3.4 Population space $\mathcal{P}^1$ at the first iteration

No. of individual	X	y	$f(X_i)$
1	1.8885	1.8885	6.8614
2	5.7734	−1.3867	13.3446
3	−1.1877	3.8476	9.7881
4	−0.1167	−0.1167	1.0776
5	3.6071	4.5293	13.4858

To this end, the acceptance function will select the fourth individual to update the belief space. The updated belief space at the second iteration can be expressed by the following equations:

$$\mathcal{S}^2 = [-0.1167 - 0.11677], f\left(\mathcal{S}^2\right) = 1.0776 \tag{3.27}$$

$$\mathcal{N}_1^2 = \left(\left[-0.1167\ 1.80 \times 10^{308}\right], 1.0776, \infty\right) \tag{3.28}$$

$$\mathcal{N}_2^2 = \left(\left[-0.1167\ 1.80 \times 10^{308}\right], 1.0776, \infty\right) \tag{3.29}$$

where the objective function value of the situational vector is equal to 1.0776. As it can be seen from the above equations, although the lower normative bounds have changed, the upper normative bounds remain unchanged.

The influence function in the fourth step will generate new offspring solutions as listed in Table 3.5. From this table, it can be seen that no success has been achieved at this iteration in finding new individuals with better objective function values compared to the situational vector.

Repeating steps 2–4 will result in updated belief and population spaces. For illustrative purposes, the population spaces for iterations 3, 5, and 10 are presented in Table 3.6, Table 3.7, and Table 3.8, respectively. The belief spaces for the different iterations are also as follows:

- Belief space at the third iteration:

Table 3.5 Population space $\mathcal{P}^2$ at the second iteration

No. of individual	x	Y	$f(X_i)$
1	−1.6691	−1.6691	7.7800
2	1.8989	−5.0516	10.9928
3	0.4475	1.1937	5.2768
4	2.7642	2.7642	10.1188
5	0.6425	0.6425	4.5949

Table 3.6 Population space $\mathcal{P}^3$ at the third iteration

No. of individual	x	y	$f(X_i)$
1	−1.0703	−1.2298	5.1621
2	1.4997	−4.8810	12.1288
3	−0.3549	0.4541	3.8280
4	2.4000	1.5840	8.9677
5	0.1415	0.1187	1.2554

Table 3.7 Population space $\mathcal{P}^5$ at the third iteration

No. of individual	x	y	$f(X_i)$
1	−0.7441	−0.9129	4.2824
2	0.8863	−4.6565	11.3818
3	0.0148	0.0961	0.5084
4	2.3083	1.2428	8.0589
5	0.1532	0.0933	1.2010

Table 3.8 Population space $\mathcal{P}^{10}$ at the tenth iteration

No. of individual	x	y	$f(X_i)$
1	−0.0954	−0.7923	3.1297
2	0.1933	−4.3609	11.0670
3	0.1087	0.1833	1.5055
4	1.2968	1.0395	5.5025
5	0.0178	0.0376	0.1631

$$\mathcal{S}^3 = [-0.1167 - 0.11677],\ f\left(\mathcal{S}^5\right) = 1.0776 \tag{3.30}$$

$$m_1^3 = ([-0.1167\ 0.6425],\ 1.0776,\ 4.5949) \tag{3.31}$$

$$m_2^3 = ([-0.1167\ 0.6425],\ 1.0776,\ 4.5949) \tag{3.32}$$

- Belief space at the fifth iteration:

$$\mathcal{S}^5 = [-0.0641\, 0.08197],\ f\left(\mathcal{S}^5\right) = 0.5626 \tag{3.33}$$

$$m_1^5 = ([-0.0641\, 0.1415],\ 0.5626,\ 1.2554) \tag{3.34}$$

$$m_2^5 = ([0.0819\ 0.1187],\ 0.5626,\ 1.2554) \tag{3.35}$$

- Belief space at the tenth iteration:

$$\mathcal{S}^{10} = [-0.0092\;0.0644],\; f\left(\mathcal{S}^{10}\right) = 0.2932 \tag{3.36}$$

$$\mathcal{N}_1^{10} = ([-0.0092\;0.1415],\; 0.2932,\; 1.2554) \tag{3.37}$$

$$\mathcal{N}_2^{10} = ([0.0644\;0.1187],\; 0.2932,\; 1.2554) \tag{3.38}$$

As it can be seen from the above equations, the normative intervals are getting narrower over time. The upper normative bounds have been updated at the third iteration for the first time. Table 3.8 shows that the algorithm found the best solution with the objective function value of 0.1631 after 10 iterations. If the solution-finding process continues, further improvements in the best objective function value and additional changes in normative bounds would be expected.

3.9 Summary

The chapter presented the basic CAs for the optimisation problems. CAs belong to the category of EAs which have been developed based on the biocultural evolutionary theory. Due to their characteristic features, CAs differ from conventional EAs. The conventional EAs work on the population level and the changes in individuals happen based on biological evolutionary operators. The learning process in biological evolution is based on a vertical pattern, in which the genetic information embedded within the genes is directly transferred from parents to offspring. Conventional EAs were developed based on this form of vertical information transmission. These algorithms generate new solutions for the problem by modelling the inheritance process of favourable traits between the parents and offspring.

In the eyes of the biocultural evolutionary theory, human evolution is a product of two interacting evolutionary processes, genetic evolution and cultural evolution. This theory states that genes and culture are two interacting forms of inheritance. In addition to the improvements in biological traits, humans culturally evolve through social learning and cultural information transmission over different generations. They can transmit their technology, skills, beliefs, traditions, and other cultural variants to the next generations. In cultural evolution, individuals not only learn and acquire information from their parents but also can socially learn from other non-kin individuals through different forms of social interactions.

CAs model the biocultural evolutionary theory to perform the search process for global optima, in which both types of vertical and horizontal learning behaviours of individuals are modelled. The main framework of CAs consists of population and belief spaces. As a genetic component, the population space contains a set of individuals representing potential solutions for the problem. The belief space represents the cultural component of CAs which includes a network of cultural information or knowledge that is acquired by the individuals over different generations.

These knowledge components are used by individuals within the population space to model the different social learning strategies. The chapter provided the formulations for different components of CAs and discussed their main algorithmic steps for a minimisation problem. The formulation provided in this chapter will be used in the next chapters to explain the different variants of CAs.

References

Abdolrazzagh-Nezhad M, Radgohar H, Salimian SN (2020) Enhanced cultural algorithm to solve multi-objective attribute reduction based on rough set theory. Math Comput Simul 170:332–350. https://doi.org/10.1016/j.matcom.2019.11.005

Alami J, el Imrani A, Bouroumi A (2007) A multipopulation cultural algorithm using fuzzy clustering. Appl Soft Comput 7:506–519. https://doi.org/10.1016/j.asoc.2006.10.010

Ali MZ, Awad NH (2014) A novel class of Niche hybrid cultural algorithms for continuous engineering optimization. Inf Sci 267:158–190. https://doi.org/10.1016/j.ins.2014.01.002

Ali MZ, Salhieh A, Snanieh RTA, Reynolds RG (2012) Boosting cultural algorithms with a heterogeneous layered social fabric influence function. Comput Math Organ Theory 18:193–210

Ali MZ, Awad NH, Suganthan PN, Reynolds RG (2016a) A modified cultural algorithm with a balanced performance for the differential evolution frameworks. Knowl-Based Syst 111:73–86. https://doi.org/10.1016/j.knosys.2016.08.005

Ali MZ, Awad NH, Suganthan PN et al (2016b) A novel hybrid cultural algorithms framework with trajectory-based search for global numerical optimization. Inf Sci 334–335:219–249. https://doi.org/10.1016/j.ins.2015.11.032

Ali MZ, Suganthan PN, Reynolds RG, Al-Badarneh AF (2016c) Leveraged neighborhood restructuring in cultural algorithms for solving real-world numerical optimization problems. IEEE Trans Evol Comput 20:218–231. https://doi.org/10.1109/TEVC.2015.2450018

Ali MZ, Reynolds RG (2014) Cultural algorithms: a Tabu search approach for the optimization of engineering design problems. Soft Comput 18. https://doi.org/10.1007/s00500-013-1169-5

Ali M, Reynolds R, Ali R, Salhieh A (2011a) Knowledge-based constrained function optimization using cultural algorithms with an enhanced social influence metaphor. In: Computational intelligence. Springer, pp 103–119

Ali MZ, Salhieh A, Snanieh RTA, Reynolds RG (2011b) Boosting cultural algorithms with an incongruous layered social fabric influence function. In: 2011b IEEE congress of evolutionary computation (CEC). IEEE, pp 1225–1232

Ali MZ, Awad N, Reynolds RG (2014a) Balancing search direction in cultural algorithm for enhanced global numerical optimization. In: 2014a IEEE symposium on swarm intelligence. IEEE, pp 1–7

Ali MZ, Morghem A, Albadarneh J, et al (2014b) Cultural algorithms applied to the evolution of robotic soccer team tactics: a novel perspective. In: 2014b IEEE congress on evolutionary computation (CEC). IEEE, pp 2180–2187

Ali MZ, Awad NH, Reynolds RG, Suganthan PN (2018) A balanced fuzzy cultural algorithm with a modified Levy flight search for real parameter optimization. Inf Sci 447. https://doi.org/10.1016/j.ins.2018.03.008

Arpaia P, Lucariello G, Zanesco A (2007) Automatic fault isolation by cultural algorithms with differential influence. IEEE Trans Instrum Meas 56:1573–1582. https://doi.org/10.1109/TIM.2007.903604

Awad NH, Ali MZ, Suganthan PN, Reynolds RG (2017) CADE: a hybridization of cultural algorithm and differential evolution for numerical optimization. Inf Sci 378:215–241. https://doi.org/10.1016/j.ins.2016.10.039

Awad NH, Ali MZ, Duwairi RM (2013) Cultural algorithm with improved local search for optimization problems. In: 2013 IEEE congress on evolutionary computation. IEEE, pp 284–291

Becerra RL, Coello CAC (2004) A cultural algorithm with differential evolution to solve constrained optimization problems. In: Ibero-American conference on artificial intelligence, pp 881–890

Becerra RL, Coello CAC (2005) A cultural algorithm for solving the job shop scheduling problem. In: Knowledge incorporation in evolutionary computation. Springer, pp 37–55

Best C, Che X, Reynolds RG, Liu D (2010) Multi-objective cultural algorithms. In: IEEE congress on evolutionary computation. IEEE, pp 1–9

Bhattacharya B, Mandal K, Chakraborty N (2012a) Knowledge based evolutionary programming: cultural algorithm approach for constrained optimization. In: Proceedings of the International conference on information systems design and intelligent applications 2012a (INDIA 2012a) held in Visakhapatnam, India, January 2012a, pp 93–101

Bhattacharya B, Mandal KK, Chakraborty N (2012b) Reactive power optimization using hybrid cultural algorithm. In: International conference on swarm, evolutionary, and memetic computing, pp 106–115

bin Peng, Reynolds RG (2004) Cultural algorithms: knowledge learning in dynamic environments. In: Proceedings of the 2004 congress on evolutionary computation (IEEE Cat. No.04TH8753). IEEE, pp 1751–1758

Cabrera G, Rubio JM, Diaz D, et al (2011) A cultural algorithm applied in a bi-objective uncapacitated facility location problem. In: International conference on evolutionary multi-criterion optimization, pp 477–491

Cai Z, Zhang M, Liu Y (2018) Sea-surface weak target detection scheme using a cultural algorithm aided time-frequency fusion strategy. IET Radar Sonar Navig 12:711–720. https://doi.org/10.1049/iet-rsn.2018.0004

Chen C-H, Yang S-Y (2015) Efficient DE-based symbiotic cultural algorithm for neuro-fuzzy system design. Appl Soft Comput 34:18–25. https://doi.org/10.1016/j.asoc.2015.04.026

Chen L, Pan ZL, Zhang GZ, Wu PH (2013) A novel design approach of high temperature superconducting magnets by cultural evolution algorithms. J Low Temp Phys 170:359–365

Chen C-H, Liu Y-C, Lin C-J, Lin C-T (2008) A hybrid of cooperative particle swarm optimization and cultural algorithm for neural fuzzy networks. In: 2008 IEEE International conference on fuzzy systems (IEEE World congress on computational intelligence). IEEE, pp 238–245

Chung C, Reynolds RG (1997) Function optimization using evolutionary programming with self-adaptive cultural algorithms, pp 17–26

Chung C-J, Reynolds RG (1998) CAEP: an evolution-based tool for real-valued function optimization using cultural algorithms. Int J Artif Intell Tools 7:239–291. https://doi.org/10.1142/S0218213098000135

Coello Coello CA, Becerra RL (2004) Efficient evolutionary optimization through the use of a cultural algorithm. Eng Optim 36:219–236. https://doi.org/10.1080/03052150410001647966

Cortés Rivera D, Landa Becerra R, Coello Coello CA (2007) Cultural algorithms, an alternative heuristic to solve the job shop scheduling problem. Eng Optim 39:69–85. https://doi.org/10.1080/03052150600956811

Crawford B, Lagos C, Castro C, Paredes F (2007) A cultural algorithm for solving the set covering problem. In: Analysis and design of intelligent systems using soft computing techniques. Springer, pp 408–415

Crawford B, Soto R, Monfroy E (2013) Cultural algorithms for the set covering problem. In: International conference in swarm intelligence, pp 27–34

Digalakis JG, Margaritis KG (2002) A multipopulation cultural algorithm for the electrical generator scheduling problem. Math Comput Simul 60:293–301. https://doi.org/10.1016/S0378-4754(02)00021-6

dos Santos Coelho L, Alotto P (2009) Electromagnetic optimization using a cultural self-organizing migrating algorithm approach based on normative knowledge. IEEE Trans Magn 45:1446–1449. https://doi.org/10.1109/TMAG.2009.2012668

dos Coelho LS, Souza RCT, Mariani VC (2006) Improved differential evolution approach based on cultural algorithm and diversity measure applied to solve economic load dispatch problems. Math Comput Simul 79:3136–3147. https://doi.org/10.1016/j.matcom.2009.03.005
Engelbrecht AP (2007) Computational intelligence: an introduction. Wiley
Gao H, Diao M (2011) Cultural firework algorithm and its application for digital filters design. Int J Model Ident Control 14:324. https://doi.org/10.1504/IJMIC.2011.043157
Gao H, Xu C (2013) Cultural quantum-inspired shuffled frog leaping algorithm for direction finding of non-circular signals. Int J Comput Sci Math 4:321. https://doi.org/10.1504/IJCSM.2013.058066
Gao M, Zhu Y, Cao C, Zhu Y (2020) A hybrid cultural harmony search algorithm for constrained optimization problem of diesel blending. IEEE Access 8:6673–6690. https://doi.org/10.1109/ACCESS.2019.2963244
Gao H, Diao M (2010) Differential cultural algorithm for digital filters design. In: 2010 Second International conference on computer modeling and simulation. IEEE, pp 459–463
Gao F, Cui G, Liu H (2006) Integration of genetic algorithm and cultural algorithms for constrained optimization, pp 817–825
Gao F, Zhao Q, Liu H, Cui G (2007) Cultural particle swarm algorithms for constrained multi-objective optimization, pp 1021–1028
Gao HY, Chi PF, Du YN, et al (2019) Digital watermarking based on wavelet transform and cultural invasive weed algorithm, pp 1620–1627
Goli A, Zare HK, Tavakkoli-Moghaddam R, Sadeghieh A (2019) An improved artificial intelligence based on Gray Wolf optimization and cultural algorithm to predict demand for dairy products: a case study. Int J Interact Multimedia Artif Intell 5:15. https://doi.org/10.9781/ijimai.2019.03.003
Goudarzi A, Ahmadi A, Swanson AG, van Coller J (2016) Non-convex optimisation of combined environmental economic dispatch through cultural algorithm with the consideration of the physical constraints of generating units and price penalty factors. SAIEE Africa Res J 107:146–166. https://doi.org/10.23919/SAIEE.2016.8532239
Goudarzi A, Swanson AG, Tooryan F, Ahmadi A (2017) Non-convex optimization of combined environmental economic dispatch through the third version of the cultural algorithm (CA3). In: 2017 IEEE Texas power and energy conference (TPEC). IEEE, pp 1–6
Gu W, Wu Y (2010) Application of multi-objective cultural algorithm in water resources optimization. In: 2010 Asia-Pacific power and energy engineering conference. IEEE
ben Guedria N, Hassine H (2019) Efficient cultural algorithm for structural damage detection problem based on modal data
Guo Y, Cheng J, Cao Y, Lin Y (2011) A novel multi-population cultural algorithm adopting knowledge migration. Soft Comput 15:897–905. https://doi.org/10.1007/s00500-010-0556-4
Guo Y, Zhang P, Cheng J et al (2018) Interval multi-objective quantum-inspired cultural algorithms. Neural Comput Appl 30:709–722
Guo Y, Liu D, Cheng J (2012) Multi-population cooperative cultural algorithms, pp 199–206
Guo Y, Chen M, Wang C (2013a) Multi-objective quantum cultural algorithm and its application in the wireless sensor networks' energy-efficient coverage optimization, pp 161–167
Guo Y, Liu D, Liu Y, Chen M (2013b) The coverage optimization for wireless sensor networks based on quantum-inspired cultural algorithm, pp 87–96
Haikal A, El-Hosseni M (2011) Modified cultural-based genetic algorithm for process optimization. Ain Shams Eng J 2:173–182. https://doi.org/10.1016/j.asej.2011.09.002
Haldar V, Chakraborty N (2014) Minimisation of power loss by network reconfiguration and capacitor installation using modified cultural algorithm. Int J Swarm Intell 1:226. https://doi.org/10.1504/IJSI.2014.066553
Haldar V, Chakraborty N (2015) Power loss minimization by optimal capacitor placement in radial distribution system using modified cultural algorithm. Int Trans Electr Energy Syst 25:54–71. https://doi.org/10.1002/etep.1820
He J, Xu F (2011) Chaotic-search-based cultural algorithm for solving unconstrained optimization problem. Model Simul Eng 2011:1–6. https://doi.org/10.1155/2011/239743

Jafari M, Salajegheh E, Salajegheh J (2019) An efficient hybrid of elephant herding optimization and cultural algorithm for optimal design of trusses. Eng Comput 35. https://doi.org/10.1007/s00366-018-0631-5
Jalili S, Hosseinzadeh Y (2015) A cultural algorithm for optimal design of truss structures. Latin Am J Solids Struct 12. https://doi.org/10.1590/1679-78251547
Jalili S, Hosseinzadeh Y, Rabczuk T (2019) Simultaneous size and shape optimization of dome-shaped structures using improved cultural algorithm
Jin X, Reynolds RG (1999a) Using knowledge-based system with hierarchical architecture to guide the search of evolutionary computation. In: proceedings 11th international conference on tools with artificial intelligence. IEEE Comput Soc pp 29–36
Jin X, Reynolds RG (1999b) Using knowledge-based evolutionary computation to solve nonlinear constraint optimization problems: a cultural algorithm approach. In: proceedings of the 1999 congress on evolutionary computation-CEC99 (Cat. No. 99TH8406). IEEE, pp 1672–1678
Kim Y, Cho SB (2009) A hybrid cultural algorithm with local search for traveling salesman problem. In: 2009 IEEE international symposium on computational intelligence in robotics and automation—(CIRA), 2009, pp 188–192. https://doi.org/10.1109/CIRA.2009.5423211
Khan SU, Qureshi IM, Zaman F et al (2014) Correction of faulty sensors in phased array radars using symmetrical sensor failure technique and cultural algorithm with differential evolution. Sci World J 2014:1–10. https://doi.org/10.1155/2014/852539
Khan SU, Qureshi IM, Zaman F, Khan W (2017) Detecting faulty sensors in an array using symmetrical structure and cultural algorithm hybridized with differential evolution. Front Inf Technol Electron Eng 18:235–245. https://doi.org/10.1631/FITEE.1500315
Khodabakhshian A, Hemmati R (2013) Multi-machine power system stabilizer design by using cultural algorithms. Int J Electr Power Energy Syst 44:571–580. https://doi.org/10.1016/j.ijepes.2012.07.049
Kulu Amadu A, Birninkudu SI, Bukata BB et al (2020) Cultured bat algorithm for optimized MPPT tracking under different shading conditions. In: 2020 International conference in mathematics, computer engineering and computer science (ICMCECS). IEEE, pp 1–8
Lagos C, Crawford B, Cabrera E et al (2014) Comparing evolutionary strategies on a biobjective cultural algorithm. Sci World J 2014:1–10. https://doi.org/10.1155/2014/745921
Lin CJ, Weng CC, Lee CL, Lee CY (2009b) Using an efficient hybrid of cooperative particle swarm optimization and cultural algorithm for neural fuzzy network design. In: 2009 international conference on machine learning and cybernetics. IEEE, pp 3076–3082
Liu W-Y, Lin C-C (2015) Spatial forest resource planning using a cultural algorithm with problem-specific information. Environ Model Softw 71. https://doi.org/10.1016/j.envsoft.2015.06.002
Lu Y, Zhou J, Qin H et al (2011) A hybrid multi-objective cultural algorithm for short-term environmental/economic hydrothermal scheduling. Energy Convers Manage 52:2121–2134. https://doi.org/10.1016/j.enconman.2010.12.003
Maheri A, Jalili S, Hosseinzadeh Y et al (2021) A comprehensive survey on cultural algorithms. Swarm Evol Comput 62:100846. https://doi.org/10.1016/j.swevo.2021.100846
Mao Z, Xiang Y, Zhang Y, Liu M (2020) A novel multi-objective cultural algorithm embedding five-element cycle optimization. In: 2020 IEEE congress on evolutionary computation (CEC). IEEE, pp 1–10
Mohammadhosseini M, Toroghi Haghighat A, Mahdipour E (2019) An efficient energy-aware method for virtual machine placement in cloud data centers using the cultural algorithm. J Supercomput 75:6904–6933. https://doi.org/10.1007/s11227-019-02909-3
Mojab SZM, Ebrahimi M, Reynolds R, Lu S (2019) Icats: scheduling big data workflows in the cloud using cultural algorithms. In: 2019 IEEE fifth international conference on big data computing service and applications (BigDataService), IEEE, 2019, pp 99–106
Muhamediyeva DT (2020) Fuzzy cultural algorithm for solving optimization problems. J Phys: Conf Ser 1441:012152. https://doi.org/10.1088/1742-6596/1441/1/012152

Murugadass G, Sivakumar P (2020) A hybrid elephant herding optimization and cultural algorithm for energy-balanced cluster head selection scheme to extend the lifetime in WSNs. Int J Commun Syst e4538. https://doi.org/10.1002/dac.4538

Nguyen TT, Yao X (2006) Hybridizing cultural algorithms and local search, pp 586–594

Oliveira De Freitas CA, Limao de Oliveira RC, Azevedo Da Silva DJ, et al (2018) Solution to economic—emission load dispatch by cultural algorithm combined with local search: case study. IEEE access 6:64023–64040. https://doi.org/10.1109/ACCESS.2018.2877770

Oloruntoba O, Cosma G, Liotta A (2019) Clan-based cultural algorithm for feature selection. In: 2019 International Conference on Data Mining Workshops (ICDMW). IEEE, pp 465–472

Pan Z-L, Chen L, Zhang G-Z (2010) Cultural algorithm for minimization of binary decision diagram and its application in crosstalk fault detection. Int J Autom Comput 7:70–77. https://doi.org/10.1007/s11633-010-0070-2

Pierezan J, Maidl G, Massashi Yamao E et al (2019) Cultural coyote optimization algorithm applied to a heavy duty gas turbine operation. Energy Convers Manage 199:111932. https://doi.org/10.1016/j.enconman.2019.111932

Reynolds RG, Peng B (2005) Cultural algorithms: computational modeling of how cultures learn to solve problems: an engineering example. Cybern Syst 36:753–771. https://doi.org/10.1080/01969720500306147

Reynolds RG, Zhu S (2001) Knowledge-based function optimization using fuzzy cultural algorithms with evolutionary programming. IEEE Trans Syst Man Cybern Part B (cybern) 31:1–18. https://doi.org/10.1109/3477.907561

Reynolds RG, Peng B, Ali MZ (2008) The role of culture in the emergence of decision-making roles: an example using cultural algorithms. Complexity 13:27–42. https://doi.org/10.1002/cplx.20196

Reynolds RG, Ali MZ (2008) Cultural algorithms: knowledge-driven engineering optimization via weaving a social fabric as an enhanced influence function. In: 2008 IEEE congress on evolutionary computation (IEEE World congress on computational intelligence). IEEE, pp 4192–4199

Reynolds R, Al-Shehri H (1997) The use of cultural algorithms with evolutionary programming to control the data mining of large-scale spatio-temporal databases. In: 1997 IEEE International conference on systems, man, and cybernetics. computational cybernetics and simulation. IEEE, pp 4098–4103

Reynolds RG, Saleem SM (2005) The impact of environmental dynamics on cultural emergence. In: Perspectives on adaptions in natural and artificial systems, pp 253–280

Reynolds RG (1979) An adaptive computer model of the evolution of agriculture for hunter-gatherers in the valley of Oaxaca, Mexico

Selvarajah K, Kobti Z, Kargar M (2020) Cultural algorithms for cluster hires in social networks. Procedia Comput Sci 170:514–521. https://doi.org/10.1016/j.procs.2020.03.117

Shah P, Kobti Z (2020) Multimodal fake news detection using a cultural algorithm with situational and normative knowledge. In: 2020 IEEE congress on evolutionary computation (CEC). IEEE, pp 1–7

Soza C, Landa R, Riff MC, Coello C (2007) A cultural algorithm with operator parameters control for solving timetabling problems. Foundations of fuzzy logic and soft computing. Springer Berlin Heidelberg, Berlin, Heidelberg, pp 810–819

Soza C, Becerra RL, Riff MC, Coello Coello CA (2011) Solving timetabling problems using a cultural algorithm. Appl Soft Comput 11:337–344. https://doi.org/10.1016/j.asoc.2009.11.024

Srinivasan S, Ramakrishnan S (2012) Multi objective optimization of classification rules using cultural algorithms. Procedia Eng 30:457–465

Stanley SD, Kattan K, Reynolds R (2019) CAPSO: A parallelized multiobjective cultural algorithm particle swarm optimizer. In: 2019 IEEE congress on evolutionary computation (CEC). IEEE, pp 3060–3069

Sun X-M, Lv X-Y, Duan X-M (2009) Novel QoS routing algorithm based on cultural-simulated annealing algorithm. In: 2009 Second International conference on intelligent networks and intelligent systems. IEEE, pp 209–212

Sun Y, Zhang L, Gu X (2010) Co-evolutionary cultural based particle swarm optimization algorithm, pp 1–7
Terán J, Aguilar J, Cerrada M (2017) Integration in industrial automation based on multi-agent systems using cultural algorithms for optimizing the coordination mechanisms. Comput Ind 91:11–23. https://doi.org/10.1016/j.compind.2017.05.002
Tremayne M, Chong SY, Bell D (2009) Optimisation of algorithm control parameters in cultural differential evolution applied to molecular crystallography. Front Comput Sci China 3:101–108. https://doi.org/10.1007/s11704-009-0009-3
Wang W, Song Y, Xue Y et al (2015) An optimal vibration control strategy for a vehicle's active suspension based on improved cultural algorithm. Appl Soft Comput 28:167–174. https://doi.org/10.1016/j.asoc.2014.11.047
Wang X, Hao W, Li Q (2017) An adaptive cultural algorithm with improved quantum-behaved particle swarm optimization for sonar image detection. Sci Rep 7:17733. https://doi.org/10.1038/s41598-017-17945-3
Wu Y, Gao X-Z, Huang X-L, Zenger K (2010) A hybrid optimization method of particle swarm optimization and cultural algorithm. In: 2010 Sixth International conference on natural computation. IEEE, pp 2515–2519
Xie W, Ji C, Li X (2009) Particle swarm optimization based on cultural algorithm for short-term optimal operation of cascade hydropower stations. In: 2009 Fifth International conference on natural computation. IEEE, pp 289–293
Xu W, Wang R, Zhang L, Gu X (2012) A multi-population cultural algorithm with adaptive diversity preservation and its application in ammonia synthesis process. Neural Comput Appl 21:1129–1140. https://doi.org/10.1007/s00521-011-0749-5
Xu W, Zhang L, Gu X (2010) A novel cultural algorithm and its application to the constrained optimization in ammonia synthesis. In: Life system modeling and intelligent computing. Springer, pp 52–58
Xue Z, Guo Y (2007) Improved cultural algorithm based on genetic algorithm. In: 2007 IEEE International conference on integration technology. IEEE, pp 117–122
Xue X, Yao M, Cheng R (2011) A novel selection operator of cultural algorithm, pp 71–77
Xue H (2020) Adaptive cultural algorithm-based cuckoo search for time-dependent vehicle routing problem with stochastic customers using adaptive fractional Kalman speed prediction. Math Probl Eng 2020. https://doi.org/10.1155/2020/7258780
Ya-Li W, Li-qing X (2010) An improved cultural particle swarm optimization algorithm based on feedback mechanism for economic load dispatch problems. In: 2010 International conference on computational aspects of social networks. IEEE, pp 319–322
Yan X, Song T, Wu Q (2017b) An improved cultural algorithm and its application in image matching. Multimedia Tools and Applications 76:14951–14968. https://doi.org/10.1007/s11042-016-4313-2
Yan X, Gong W, Wu Q (2017a) Contaminant source identification of water distribution networks using cultural algorithm. Concurrency Comput: Pract Exp 29. https://doi.org/10.1002/cpe.4230
Yang Y, Gu X (2014) Cultural-based genetic Tabu algorithm for multiobjective job shop scheduling. Math Probl Eng 2014:1–14. https://doi.org/10.1155/2014/230719
Yuan X, Nie H, Yuan Y et al (2009) Hydrothermal systems generation scheduling using cultural algorithm. J Hydroinf 11:65–78. https://doi.org/10.2166/hydro.2009.056
Yuan X, Nie H, He L, et al (2008) A cultural algorithm for scheduling of hydro producer in the power market. In: 2008 Second International conference on genetic and evolutionary computing. IEEE, pp 364–367
Zadeh PM, Kobti Z (2015) A multi-population cultural algorithm for community detection in social networks. Procedia Comput Sci 52:342–349. https://doi.org/10.1016/j.procs.2015.05.105
Zhang Y, Zhu X (2012) A cultural particle swarm optimization algorithm for flow shop scheduling problem, pp 1575–1583
Zhou Y, Zhang S, Luo Q, Abdel-Baset M (2019) CCEO: cultural cognitive evolution optimization algorithm. Soft Comput 23:12561–12583. https://doi.org/10.1007/s00500-019-03806-w

Part II
Application of Cultural Algorithms

Chapter 4
Applications of Cultural Algorithms in Engineering

Abstract Since the emergence of CAs, they have been successfully extended and widely employed to solve a wide variety of problems in different engineering disciplines, including civil engineering, mechanical engineering, chemical engineering, electrical engineering, and industrial engineering. This chapter focuses on the applications of CAs in these engineering disciplines and provides the overall mathematical formulations for different optimisation problems.

Keywords Applications · Engineering · Optimisation · Cultural algorithms

4.1 Introduction

Since the emergence of CAs, they have been applied to a wide range of problems in different branches of engineering science. Literature reveals that CAs are suitable tools to find promising results for different problem types with different characteristic features. Since most of the problems in literature are inherently interdisciplinary problems, providing a clear taxonomy of application areas is not an easy task. Despite this difficulty, this chapter tries to categorise the application areas and briefly explain the details for each problem. The chapter provides useful information to readers about the taxonomy of application areas in different engineering disciplines and formulations for each category of problems. The problems are explained briefly and they are formalised based on a unified approach. In the formulations, the decision variables, objective functions, constraints, as well as other key assumptions are presented. In this chapter, the application areas of CAs are categorised into civil engineering, mechanical engineering, chemical engineering, electrical engineering, and industrial engineering. Figure 4.1 shows the subjects in each engineering branch covered in this chapter.

S. Jalili, *Cultural Algorithms*, Engineering Optimization: Methods and Applications,
https://doi.org/10.1007/978-981-19-4633-2_4

Fig. 4.1 Application areas of CAs in different branches of engineering science

4.2 Civil Engineering

CAs have found some applications in civil engineering. The application studies of CAs in the civil engineering area can be categorised into five sub-disciplines as follows: (i) transportation engineering, (ii) structural engineering, (iii) environmental engineering, (iv) urban engineering, and (v) water engineering. Table 4.1 lists the references in which the CAs are applied to solve problems in different sub-disciplines of civil engineering. In the following subsections, each reference are investigated in more detail.

Table 4.1 Applications of CAs and their variants in civil engineering

Application area	References
Transportation engineering	Reyes et al. (2010), Xue (2020), Liu et al. (2012), Arpaia et al. (2007), Maldonado et al. (2013)
Structural engineering	Jalili and Hosseinzadeh (2015), Jalili et al. (2019), Jafari et al. (2019), Ali et al. (2018)
Environmental engineering	Liu and Lin (2015)
Urban engineering	Jayyousi and Reynolds (2014)
Water engineering	Yan et al. (2017a), Gu and Wu (2010), Jafari-Asl et al. (2020)

4.2.1 Transportation Engineering

In recent years, the application of computational intelligence methods for traffic control has attracted much attention (Araghi et al. 2015; Nallaperuma et al. 2020; Jalili et al. 2021b). As intelligent optimisation algorithms, CAs are also applied to different problems in the transportation engineering area. Vehicle routing, fault isolation, and accident prevention are the problems that are explained in the following subsections.

4.2.1.1 Vehicle Routing

The CA has been employed by Reyes et al. (2010) for the optimum routing problem of public buses. Let us assume a directed graph $G = (V, E)$ representing the public transportation network, in which $V = (v_0, v_1, \ldots, v_n)$ indicates a set of vertexes representing the bus stations and $E = \left(e_0, e_1, \ldots, e_{n_R}\right)$ includes a set of edges in the network which connect the stations. Let n_R be the number of edges between the origin station v_0 and final station v_n. Reyes et al. (2010) investigated the following public bus routing problem:

$$\text{Find: } E = \left(e_0, e_1, \ldots, e_{n_{\mathrm{R}}}\right)$$

$$\text{To minimise: } f_1(E) = \sum_{i=0}^{n} t_i \text{ and } f_2(E) = n_{\mathrm{R}} \tag{4.1}$$

where f_1 is the first objective function representing the travel time of the selected route, f_2 is the second objective function representing the number of routes, and t_i is the travel time between the i th and $i+1$ th stops. Reyes et al. (2010) investigated the performance of CA in routing optimisation of the public buses in Leon city, Mexico. The numerical results revealed that the CA is capable of generating better optimal routes in comparison to the traditional handbook estimates (Reyes et al. 2010).

In another research, Xue (2020) studied the time-dependent vehicle routing problem with stochastic customers. The study defined the road network as a directed time-dependent graph represented by $G = \left(C, E, v_{ij}^{k}\right)$, where C represents a set of customer nodes, E is a set of edges that connect the nodes, and v_{ij}^{k} indicates the travel speed from the i th node to the j th node. The study employed an adaptive fractional-order Kalman filter to predict the travel speed v_{ij}^{k}. The minimisation of travel and service times is assumed as the objective function, in which the stochastic customer demand, time-varying speed, and real-time traffic conditions on roads are considered. The authors applied a CA-based Cuckoo Search (CS) to find optimal routes for a road network with different business districts in Xiamen, China.

In another study, Liu et al. (2012) employed CA to find the shortest path between two points based on the GIS data, in which the weather, road quality, traffic, and other uncertain factors were considered. The obtained optimum routes revealed the efficiency of the CA.

4.2.1.2 Fault Isolation

Arpaia et al. (2007) employed CA for automatic fault isolation in the transport of dangerous goods. The study assumes that each fleet vehicle is monitored by two master–slave units, including the Vehicle Monitoring Unit of Tractor UMAT and Vehicle Monitoring Unit of Tank UMAB. The data measured by the UMAT and UMAB are delivered via the GPRS-GSM channel to the monitoring unit. In fault isolation, the monitoring unit usually faces a set of multiple types of faults in the system. However, only a set of faults that adequately represent the abnormal behaviours of the system should be detected. The study considered various types of faults with different predefined probabilities for the fault isolation problem, such as tanker wheel slip, road accident, carried substance loss, fuel theft, starter breaking, radiator broken, fifth wheel irregular vibrations, fuel metre uncalibration, gas exhaust pollution, and road traffic (Arpaia et al. 2007). The authors designed the DE-based influence functions, in which the topographical and historical knowledge components are used.

The efficiency of CA was investigated in the fleet vehicle case study under two fault scenarios, including the tanker wheel slip and the fuel metre uncalibration faults. The result comparisons showed that the CA accurately detects the real fault of the system with better performance than the GA.

4.2.1.3 Accident Prevention

Maldonado et al. (2013) created a software based on the CA and data mining to urge drivers on traffic accident risk in Ciudad Juárez city, Mexico, which works on the Android mobile operating system. The software uses the historical accident data and CA to predict the risk of a traffic accident in each part of the city. The authors reported that the long-term application of this software can reduce the number of traffic accidents and prevent traffic congestion.

4.2.2 Structural Engineering

Structural optimisation is one of the fundamental problems in the field of structural engineering. Nowadays, the desire for economic and safe structures has intensified the engineers' effort to attain optimum designs for complex real-world structures under international code requirements. Most structural optimum design problems assume the construction costs as the objective function. However, due to a lack of data, it is quite common in the literature to consider the structural weight as the objective function. The general structural design optimisation problem can be mathematically expressed as follows:

$$\text{Find: } A = [a_1, a_2, \ldots, a_n]$$

$$\text{To minimise: } W(A) = \sum_{i=1}^{m} \gamma_i a_i L_i$$

$$\text{Subject to: } g(A) = \frac{\sigma_i}{\sigma_i^c} - 1 \geq 0, \quad i = 1, 2, \ldots, m$$

$$h(A) = \frac{\sigma_i}{\sigma^t} - 1 \leq 0, \quad i = 1, 2, \ldots, m$$

$$I(A) = \frac{\delta_j}{\delta_{\max}} - 1 \leq 0, \quad j = 1, 2, \ldots, n \tag{4.2}$$

where A represents the vector of cross-sectional areas for different element groups; m is the number of element groups in the structure; $W(.)$ indicates the weight of the structure; γ_i is the material density of elements in group i; a_i is the cross-sectional area for element group i; L_i is the length of elements in group i; $g(A)$, $h(A)$, and $I(A)$ represent the compressive stress, tension stress, and nodal displacement constraints, respectively; σ_i, σ_i^c, and σ^t are the stress, allowable compressive stress, and allowable tension stress in the i th element group; δ_j is the displacement of the j th node; $\delta_{\max}$ indicates the allowable nodal displacement; n is the number of nodes in the structure.

In the field of structural engineering, Jalili and Hosseinzadeh (2015) applied CA for the optimal design of truss structures under stress and displacement constraints. The authors verified the efficiency of CA by optimising four real-world truss structures. Comparison results between CA and other meta-heuristics, such as BB-BC, PSO, and GA, illustrated that CA can provide relatively better structural designs.

In another research, Jalili et al. (2019) introduced an Improved CA (ICA) for simultaneous size and shape optimisation of dome-shaped structures under static and frequency constraints. Structural optimisation under frequency constraints is a highly nonlinear optimisation problem. The authors investigated the size and layout optimisation of a 276-bar lamella dome structure under stress and displacement constraints and a 708-bar lamella dome structure under frequency constraints.

Jafari et al. (2019) developed a hybrid CA and Elephant Herding Optimisation (EHO) for optimum design of truss structures. Authors applied their algorithm to solve some benchmark structural design problems, including 10-bar truss, 25-bar truss, 72-bar truss, and 120-bar dome structures. The results indicated that the standard CA was able to provide better structural designs than the EHO algorithm for all design examples.

Ali et al. (2018) presented a balanced fuzzy CA with a modified Levy flight search to solve real-world parameter optimisation. The authors tested the performance of the approach on a set of benchmark functions and a 120-bar dome structure.

Figures 4.2 and 4.3 show some of the structural design optimisation benchmarks which are investigated in the literature to test the effectiveness of CAs.

4.2.3 *Environmental Engineering*

In the field of environmental engineering, Liu and Lin 2015 proposed a CA-based approach to solve the spatial forest resource planning problem. The forest lands consist of several smaller sub-lands, called *polygons*. The vector $X = [x_1, x_2, \ldots, x_n]$ represents the vector containing the IDs of different polygons in the forest land that need to be harvested in different periods. In the study, the optimisation vector X is divided into different partitions, each representing harvested polygons in a specific period as follows:

$$X = \left[x_1^1, x_2^1, \ldots, x_{n_1}^1 \middle| x_1^2, x_2^2, \ldots, x_{n_2}^2 \middle| \ldots \middle| x_1^m, x_2^m, \ldots, x_{n_m}^m \middle| x_1^R, x_2^R, \ldots, x_{n_R}^R\right] \tag{4.3}$$

where x_j^i $(i = 1, 2, \ldots, m, \ j = 1, 2, \ldots, n_i)$ represents the j th polygon harvested during the i th period, m is the number of periods, n_i indicates the number of polygons harvested during the i th period, x_i^R $(i = 1, 2, \ldots, n_R)$ is the i th polygon that will not be harvested, and n_R is the total number of polygons which will not be harvested.

As it can be seen from Eq. (4.3), the last partition of the vector is the residual part that includes polygons that should not be harvested. With the definition of problem variables, the main objective function of the spatial forest resource planning problem is the total timber volume maximisation of all polygons harvested under some constraints on the harvest age, adjacency, and even harvest volume for each period. The authors performed a statistical comparison between the CA and SA by assuming different values for maximum iterations and population sizes. Their statistical comparisons and the computed p-values revealed that CA is generally able to perform better than the SA algorithm in terms of efficiency and stability of the results.

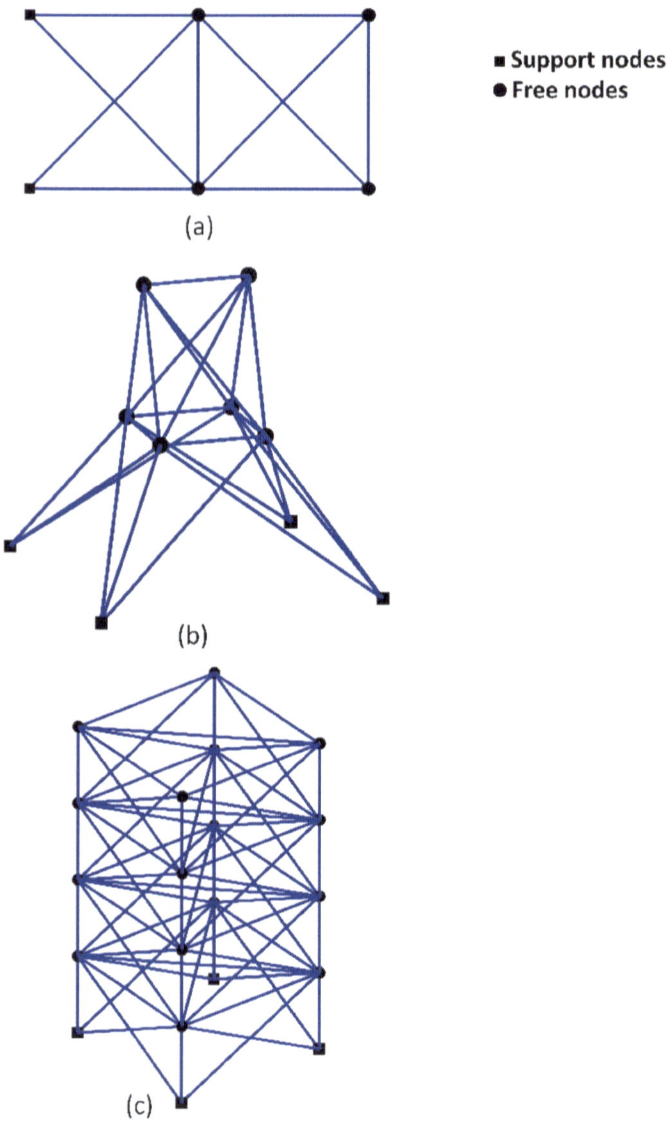

Fig. 4.2 Structural types investigated in the structural optimisation field: **a** 10-bar truss, **b** 25-bar truss, and **c** 72-bar

4.2.4 Urban Engineering

CA was also applied to urban engineering. Jayyousi and Reynolds (2014) applied CA to produce high-level structural–functional models of cities, in which each individual of CA competes to build up a high-level model of the site from a set of building blocks.

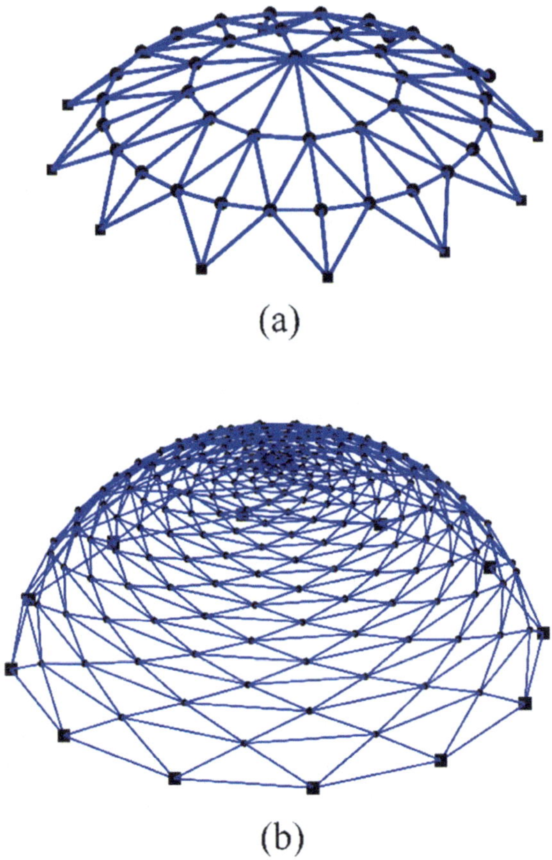

Fig. 4.3 Structural types investigated in the structural optimisation field: **a** 120-bar dome and **b** 708 lamella dome

Their study proved that the application of CA to urban planning can provide new insights into the plans produced by human experts in the field.

4.2.5 Water Engineering

CAs have been applied to optimise the water supply networks to improve their performance. In the following subsection, the contaminant source identification and leakage optimisation problems are described.

4.2.5.1 Contaminant Source Identification

In the field of water engineering, Yan et al. (2017a) employed CA to solve the contaminant source identification problem in water distribution networks. The water supply networks are susceptible to sudden contaminations which can rapidly propagate across the network and disturb urban water supply systems. Nowadays, the water supply pipe networks are equipped with a set of sensors that can monitor the network performance against possible pollution scenarios in real-time and take actions against possible contamination scenarios. However, identification of the types and exact locations of the pollution in the water network based on the information provided by the sensors is not an easy task. Hence, the contaminant source identification is an optimisation problem that looks for the locations of contaminants sources, the contaminants' mass loading profile, and the start time of the injection at different contaminant sources to minimise simulated contaminant concentration error.

The contaminant source identification problem can be mathematically written as follows (Zechman and Ranjithan 2009):

$$\begin{aligned}
&\text{Find: } X = [M, n, t_{\text{I}}],\ M \in \mathbb{C}^K,\ X \in \mathbb{C}^{K+2} \\
&\text{To minimise: } E(X) = \sum_{i=1}^{n_s} \sum_{t=1}^{T_s} \left(\tilde{c}_j^t - c_j^t\right)^2 \\
&\text{Subject to: } m_i \geq 0, \quad i = 1, 2, \ldots, K \\
&\qquad n \in \{1, 2, \ldots, N\} \\
&\qquad t_I \leq T_s
\end{aligned} \tag{4.4}$$

where X represents the vector of decision variables, $M = [m_1, m_2, \ldots, m_K]$ is the mass loading profile vector of the injected contaminant into the network, n is the node ID at which the contaminant source is introduced, t_{I} is the contaminant starting time, $E(.)$ indicates the simulated contaminant concentration error, n_s is the number of sensor nodes in the network, T_s represents the simulation duration of the whole water network, $\tilde{c}_j^t$ the contaminate concentration obtained from the simulation at the j th sensor, c_j^t is the observed contaminate concentration at the j th sensor, and N is the number of nodes in the network.

By assuming different sizes of water supply networks, Yan et al. (2017a) evaluated the performance of CA based on three water supply networks with 129, 430, and 12,527 nodes. The results revealed that the CA is capable to identify the source of the contaminant with an appropriate level of accuracy. However, it seems that its performance needs to be improved to identify the pollutant sources more accurately in larger scales of water distribution networks.

4.2.5.2 Leakage Optimisation

Water loss is one of the important issues in water supply networks that can increase the water supply costs and reduce the performance of the network. Jafari-Asl et al. (2020) employed CA to solve the leakage minimisation problem in a water supply network under mass and energy conversation constraints. The study formulated the leakage optimisation as follows (Jafari-Asl et al. 2020):

$$f(X) = \frac{1}{N_{\mathrm{L}}} \sum_{k=1}^{N_{\mathrm{L}}} \sum_{i=1}^{N} W_k C_{\mathrm{L}} L_{\mathrm{t},i} \left(P_{i,k}\right)^{\gamma}$$
$$\text{Subject to: } H_{i,k} \geq H_{\min,i}, \quad i = 1, 2, \ldots, N \tag{4.5}$$

where N_L is the total number of load conditions, N is the total number of nodes within the water supply network, W_k is the weight associated with the k th load condition, C_L is the coefficient for the leakage per unit length of pipe, $L_{t,i}$ represents the total length of pipes connected to the i th node, $P_{i,k}$ indicates the pressure at the i th node for the k th load, γ is a leakage exponent coefficient, $H_{i,k}$ is the head at the i th node for the k th load condition, and $H_{\min,i}$ is the minimum allowable head value at the i th node.

4.3 Mechanical Engineering

CA has also been employed to find solutions for the optimisation problems in mechanical engineering. Table 4.2 presents the application areas and references in the field of mechanical engineering. Subsequent sections provide the details for each study.

4.3.1 Damage Detection

The structural systems can experience serious damages during their design lifetime that can significantly affect their behaviour. In most cases, these damages are not detectable by traditional visual expert inspections which can put the safety of the structure at risk. The propagation and accumulation of undetected damages can cause sudden failure of the structural system with serious repair costs or loss of human lives.

Table 4.2 Applications of CAs and their variants in mechanical engineering

Application area	References
Damage detection	Guedria and Hassine (2019)
Optimum design	Ali and Reynolds (2014), Reynolds and Ali (2008a, b), dos Santos Coelho and Mariani (2006), Reynolds and Peng (2005), Coello and Becerra (2004)
Composite structures	Jalili et al. (2022)
Optimal vibration control	Wang et al. (2015a)
Vehicle safety	Kobti et al. (2006b)
Flight control	Meng et al. (2011)
Gas turbine operation	Pierezan et al. (2019)
Renewable energy	Amalo et al. (2020)

One practical way to deal with this challenge is the application of damage detection techniques. These techniques try to detect the possible damages to the structure by monitoring the changes in the physical properties of structural elements. Vibration-based damage detection is one of the popular approaches which employ the different types of vibrational data to identify the existence, location, and extents of damages, such as time, frequency responses, and modal information (Avci et al. 2021; Chang and Kim 2016; Kopsaftopoulos and Fassois 2013; Seyedpoor 2012; Vamvoudakis-Stefanou et al. 2018). In recent years, researchers are focused on the application of Artificial Intelligence (AI) and evolutionary computation techniques to solve damage detection problems (Alkayem et al. 2018; Avci et al. 2021; Neves et al. 2017).

Guedria and Hassine (2019) employed the CA for the vibrational damage detection based on the modal data. The main objective of the problem is to minimise the residual difference between the flexibility matrixes obtained from the experimental and numerical results as follows (Guedria and Hassine, 2019):

$$\begin{aligned}&\text{Find } X = [x_1, x_2, \ldots, x_n],\ X \in \mathbb{C}^n\\&\text{To Minimise: } f(X) = \left\| F_{\mathrm{E}} - \widetilde{F}(X) \right\|^2,\ F_{\mathrm{E}} \in \mathbb{C}^{n_{\mathrm{F}} \times n_{\mathrm{F}}},\ \widetilde{F}(X) \in \mathbb{C}^{n_{\mathrm{F}} \times n_{\mathrm{F}}}\\&\text{Subject to: } 0 \leq x_i \leq 1\end{aligned} \tag{4.6}$$

where X is the vector of the damage indexes, in which each variable x_i represents the damage index for the i th element of the structure, n represents the number of structural elements, F_E is the flexibility matrix obtained from the experimental data, $\widetilde{F}(X)$ is the flexibility matrix yielded by considering the damage indexes, $\|.\|$ represents the Euclidean norm, and n_F is the number of degrees of freedom in the structural model. The flexibility matrix F is a function of free modal shapes of the structure as follows:

$$\boldsymbol{F} = \boldsymbol{\Phi}\lambda\boldsymbol{\Phi}^{\mathrm{T}} \quad \boldsymbol{\Phi}, \lambda \in \mathbb{C}^{n_F \times n_F} \tag{4.7}$$

in which:

$$\boldsymbol{\Phi} = \begin{bmatrix} \phi_{1,1} & \phi_{2,1} & \cdots & \phi_{n_F,1} \\ \phi_{1,2} & \phi_{2,2} & \cdots & \phi_{n_F,2} \\ \vdots & \vdots & \vdots & \vdots \\ \phi_{1,n_F} & \phi_{2,n_F} & \cdots & \phi_{n_F,n_F} \end{bmatrix} \tag{4.8}$$

and

$$\boldsymbol{\lambda} = \begin{bmatrix} \omega_1^2 & 0 & \cdots & 0 \\ 0 & \omega_2^2 & \cdots & 0 \\ \vdots & \vdots & \vdots & \vdots \\ 0 & 0 & \cdots & \omega_{n_F}^2 \end{bmatrix} \tag{4.9}$$

where $\boldsymbol{\Phi}$ represents the modal matrix, in which the i th column is the i th modal shape of the structure, and $\boldsymbol{\lambda}$ is a diagonal matrix, in which each diagonal element ω_i^2 indicates the square frequency corresponding to the modal shape Φ_i. As it can be seen from Eq. (4.7), the flexibility matrix $\boldsymbol{F}$ is a function of modal matrix $\boldsymbol{\Phi}$. The flexibility matrix obtained based on the experimental modal shapes represented by $\boldsymbol{\Phi}_E$ is called the experimental flexibility matrix $\boldsymbol{F}_{\mathrm{E}}$.

The flexibility matrix $\widetilde{\boldsymbol{F}}(X)$ yielded by considering the damage indexes X should be calculated based on the damaged modal shapes matrix $\boldsymbol{\Phi}_{\mathrm{D}}$, which is calculated as follows:

$$\boldsymbol{K}_{\mathrm{D}}\boldsymbol{\Phi}_{\mathrm{D}} = \boldsymbol{\lambda}\boldsymbol{M}\boldsymbol{\Phi}_{\mathrm{D}}, \boldsymbol{K}_{\mathrm{D}}, \boldsymbol{M} \in \mathbb{C}^{n_{\mathrm{F}} \times n_{\mathrm{F}}} \tag{4.10}$$

where $\boldsymbol{K}_D$ and $\boldsymbol{M}$ are the damaged stiffness matrix and mass matrix obtained from the finite element model of the structure. The damaged stiffness matrix $\boldsymbol{K}_D$ is a function of damage indexes X (or decision variables) as follows (Guedria and Hassine 2019):

$$\boldsymbol{K}_D = \sum_{i=1}^{n} (1 - x_i)\boldsymbol{k}_i \tag{4.11}$$

where x_i and $\boldsymbol{k}_i$ are the damage index and element stiffness for the ith element in the structure, respectively. The flexibility matrix is calculated by using Eq. (4.7) and the damaged modal shape $\boldsymbol{\Phi}_D$ obtained from Eq. (4.10) will be the damaged flexibility matrix $\widetilde{\boldsymbol{F}}(X)$. The damage index for each element can be any value between 0 and 1. For the elements without any damages, the damages index will be equal to 0, while the damage index equal to 1 means the element is entirely damaged. The interested readers are referred to finite element analysis books for the details about the mass and stiffness matrixes discussed in this section (Bathe 2006).

Guedria and Hassine (2019) investigated the performance of CA against other algorithms, such as DE and PSO, in damage detection of a cantilever beam as shown in Fig. 4.4 under different damage scenarios. The results revealed that the CA can detect the location of damages with a higher level of accuracy in comparison to other algorithms.

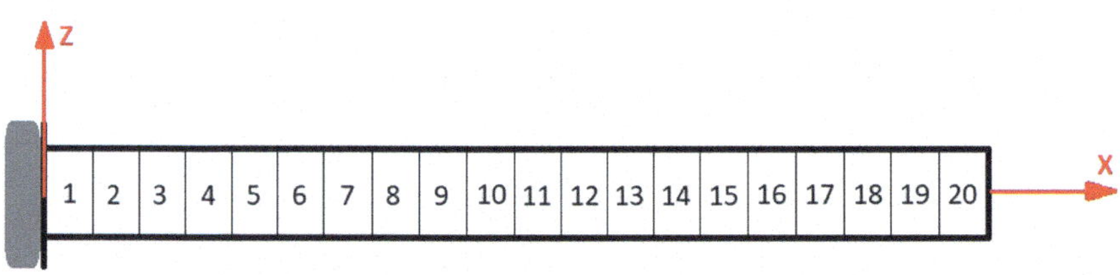

Fig. 4.4 Finite element model of the cantilever beam (Guedria and Hassine 2019)

4.3.2 *Optimum Design*

In the field of optimum design, a range of mechanical engineering benchmark design problems under different design constraints has been mathematically formulated. The studies for design optimisation are performed by different researchers. Reynolds and Ali (2014; 2008a, b) and Reynolds and Peng (2005) applied CA to solve different engineering design problems, including spring and gearbox design problems. dos Santos Coelho and Mariani (2006) proposed a hybrid CA and Gaussian Particle Swarm Optimisation (GPSO-CA) for the pressure vessel design problem. In the following subsections, the formulations for some of the commonly used mechanical engineering benchmark design problems are briefly presented.

4.3.2.1 Spring Design

The spring under tension/compression is one of the most commonly investigated engineering design problems in literature as illustrated in Fig. 4.5. The main aim of the problem is to minimise the weight of the spring by considering the wire diameter $x_1 = d$, the average coil diameter $x_2 = D$, and the number of active coils $x_3 = N$ as decision variables. The optimum design should satisfy the deflection, surge frequency, shear stress, and outside diameter constraints (Arora 2012). The optimum design problem can be mathematically expressed as follows:

$$
\begin{aligned}
&\text{Find: } X = [x_1, x_2, x_3] \\
&\text{To minimise: } W(X) = (x_3 + 2)x_2 x_1^2 \\
&\text{Subject to: } g_\delta(X) = 1 - \frac{x_2^3 x_3}{71,785 x_1^4} \le 0 \\
&\qquad g_\tau(X) = \frac{4x_2^2 - x_1 x_2}{12,566\left(x_2 x_1^3 - x_1^4\right)} + \frac{1}{5108 x_1^2} - 1 \le 0 \\
&\qquad g_\omega(X) = 1 - \frac{140.45 x_1}{x_2^2 x_3} \le 0 \\
&\qquad g_{\mathrm{D}}(X) = \frac{x_2 + x_1}{1.5} - 1 \le 0 \\
&\qquad 0.05 \le x_1 \le 2 \\
&\qquad 0.25 \le x_2 \le 1.3 \\
&\qquad 2 \le x_3 \le 15
\end{aligned}
\tag{4.12}
$$

where $g_\delta(.)$ represents the deflection constraint, $g_\tau(.)$ is the shear stress constraint, $g_\omega(.)$ indicates the frequency of surge waves constraint, and $g_D(.)$ is the outer diameter constraint.

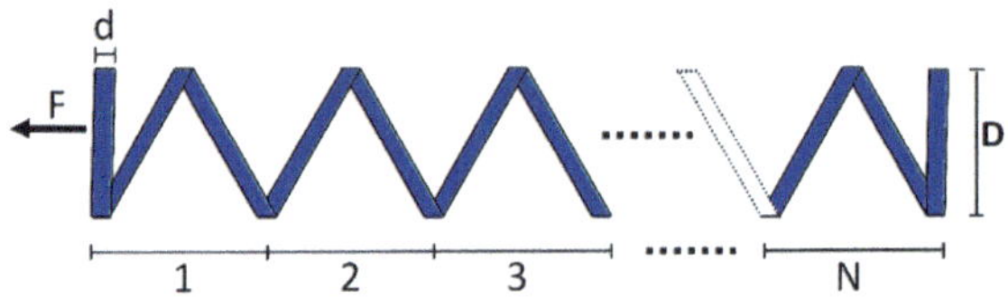

Fig. 4.5 Spring under tension/compression design problem (Ali and Reynolds 2014)

4.3.2.2 Pressure Vessel Design

The pressure vessel design problem shown in Fig. 4.6 consists of a cylindrical section that is capped at both ends by hemispherical heads. In this design problem, the total cost of the material, formation and welding is assumed as the objective function. The design variables are as follows: (i) shell thickness ($x_1 = t_s$), (ii) the thickness of the head section ($x_2 = t_h$), (iii) the inner radius of the cylindrical section ($x_3 = R$), and the length of the cylindrical section ($x_4 = L$). This is an optimisation problem with mixed continuous and discrete variables. The length L and radius R of cylindrical section are continuous variables, while the thickness parameters should be selected as integer multiples of 0.0625 in. The optimum design problem for the pressure vessel can be written as follows:

$$\begin{aligned}
&\text{Find } X = [x_1, x_2, x_3, x_4] \\
&\text{To minimise}: f(X) = 0.6224x_1x_3x_4 + 1.7781x_2x_3^2 \\
&\quad + 3.1661x_1^2x_4 + 19.84x_1^2x_3 \\
&\text{Subject to: } g_{\sigma_h}(X) = 0.0193x_3 - x_1 \le 0 \\
&\qquad g_{\sigma_l}(X) = 0.00954x_3 - x_2 \le 0 \\
&\qquad g_V(X) = -\pi x_3^2x_4 - \frac{4}{3}\pi x_3^3 + 1,296,000 \le 0 \\
&\qquad g(X) = x_4 - 240 \le 0 \\
&\qquad 0 \le x_1 \le 99 \\
&\qquad 0 \le x_2 \le 99 \\
&\qquad 10 \le x_3 \le 200 \\
&\qquad 10 \le x_4 \le 200
\end{aligned} \tag{4.13}$$

where $g_{\sigma_h}(.)$ represents the Hoop stress constraint, $g_{\sigma_l}(.)$ indicates the longitudinal stress constraint, and $g_V(.)$ is the volume constraint.

4.3.2.3 Cantilever Welded Beam Design

The cantilever welded beam shown in Fig. 4.7 is another popular mechanical design problem that has been widely used by different researchers to assess the performance of meta-heuristic algorithms. The length of the beam L is equal to 14 in. The Young's

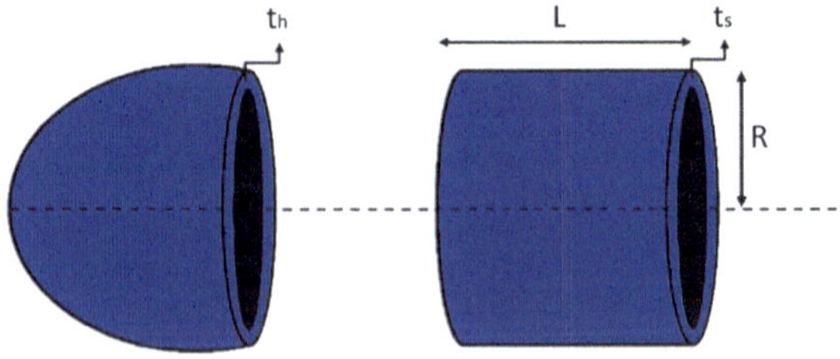

Fig. 4.6 Pressure vessel design problem (Ali and Reynolds 2014)

modulus E and shear modulus G of the beam are 30×10^6 psi and 12×10^6 psi, respectively. The weld section of the beam is subjected to shear force and moment resulting from a vertical load F as illustrated in Fig. 4.7. Hence, the shear and bending stresses at the welded section should not exceed the allowable values, which are the stress constraints in this problem. The deflection of the beam should also not exceed a given value. The design variables are the weld throat size $x_1 = t_w$ and the weld length $x_2 = l_w$, the height $x_3 = h$, and width $x_4 = b$ of the beam. The main objective of the problem is to minimise the cost of the beam under some design constraints on the shear and bending stresses as well ass deflection, which can be mathematically expressed as follows:

$$
\begin{aligned}
&\text{Find } X = [x_1, x_2, x_3, x_4] \\
&\text{To minimise: } f(X) = 1.10471x_1^2x_2 + 0.04811x_3x_4(14 + x_2) \\
&\text{Subject to:} \\
&g_\tau(X) = \tau(X) - \tau_{\text{all}} \leq 0 \\
&g_\sigma(X) = \sigma(X) - \sigma_{\text{all}} \leq 0 \\
&g_\delta(X) = \delta(X) - \delta_{\max} \leq 0 \\
&g_{\text{b}}(X) = F - P_b(X) \leq 0 \\
&g_1(X) = x_1 - x_4 \leq 0 \\
&g_2(X) = 0.10471x_1^2 + 0.04811x_3x_4(14 + x_2) - 5 \leq 0 \\
&g_3(X) = 0.125 - x_1 \leq 0 \\
&0.1 \leq x_1 \leq 2 \\
&0.1 \leq x_2 \leq 10 \\
&0.1 \leq x_3 \leq 10 \\
&0.1 \leq x_4 \leq 2
\end{aligned}
\qquad (4.14)
$$

where $g_\tau(.)$ represents the shear stress constraint, $\tau(.)$ indicates the existing shear stress, τ_{all} is the allowable shear stress, $g_\sigma(.)$ is the normal bending stress constraint, $\sigma(.)$ is existing normal stress resulting from the bending, σ_{all} is the allowable normal stress, $g_\delta(.)$ is the deflection constraint, $\delta(.)$ is the existing deflection of the beam, $\delta_{\max}$ is the maximum allowable displacement value, $g_{\text{b}}(.)$is the buckling constraint, $P_b(X)$ is the buckling load capacity, $g_1(.)$, $g_2(.)$, and $g_3(.)$ are the design constraints

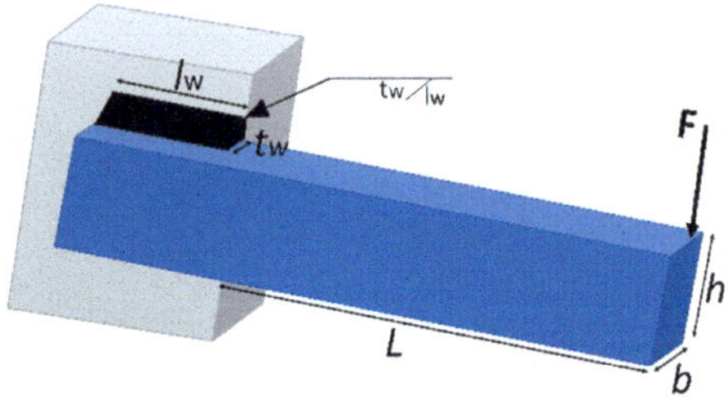

Fig. 4.7 Cantilever welded beam design problem (Ali and Reynolds 2014)

on the dimensions of the weld and beam. The values of existing normal and shear stresses are obtained from the following equations:

$$\sigma(X) = \frac{6FL}{x_4 x_3^2} \tag{4.15}$$

$$\tau(X) = \sqrt{\left(\frac{F}{\sqrt{2}x_1x_2}\right)^2 + 2\left(\frac{F}{\sqrt{2}x_1x_2}\right)\left(\frac{MR}{J}\right)\frac{x_2}{2R} + \left(\frac{MR}{J}\right)^2} \tag{4.16}$$

where:

$$M = F\left(L + \frac{x_2}{2}\right) \tag{4.17}$$

$$R = \sqrt{\left(\frac{x_1 + x_3}{2}\right)^2 + \frac{x_2^2}{4}} \tag{4.18}$$

$$J = 2\sqrt{2}x_1x_2\left[\frac{x_2^2}{12} + \left(\frac{x_1 + x_3}{2}\right)^2\right] \tag{4.19}$$

The buckling load capacity is calculated by the following formula:

$$P_b(X) = \frac{4.013E\sqrt{\frac{x_3^2x_4^6}{36}}}{L^2}\left(1 - \frac{x_3}{2L}\sqrt{\frac{E}{4G}}\right) \tag{4.20}$$

For further details about the assumptions on the objective function and stress calculations, interested readers are referred to Rao 2019.

4.3.3 *Composite Structures*

Composite materials can be described as multi-phase materials with improved properties that are constructed by the combination of two or more distinct materials with different features (Chung 2010). Laminated composite plates have been widely applied as structural elements in different engineering industries, such as aerospace, building, renewable energy, and shipping sectors. Their high strength-to-weight ratios make them attractive options for engineers to reduce the design weights and improve structural performance. Moreover, the changeable design parameters of laminated composite plates make it possible to optimise their performance to achieve the most cost-effective designs for a given application. In recent years, researchers have been focused on the application of meta-heuristic algorithms for the optimum design of laminated composite structures (Nikbakt et al. 2018; Hosseinzadeh et al. 2020; Jalili et al. 2021a).

Recently, Jalili et al. (2022) applied CA for the optimum design of laminated composite plates by considering the buckling load capacity as the objective function. Let us assume a laminated composite plate under in-plane compressive loads as shown in Fig. 4.8. The width and length of the plate are represented by a and b, respectively. In Fig. 4.8, N_x and N_y indicate the in-plane compressive loads in x and y directions, respectively. The study employed the classical laminated plate theory to calculate the buckling load capacity of the structure. Based on this assumption, the buckling load factor of the laminated plate can be written as follows (Soremekun et al. 2001):

$$\lambda_b(p,q) = \pi^2\left[\frac{p^4 D_{11} + 2(D_{12} + 2D_{66})(\mathrm{rpq})^2 + (rq)^4 D_{22}}{(ap)^2 N_x + (\mathrm{raq})^2 N_y}\right] \tag{4.21}$$

In the above equation, λ_b represents the buckling load factor of the laminated composite, r indicates the ratio of the length of the laminate to its width, p and q are the half-waves in the x and y directions, respectively, and D_{ij} are the parameters representing the bending stiffness of composite plate which are functions of material types, stacking sequences, and structure geometry. It is worth mentioning that Eq. (4.21) provides the exact buckling load factor for specially orthotropic laminates, while it will yield approximate results for the symmetrically balanced laminates. Different

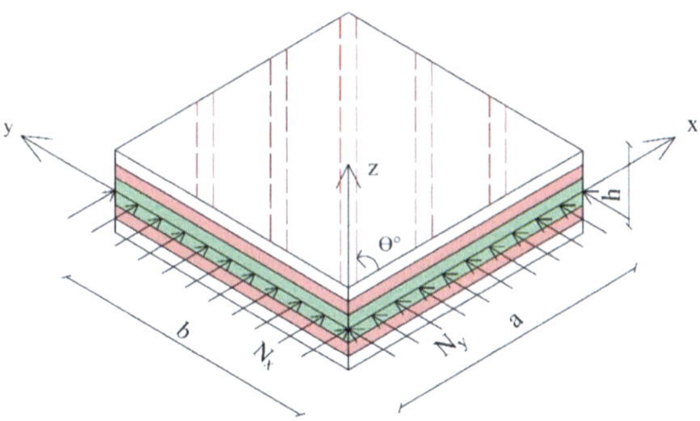

Fig. 4.8 Laminated composite plate structure (Jalili et al. 2022)

buckling load capacities can be calculated by Eq. (4.21) for different values of parameters p and q. However, the minimum buckling load factor, which is typically referred to as the critical buckling load factor, must be considered for design purposes.

Jalili et al. (2022) formulated the buckling maximisation problem of laminated composite plates as follows:

$$\begin{aligned} &\text{Find } X = [x_1, x_2, \ldots, x_n] \\ &\text{To maximise : } f(X) = \lambda_b \\ &\text{Subjected to : } x_i \in S = \left[s_1, s_2, \ldots, s_p\right] \end{aligned} \tag{4.22}$$

where X represents the vector containing the orientation variables of different layers of the composite structure, x_i indicates the orientation of the ith layer, and n is the number of design variables. The design variables in Eq. (4.22) are discrete variables that should be selected from a given set of allowable fibre orientations determined based on the manufacturing constraints. The vector S in Eq. (4.22) consists of allowable orientations that can be assigned for different layers of the laminated composite structure. The study assumed a wide range of orientations within the vector S in order to challenge the performance of CA in finding the optimum buckling load capacity for the laminated composite structure.

4.3.4 Optimal Vibration Control

Wang et al. (2015a) applied CA to control the vibration of the vehicle suspension system. The authors modelled the quarter of the vehicle and considered one wheel in their vibration control approach. Figure 4.9 illustrates the free body diagram of a wheel, in which m_t and m_g represent the sprung and unsprung masses, respectively, k_t indicates the stiffness of the tyre, c_0 is the linear damping coefficient, k_s is the stiffness provided by the suspension system, and u is the active control force generated by the actuator. The authors applied the point contact theory to define the contact between the tyre and the surface of the road. The system shown in Fig. 4.9 is a three degree of freedom dynamic system with displacement variables of x_0, x_1 and x_2 representing the surface of the road, vertical displacements of wheel and bodywork, respectively. The differential motion equations for this dynamic system can be written as follows (Wang et al. 2015):

$$m_2\ddot{x}_2 + c_0(\dot{x}_2 - \dot{x}_1) + k_s(x_2 - x_1) + u = 0 \tag{4.23}$$

$$m_1\ddot{x}_1 + c_0(\dot{x}_1 - \dot{x}_0) + k_t(x_1 - x_0) + k_s(x_1 - x_2) - u = 0 \tag{4.24}$$

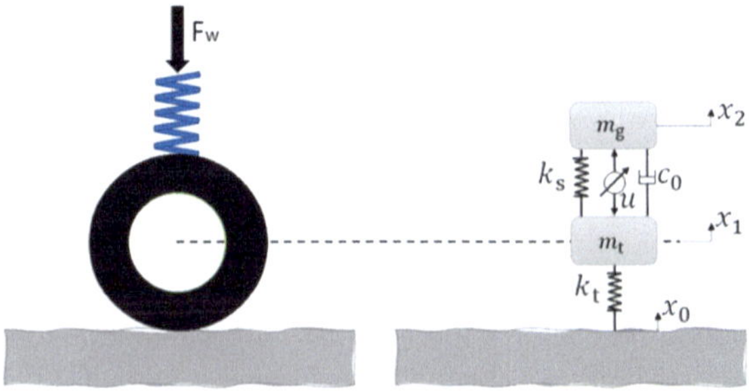

Fig. 4.9 Dynamic model of car wheel (Wang et al. 2015a)

By solving the above equations, the displacements and their derivatives with respect to the time can be obtained. Wang et al. (2015a) defined a Fuzzy-PID control strategy for the active suspension system and defined the following problem to optimise the fuzzy control variables:

$$\text{Find: } X = [x_1, x_2, \ldots, x_n] \tag{4.25}$$

$$\text{To minimise : } f(X) = \sqrt{\frac{1}{N}\sum_{t=1}^{N} |\ddot{x}_2|^2} \tag{4.26}$$

where n represents the number of fuzzy control variables, $f(.)$ represents the objective function which is defined as the root-mean-square value of bodywork vertical vibration acceleration, and N is the number of time steps that the accelerations are recorded.

4.3.5 *Vehicle Safety*

Kobti et al. (2006b) applied CA to guide the drivers on using the child vehicle safety restraint. The study employed CA to provide a social knowledge exchange mechanism between the different drivers to enhance the safety requirements for child vehicle safety restraint. The results revealed that the learning mechanism has a significant impact on the injury level in children.

4.3.6 *Flight Control*

Meng et al. (2011) applied CA based on PSO (CAPSO) to solve the flight control laws problem. The flight control law designers try to develop algorithms that can transform the movement produced by an aircraft pilot into movements of the aircraft control surfaces. However, due to the high nonlinearity and uncertainty of various flight parameters, the validation of flight control laws is a challenging problem. The flight control laws problem is a complex and nonlinear problem that looks for the worst combination of flight parameters that satisfy a set of clearance criteria (Joos 2009). The authors selected the aircraft mass m, aircraft inertia around the y-axis I_y, pitching moment due to pitch rate C_{mq}, pitching moment due to angle of attack (AoA) C_{ma}, and pitching moment due to elevator C_{meta} as decision variables. The study considered two clearance criteria, the unstable eigenvalue and stability margin. The authors investigated a flight condition with Mach 0.7765 and an attitude of 10,000 m to evaluate the performance of the CAPSO algorithm.

4.3.7 *Gas Turbine Operation*

Gas turbines are continuous and internal combustion engines that are widely used in the energy generation industry. They offer a variety of advantages, such as high power-to-weight ratios, low emission, and fuel flexibility. However, the maintenance difficulties, emission pollutions, pressure oscillations as well as complexity of mathematical formulation are serious challenges in the gas turbine operation life cycle. Pierezan et al. (2019) applied the Cultural Coyote Optimisation Algorithm (CCOA) to optimise the operation of heavy-duty gas turbines.

4.3.8 *Renewable Energy*

Tazay et al. (2020) applied CA for the optimum design of Hybrid Renewable Energy Systems (HRES) of buildings. The HRES for the building is a combination of Photo-Voltaics (PVs), wind turbines, battery banks, H_2 tanks, fuel cells, and electrolysers as shown in Fig. 4.10. The study considered the cost minimisation of HRES under a set of renewable resource constraints as follows:

$$
\begin{aligned}
&\text{Find: } X = [N_{\text{PV}}, N_B, N_{\text{WT}}, N_{\text{FC}}, N_E, N_H] \\
&\text{To minimise: } f(X) \\
&\text{Subject to: } N_{\text{PV}}^{\min} \leq N_{\text{PV}} \leq N_{\text{PV}}^{\max} \\
&\qquad N_B^{\min} \leq N_B \leq N_B^{\max} \\
&\qquad N_{\text{WT}}^{\min} \leq N_{\text{WT}} \leq N_{\text{WT}}^{\max} \\
&\qquad N_{\text{FC}}^{\min} \leq N_{\text{FC}} \leq N_{\text{FC}}^{\max} \\
&\qquad N_E^{\min} \leq N_E \leq N_E^{\max} \\
&\qquad N_H^{\min} \leq N_H \leq N_H^{\max} \\
&\qquad 0 \leq L \leq L_{\max}
\end{aligned}
\tag{4.27}
$$

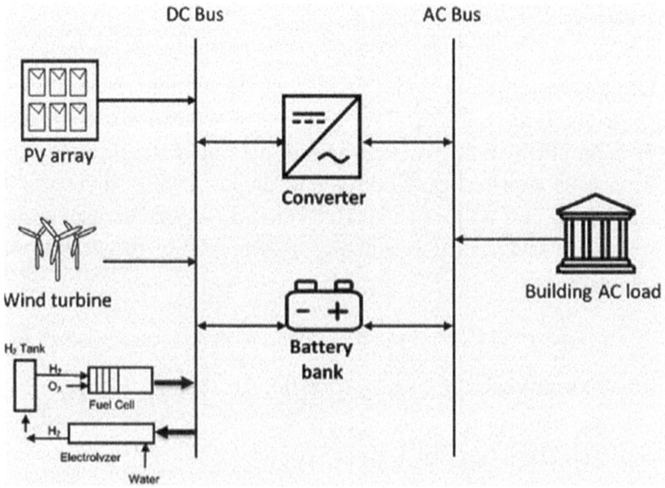

Fig. 4.10 Hybrid renewable energy system for a building (Tazay et al. 2020)

where N_{PV}, N_B, N_{WT}, N_{FC}, N_E, and N_H are the number of PVs, battery banks, wind turbines, fuel cells, electrolysers, and hydrogen tanks, respectively, $f(.)$ is the total net cost function of HRES, L represents the Loss of Power Supply Probability (LPSP), the variables with min and max superscripts show the lower and upper bounds for the corresponding variables, respectively, and L_{max} is the maximum allowable LPSP value.

Tazay et al. (2020) compared the performance of CA against PSO and JAYA algorithms for HRES optimisation of a University Building in Saudi Arabia and reported the competitive performance of CA.

4.4 Chemical Engineering

CA has also been employed to find solutions for the optimisation problems in chemical engineering as listed in Table 4.3. The following subsection focuses on the application of CAs in chemical engineering problems.

Table 4.3 Applications of CAs and their variants in chemical engineering

Application area	References
Process optimisation	Haikal and El-Hosseni (2011)
Ammonia synthesis	Xu et al. (2010), Xu et al. (2012)
Molecular crystallography	Tremayne et al. (2009)
Diesel blending	Gao et al. (2020)
Wax precipitation	Amiri-Ramsheh et al. (2022)

4.4.1 Penicillin Production

Haikal and El-Hosseni (2011) proposed a cultural-based genetic algorithm to solve the Penicillin production problem. The problem is a fed-batch fermentor for the production of penicillin which has been investigated by different researchers using different optimisation approaches (Cuthrell and Biegler 1989; Lim et al. 1986; Luus 1993). The problem looks for the optimum feed rate to a fed-batch reactor to maximise Penicillin production under various equality and inequality constraints, which can be formulated as follows:

$$\begin{aligned}
&\text{Find}: X = \left[u(t), t_f\right] \\
&\text{To minimise}: f(t) = x_2(t)x_4(t), t \in \left[t_0, t_f\right] \\
&\frac{dx_1(t)}{dt} = h_1 x_1 - u\left(\frac{x_1(t)}{500x_4(t)}\right) \\
&\frac{dx_2(t)}{dt} = h_2 x_1(t) - 0.001x_2(t) - u\left(\frac{x_2(t)}{500x_4}\right) \\
&\frac{dx_3(t)}{dt} = \frac{h_1 x_1(t)}{0.47} - h_2\frac{x_2(t)}{1.2} \\
&\quad - x_1(t)\left(\frac{0.029x_3(t)}{0.0001 + x_3(t)}\right) + \frac{u}{x_4(t)}\left(1 - \frac{x_3(t)}{500}\right) \\
&\frac{dx_4(t)}{dt} = \frac{u(t)}{500} \\
&h_1 = 0.11\left(\frac{x_3(t)}{0.006x_1(t) + x_3(t)}\right) \\
&h_2 = 0.0055\left(\frac{x_3(t)}{0.0001 + x_3(t)(1 + 10x_3(t))}\right) \\
&0 \le x_1(t) \le 40 \\
&0 \le x_3 \le 25 \\
&0 \le x_4(t) \le 10 \\
&0 \le u(t) \le 50
\end{aligned} \tag{4.28}$$

where $x_1(t), x_2(t), x_3(t)$ and $x_4(t)$ represent the biomass, penicillin, substrate concentration, and volume. The initial conditions for the mentioned parameters are as follows: $x_1(t_0) = 1.5$, $x_2(t_0) = 0$, $x_3(t_0) = 0$, and $x_4(t_0) = 7$. Interested readers are referred to Cuthrell and Biegler (1989) for more details.

4.4.2 Ammonia Synthesis

Ammonia is a chemical material produced from nitrogen and hydrogen, and has important applications, such as fertilising agricultural crops and the production of plastics, fibres, explosives, and nitric acid. Ammonia production technology has been evolved during the past decades (Valera-Medina and Banares-Alcantara 2020).

Xu et al. (2010, 2012) described the ammonia synthesis system of medium and small fertiliser plants in China as shown in Fig. 4.11 and formulated the ammonia production process in terms of an optimisation problem, in which the main objective is to maximise the net value of ammonia Δ_{NH_3} under some inequality constraints as follows:

$$
\begin{aligned}
&\text{Find: } X = [x_1, x_2, \ldots, x_5] \\
&\text{To maximise: } f(X) = \Delta_{NH_3}(X) \\
&x_3 - x_2 < 0 \\
&x_4 - x_3 < 0 \\
&x_4 - x_5 < 0 \\
&x_5 - x_2 < 0 \\
&55 \le x_1 \le 65 \\
&470 \le x_2 \le 500 \\
&475 \le x_3 \le 495 \\
&460 \le x_4 \le 485 \\
&470 \le x_5 \le 490
\end{aligned}
\tag{4.29}
$$

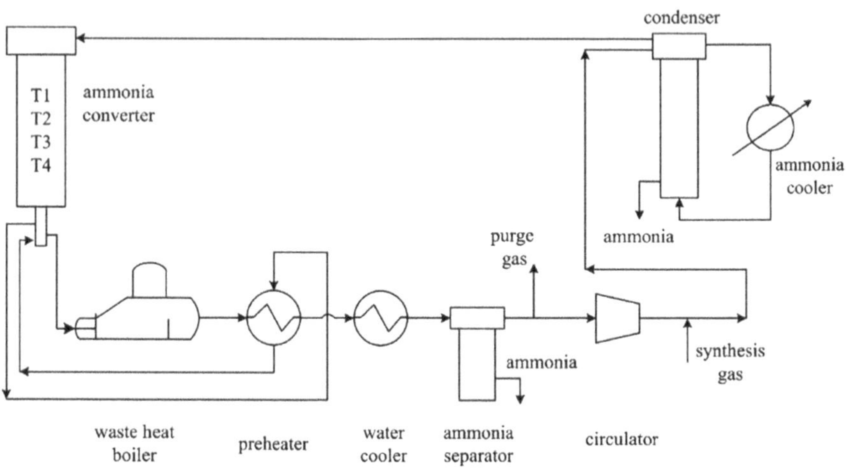

Fig. 4.11 Flowchart of the ammonia production investigated by Xu et al. (2010, 2012)

where $\Delta_{NH_3}(.)$ represents the net ammonia value, x_1 is the hydrogen concentration in recycled gas, x_2 indicates the hot-spot temperature of the axial stage (°C), and x_3, x_4, x_5 are the hot-spot temperature of different radial stages of the production process. Xu et al. (2010, 2012) trained and used the Back-Propagation Neural Networks (BPNN) to calculate the net ammonia value for a given vector of input variables.

4.4.3 *Diesel Blending*

Gao et al. (2020) applied a hybrid Cultural HS (CHS) algorithm for the diesel blending problem. In the diesel blending units, the different components of intermediate oils and additives are combined to gain new oil products with enhanced properties that meet quality and environmental regulations (Gao et al. 2020; Ounahasaree et al. 2016). Researchers have been formulated the diesel blending process as an optimisation problem (Long et al. 2021). The main objective of the problem is to maximise the amount of profit of the product under some constraints on quality and environmental requirements. The problem can be formulated as follows (Gao et al. 2020):

$$\begin{aligned}
&\text{Find: } X = [x_1, x_2, \ldots, x_n] \\
&\text{To maximise: } f(X) = \sum_{i=1}^{N} Q_i \\
&\text{Subject to:} \\
&g_i(X) = 0, \quad i = 1, 2, \ldots, n_{\text{ec}} \\
&h_i(X) < 0, \quad i = 1, 2, \ldots, n_{\text{uc}}
\end{aligned} \tag{4.30}$$

where X is a vector containing the flow rates from different units in the diesel blending process, Q_i represents the flow rate (or amount) of the ith product, N is number of products, $g_i(X)$ and $h_i(X)$ represent some constraints on flow rates in different units, operation, and quality constraints, n_{ec} is the number of equality constraints, and n_{uc} is the number of inequality constraints.

4.4.4 *Wax Precipitation*

One of the challenges in oil production is the wax deposition in the oil pipelines, which increases the costs and damages in oil production and transportation (Aiyejina et al. 2011). According to the definition, "*wax precipitation within pipelines at and below the Cloud Point or Wax Appearance Temperature (WAT) can lead to gelling that inhibits flow by causing significant non-Newtonian behaviour and increasing effective viscosities as the temperature of a waxy crude oil approaches its Pour Point*" (Pedersen and Rønningsen 2003; Aiyejina et al. 2011). This phenomenon usually happens in low temperatures and decreases the effective diameter of oil pipelines. In wax deposition terminology, the Wax Disappearance Temperature (WDT) refers to the lowest temperature at which the wax deposits start to melt and vanish which is a key parameter alongside the WAT (Amiri-Ramsheh et al. 2022).

Amiri-Ramsheh et al. (2022) developed an AI-based approach to predict the WDT based on a set of available data. The study developed an Artificial Neural Network (ANN) that is trained by different meta-heuristic algorithms, such as CA, BBO, and TLBO. In the learning model, the approach considered a data set, in which the pressure and molar mass of the oil composition are considered as input and the WDT is treated as an output.

4.5 Electrical Engineering

CAs have been applied by different researchers to deal with a variety of optimisation problems in the field of electrical engineering as summarised in Table 4.4. In the following subsections, these application studies are investigated in detail.

4.5.1 Economic/Environmental Dispatch of Power Systems

The economic/environmental dispatch problem is one of the most important problems in the field of the electric power systems. The problem looks for optimum output schedules of the electricity generation facilities to allocate system load demand and minimise the total fuel cost under some transmission and operational constraints. The economic/environmental dispatch problem has been addressed by different researchers in the field of the electrical engineering and computer science (Qu et al. 2018). Let us assume a power network consisted of n thermal power generators with the power generation capacities of $P = [p_1, p_2, \ldots, p_n]$. The economic dispatch problem considers the cost of generated power as the objective function, while the environmental dispatch problem tries to minimise the emission rate resulted from the power generation. In following subsections, the formulations for the economic and environmental dispatch problems are presented.

4.5.1.1 Economic Dispatch

The economic dispatch problem can be formulated as an optimisation problem as follows:

$$\text{Find: } P = [p_1, p_2, \ldots, p_n]$$

Table 4.4 Applications of CAs and their variants in electrical engineering

Application area	References
Economic/environmental dispatch of power systems	Goudarzi et al. (2016, 2017), Zhang et al. (2013a), Bhattacharya et al. (2012a), Coelho et al. (2009), Oliveira De Freitas et al. (2018b), Ya-Li and Li-qing (2010)
Power distribution systems	Haldar et al. (2014, 2015)
Phased array radars	Khan et al. (2014, 2017)
Wireless sensor networks	Guo et al. (2013a, b), Kulkarni et al. (2020), Murugadass and Sivakumar (2020), Nabavi and Najafi (2021)
Signal processing	Gao and Diao (2010, 2011), Gao and Xu (2013)
Maintenance scheduling	Digalakis and Margaritis (2002)
Digital watermarking	Gao et al. (2019)
Power system stabiliser design	Khodabakhshian and Hemmati (2013), Tavakoli et al. (2014)
Optimal reactive power dispatch	Bhattacharya et al. (2012b)
Optimal operation of cascade hydropower station	Wang et al. (2009), Xie et al. (2009)
Optimal design of passive power filters	Guo et al. (2008)
Hydrothermal scheduling	Yuan and Yuan (2006), Lu et al. (2011), Yuan et al. (2009)
Hydroproduction scheduling	Yuan et al. (2008)
Multi-robot tasks allocation	Yu et al. (2011)
Robot soccer players	Reynolds and Chung (1997), Ali et al. (2014b)
Sea-surface weak targets detection	Cai et al. (2018b)
Fault detection	Pan et al. (2010)
PID parameter optimisation	Xiao et al. (2010)
Frequency assignment	Alami and Imrani (2008)

$$\text{To minimise}: f_c(P) = \sum_{i=1}^{n} \left(a_i + b_i p_i - c_i p_i^2\right)$$

Subject to:

$$\sum_{i=1}^{n} p_i - p_D - p_L = 0$$
$$p_i^{\min} \le p_i \le p_i^{\max} \tag{4.31}$$

where P represents the vector containing the power output from each generator, $f_c(.)$ is the fuel cost function of power generation per hour (\$/hour), the parameters a_i, b_i and c_i are the fuel coefficients for the ith generator, p_D indicates is the total power demand, p_L is the total power loss, $p_i^{\min}$ and $p_i^{\max}$ are the minimum and maximum power outputs of the ith generator, respectively. The fuel cost function can be obtained by considering the rippling effects which are results from the valve openings. In this case, the fuel cost function is written as follows (Zhang et al. 2013b):

$$f_v(P) = \sum_{i=1}^{n} \left(a_i + b_i p_i - c_i p_i^2 + \left|d_i \sin\left[e_i\left(p_i^{\min} - p_i\right)\right]\right|\right) \tag{4.32}$$

where $f_v(.)$ represents the cost function considering the valve point effects, and the parameters d_i and e_i are the cost coefficients.

Researchers have been applied different versions of CAs to solve the economic dispatch problem. Bhattacharya et al. (2012a) applied CA to solve the economic dispatch problem of a power system with 13 generators. Coelho et al. (2009) proposed a cultural DE algorithm with a diversity measure for different economic dispatch case studies, including power systems with 13 and 40 generators.

4.5.1.2 Environmental Dispatch

As was mentioned earlier, the objective function of the environmental economic dispatch problem is to minimise the emission rate of the power generation, which is written as follows (Qu et al. 2018):

$$f_e(P) = \sum_{i=1}^{n} \left[10^{-2}\left(\alpha_i + \beta_i p_i + \gamma_i p_i^2\right) + \xi_i e^{\varphi_i p_i}\right] \tag{4.33}$$

where $f_e(.)$ represents the emission rate of the power system and the parameters α_i, β_i, γ_i, ξ_i and φ_i are the constant emission coefficients. It should be noted that the constraints for the environmental dispatch problem are similar to those in the economic dispatch problem.

Goudarzi et al. (2016, 2017) applied different versions of CAs with different influence functions to solve the combined economic/environmental dispatch problem of power systems with 5, 20, and 50 power generator units. The authors have converted the multi-objective problem of economic/environmental dispatch problem to an equivalent single objective one by introducing the following mixed objective function:

$$f(P) = f_v(P) + f_e(P) \tag{4.34}$$

The comparisons between the results obtained by the different versions of CAs and other algorithms in the literature, such as BBO, PSO, GA, Artificial Bee Colony (ABC), and Firefly Algorithm (FA), showed that the CAs are computationally more efficient, and they can produce the results with higher qualities.

Zhang et al. (2013a) employed an Enhanced Multi-Objective CA (EMOCA) to solve the dispatch problem by treating the economic and environmental objectives as two separate objective functions and investigated the Pareto front solutions. The results obtained from the power systems with 6 and 10 units revealed that the EMOCA can generate better Pareto solutions than the multi-objective version of DE and non-dominated sorting GA (NSGA-II). The study also revealed that the valve point effect can potentially affect the Pareto front distribution. In a similar study, Oliveira De Freitas et al. (2018b) applied CAs combined with TS and SA local search techniques to solve the multi-objective problem of economic and environmental dispatch problem.

4.5.2 Power Distribution Systems

Capacity allocation problem in the power distribution networks looks for the optimal location, size, and number of shunt capacitors in the system to improve the performance of the systems in several ways, such as cost minimisations as well as power and energy loss minimisations (Kavousi-Fard and Samet 2013). The optimal capacity allocation problem can be formulated as follows:

$$\begin{aligned}
&\text{Find: } X = \left[T_1, T_2, \ldots, T_{n_b}, Q_1^c, Q_2^c, \ldots, Q_{n_b}^c\right], X \in \mathbb{C}^{2n_b} \\
&\text{To minimise: } f_c(X) = K_p(P_L) + \sum_{i=1}^{n_b} K_i Q_i^c \\
&\text{Subject to: } V_{\min} \le V_i \le V_{\max}
\end{aligned} \tag{4.35}$$

where $T_1, T_2, \ldots, T_{n_b}$ represent the exact bus locations which are Boolean variables, $Q_1^c, Q_2^c, \ldots, Q_{n_b}^c$ are the capacitor values, n_b is the total number of buses in the distribution system, $f_c(.)$ denotes the combined cost due to real power loss and capacitor installation, $K_p(.)$ represents the equivalent cost per unit of power loss, P_L is the power losses of the network, K_i is the capacitor installation cost, Q_i^c is the capacitor value installed at the ith bus of the distribution system, V_i is the voltage value of the ith bus, $V_{\min}$ and $V_{\max}$ are the minimum and maximum allowable bus voltage, values, respectively.

Haldar et al. (2014, 2015) employed Modified CA (MCA) for optimal capacitor placement in radial distribution systems and investigated the cost minimisation of 9-bus, 34-bus, and 69-bus standard distribution systems. Compared to other approaches, such as GA and PSO, the results revealed that the MCA can further reduce the total cost resulting from the power loss and capacitor installation.

4.5.3 Phased Array Radars

Phased array radars are electronically controlled devices that can provide an angular scanning of the horizon without any mechanical rotation of the antenna array (Scott 2008). The design of arrays is one of the important research directions in the field of phased array radars. The arrays in phased array radars consisted of a set of sensors that have significant impact on the radar's performance and the failure of any sensor can cause the radar to fail in identifying the targets. Hence, development of efficient techniques for the sensor fault detection in phased array radars is vital. Khan et al. (2014, 2017) joined this research line by applying hybrid CA and DE for sensor fault detection in phased array radars. The authors formulated the sensor fault detection as a minimisation problem of following objective function:

$$f(X) = \sum_{i=1}^{90^{\circ}} \left| P_F(\theta_i) - \widetilde{P}(\theta_i) \right|^2 \tag{4.36}$$

where $f(.)$ represents the objective function, which is defined as sum of squared errors in prediction the location of the faults, X represents the location of faults, $P_F(\theta_i)$ is desired faulty pattern, and $\widetilde{P}(\theta_i)$ is the approximate faulty pattern obtained from the optimisation algorithm. The interested readers are referred to Khan et al. (2014, 2017) for more details about faulty patterns.

4.5.4 Wireless Sensor Networks

A Wireless Sensor Network (WSN) can be described as a network with several interconnected sensor nodes designed for data collection from the target environment or physical conditions (Patil and Chen 2017). They have important applications in different industries, such as agriculture (Hamami and Nassereddine 2020), healthcare (Baig Mohammad and Shitharth 2021), and oil and gas (Divoux et al. 2015). Each sensor in the network receives, processes, and transmits the data that require enough amount of energy. The energy for the sensors is usually supplied by batteries with a certain level of capacity, which are difficult to replace or charge (Tam et al. 2021). The network configuration, the distance between the nodes, and the amount of data can significantly affect energy consumption in the network. These challenges motivated the researchers to optimise the wireless sensor networks in terms of energy efficiency and lifetime extension.

Guo et al. (2013a, b) applied a Multi-Objective Quantum CA (MOQCA) to maximise the effective coverage ratio and minimise the redundancy ratio which represents the coverage performance of the network and the average distribution of nodes in the network, respectively. The study converted the multi-objective wireless sensor network problem to a single objective one as follows (Guo et al. 2013a, b):

$$\begin{aligned}
&\text{Find} : X = [x_1, x_2, \ldots, x_n] \\
&\text{To minimise} : f(X) = \omega_1 r_c(X) + \omega_2(1 - r_r(X)) \\
&\text{Subject to} : \omega_1 + \omega_2 = 1
\end{aligned} \tag{4.37}$$

where $f(.)$ represents the objective function, X is the vector containing the coordinates of nodes in the network, ω_1 and ω_2 are the weighting parameters, $r_c(X)$ and $r_r(X)$ represent the effective coverage and redundancy ratios of the sensor network, respectively. Let us assume A be a two-dimensional monitored area consisting of $a \times b$ grids, in which the intersections of beelines are pixel points labelled by p_j. The sensor network is consisted of N_s sensors represented by $S = [s_1, s_2, \ldots, s_{N_s}]$. Each node with coordinates x_i and y_i in two-dimensional space covers a circular sensing area with radius r, which can be represented as $s_i = \{x_i, y_i, r\}, i = 1, 2, \ldots, N_s$. The coverage probability of the point p_j by the sensor s_i depends on the distance, which can be calculated as follows (Guo et al. 2013a, b):

$$P_{ij} = \begin{cases} 1 & d_{ij} < r - r_\mathrm{e} \\ \mathrm{e}^{\left(\frac{-\alpha_1 \lambda_1 \beta_1}{\lambda_2 \beta_2} + \alpha_2\right)} & r - r_\mathrm{e} < d_{ij} < r + r_\mathrm{e} \\ 0 & \text{other} \end{cases} \tag{4.38}$$

where P_{ij} represents the coverage probability of the point p_j by the sensor s_i, r_e is the measuring reliability of the sensor nodes which is between 0 and r, $\lambda_1 = r_e - r + d_{ij}$, and $\lambda_2 = r_e + r + d_{ij}$. In the above equation, $\alpha_1, \alpha_2, \beta_1$, and β_2 are the parameters that depend on the type of the sensors.

Based on the aforementioned definitions, the ratio $r_c(X)$ is defined as the ratio of effective coverage area to the total monitoring area, and $r_r(X)$ is defined as the ratio of the difference between the coverage and efficient coverage areas to the coverage area, which are calculated as follows (Guo et al. 2013a, b):

$$r_c(X) = \frac{\sum_{j=1}^{a \times b} \sum_{i=1}^{N_S} P_{ij}}{a \times b} \tag{4.39}$$

$$r_r(X) = \frac{N_S \pi r^2 - \sum_{j=1}^{a \times b} \sum_{i=1}^{N_S} P_{ij}}{N_S \pi r^2} \tag{4.40}$$

The application results of MOQCA for optimisation of a wireless sensor network with the size of 20×20 reported by Guo et al. (2013a, b) showed that the MOQCA provides better results in comparison to the NSGA-II and other multi-objective versions of CAs.

In another study, Murugadass and Sivakumar (2020) Hybrid Elephant Herding Optimisation and CA (HEHO-CA) for energy optimisation in wireless sensor networks. The authors proposed another form of the objective function for the problem, in which the wireless networks with different clusters have been modelled. Each sensors cluster has a cluster-head sensor that is responsible for data collection and aggregation. The study performed by Nabavi and Najafi (2021) investigated the optimal cluster-head sensor in a WSN for objective functions based on the quality of service (QoS), including residual energy, connection quality, delivery rate, and end-to-end delay. The study converted the multi-objective problem into an equivalent single-objective one.

In some applications, the locations of the sensors are not constant due to the dynamic nature of the environment. Let us assume a network of sensors used to track the behaviour of animals in nature. In this case, the locations of sensors are not stationary and they should be accurately calculated. Hence, the wireless sensor network optimisation can be performed for the optimum sensor location estimation. Kulkarni and Desai (2020) joined this line of research and applied CA for error minimisation in optimum localisation of wireless sensor networks.

4.5.5 *Signal Processing*

Signal processing is one of the sub-disciplines of electrical engineering that focuses on the manipulation, interpretation, and feature extraction of different real-world signals, such as audio and video (Brice 2001). In real-world applications, the recorded signals include some unwanted features that can cause unrealistic results. As its name suggests, a filter is referred to a device or process that is used for signal restoration or elimination of unwanted features from a given signal. Finite Impulse Response (FIR) and Infinite Impulse Response (IIR) filters have important applications in signal processing. The design of FIR and IIR filters plays a key role in their performance and reliability for a given application. In recent years, researchers have been formulated the design problem of FIR and IIR filters as an optimisation problem and applied meta-heuristics to enhance their performance (Dwivedi et al. 2018).

The design optimisation problem of an FIR filter looks for optimum values of a set of coefficients that provides the best filter performance. In most cases, the error between the ideal frequency response and the designed filter response is selected as the objective function. The filter design optimisation problem can be mathematically written as follows (Gao and Diao 2010):

$$\begin{aligned} &\text{Find } X = [x_1, x_2, \ldots, x_n] \\ &\text{To minimise}: f_E(X) = \sum\nolimits_{i=1}^{M} \left| H_d\left(e^{j\omega_i}\right) - H\left(e^{j\omega_i}, X\right) \right|^2 \end{aligned} \tag{4.41}$$

where $f_E(.)$ represents the error function, M is the number of frequency intervals, $H_d(.)$ indicates the ideal frequency response, $H(.)$ is the current frequency response obtained from the designed filter with parameters $X = [x_1, x_2, \ldots, x_n]$, and ω_i is the frequency value at the ith interval. The form of the frequency response $H(.)$ depends on the filter type and is different for FIR and IIR filters (Gao and Diao 2010).

Gao et al. (2010, 2011) applied a Differential CA (DCA) and Cultural Firework (CF) algorithm for the optimum design of FIR and IIR digital filters. The results revealed that the DCA and CF algorithms can find better designs and provide further reductions in the error values in comparison to PSO, Quantum PSO (QPSO), and Adaptive Quantum PSO (AQPSO) algorithms.

4.5.6 Maintenance Scheduling

The maintenance scheduling of electrical systems is an optimisation problem, in which the objective is to improve the performance of electric generators through maintenance activities, reduction of the unexpected interruptions in energy generation, and lifetime extension of the electrical systems (Reihani et al. 2012; Assis et al. 2022). In recent years, researchers have been employed meta-heuristic approaches to solve this problem (Yare and Venayagamoorthy 2010; Balaji et al. 2016; Lakshminarayanan and Kaur 2018).

Digalakis and Margaritis (2002) applied multi-population CA to solve the electrical generator scheduling problem. Let us assume n number of electrical generators that are intended to produce the power for J number of periods. Each generator i should be under maintenance operations for m_i continuous period. The problem assumed the start time of maintenance operations for each generator as decision variables. The main objective of the optimum maintenance scheduling problem is to minimise the fuel and maintenance costs under some constraints which can be mathematically formulated as follows (Digalakis and Margaritis 2002):

$$
\begin{aligned}
&\text{Find: } X = [x_1, x_2, \ldots, x_n] \\
&\text{To minimise: } f(X) = \sum_{i=j}^{J} \left(C_j^f \sum\nolimits_{i=1}^{n} p_{i,j} \right) + \sum_{i=1}^{n} c_{i,j}(x_i) \\
&\text{Subject to: } 0 \le p_{i,j} \le C_i \left(1 - y_{i,j}\right) \\
&\sum\nolimits_{i=1}^{n} p_{i,j} = D_j \text{ for } \ j = 1, 2, \ldots, J \\
&\sum_{i=1}^{n} (1 - y_{i,j}) C_i \ge \left(D_j + R_j\right)
\end{aligned}
\tag{4.42}
$$

where C_j^f represents the fuel cost per unit power output during the jth period, $p_{i,j}$ denotes the power generated by the ith generator during the period j, $c_{i,j}(.)$ is the maintenance cost of the ith generator during the jth period which is a function of starting maintenance time x_i, C_i is the operating capacity of ith generator, Y_i is a parameter representing the maintenance period indexes for the ith unit which is a function of maintenance starting times, D_j indicates the anticipated demand of the systems over the jth period, and R_j is the reserve capacity during the jth period. As it can be seen from the above equation, the problem is subject to the unit and overall outputs and demand constraints. As the maintenance activities should be continued up to m_i for each unit, the vector Y_i is obtained as follows:

$$y_{i,j} = \begin{cases} 0 \text{ if } j = 1, 2, \ldots, x_i - 1 \\ 1 \text{ if } j = x_i, x_i + 1, \ldots, x_i + m_i - 1 \\ 0 \text{ if } j = x_i + m_i, \ldots, J \end{cases} \tag{4.43}$$

4.5.7 Digital Watermarking

Nowadays, a huge amount of multimedia products, such as images, videos, documents, and audio, are being shared or reused, in which the copyrights of the product owners are not necessarily protected. Digital watermarking is a technology that has important applications in the copyright protection of multimedia digital products (Tefas et al. 2009). Roughly speaking, digital watermarking techniques embed a set of identification information into the multimedia products to protect the ownership rights of the original creator without affecting the functionality of the product (Kessler and Hosmer 2011; Ou et al. 2016). Gao et al. (2019) applied a Cultural Invasive Weed Algorithm (CIWA) for optimising the robustness and the imperceptibility of the watermarking system.

4.5.8 Power System Stabiliser Design

The performance of electrical power systems is significantly sensitive to the low-frequency electromechanical oscillations which can affect the reliability, efficiency, stability, outages, and safety of the power system (Nogueira et al. 2018). Power system stabilisers are embedded into the power generators to control the excitation and provide damping of low-frequency oscillations (Talaq 2012).

Khodabakhshian and Hemmati (2013) employed CA for multi-machine power system stabiliser design by considering the design parameters of the power stabiliser as decision variables. The optimum parameter design for the power system stabilisers is formulated as follows (Khodabakhshian and Hemmati 2013):

$$\begin{aligned} &\text{Find: } X = \left[K_1^{\mathrm{DC}}, T_1^1, T_1^3, K_2^{\mathrm{DC}}, T_2^1, T_2^3, \ldots, K_n^{\mathrm{DC}}, T_n^1, T_n^3\right] \\ &f(X) = \sum_{j=1}^{N_P} \sum\nolimits_{\sigma_i \geq \sigma_0} \left(\sigma_0 - \sigma_{ij}\right)^2 + a \sum_{j=1}^{N_P} \sum\nolimits_{\xi_i \geq \xi_0} \left(\xi_0 - \xi_{ij}\right)^2 \\ &\text{Subject to: } K_{i,\min}^{\mathrm{DC}} \leq K_i^{\mathrm{DC}} \leq K_{i,\max}^{\mathrm{DC}} \\ &T_{i,\min}^1 \leq T_i^1 \leq T_{i,\max}^1 \\ &T_{i,\min}^3 \leq T_i^3 \leq T_{i,\max}^3 \end{aligned} \tag{4.44}$$

where X represents the vector of design parameters of the power system stabiliser, K_i^{DC} is the power stabiliser gain parameter, T_i^1 and T_i^3 are the timing parameters, σ_{ij} and ξ_{ij} represent the real part and the damping ratio of the ith eigenvalue of the jth operating condition, respectively, a is a constant parameter to transform the multi-objective problem to equivalent single objective one, N_P is the total number of operating points, σ_0 and ξ_0 are the constrained values for the eigenvalues and damping ratio which are defined for relative system stability, $K_{i,\min}^{\mathrm{DC}}$ and $K_{i,\max}^{\mathrm{DC}}$ are the lower and upper bounds for the power stabiliser gain parameter, $T_{i,\min}^1$, $T_{i,\max}^1$, $T_{i,\min}^3$, and $T_{i,\max}^3$ are the allowable bounds for the timing parameters. In a similar study, Tavakoli et al. (2014) applied CA to solve the integrated design of power stabilisers with the STATCOM device. The STATCOM is a shunt flexible AC transmission system device that enhances the damping capacity of a power system.

4.5.9 Optimal Reactive Power Dispatch

In electrical power systems, the voltage stability and power transfer within the network are improved by employing a set of control tools, such as switching VAr sources, changing generator voltages, and/or adjusting transformer tap settings (Hu et al. 2010). These control tools have a set of adjustable parameters that can affect the power redistribution and transmission losses in the power network.

The optimal reactive power dispatch problem looks for optimum values of the adjustment parameters of control tools to minimise the power loss under a set of equality and inequality operational constraints. The optimal reactive power dispatch problem has been investigated in the literature by considering different objective functions, such as power loss minimisation, voltage profile minimisation, and power transmission cost minimisation (Saddique et al. 2020).

Bhattacharya et al. (2012b) employed CA to solve the optimal reactive power dispatch problem, in which the power loss of the power system is considered as the objective function under the load flow balance equality constraints and the inequality constraints on the operational variables. The authors mathematically formulated this problem as follows (Nuaekaew et al. 2017; Shaheen et al. 2018):

$$\text{Find: } X = \left[P_s, V_l, Q_g, S\right], U = \left[V_g, T, Q_c\right]$$

$$\text{To minimise: } f(X, U) = P_L = \sum_{k=1}^{n_1} g_k\left[V_i^2 + V_j^2 - 2V_iV_j\cos\theta_{ij}\right]$$

Subject to :

$$P_{G_i} - \mathrm{P}_{D_i} - V_i\sum_{j=1}^{n_b} V_j\left(G_{ij}\cos\theta_{ij} + B_{ij}\sin\theta_{ij}\right) = 0, \quad i = 1, 2, \ldots, n_b$$

$$
\begin{aligned}
& Q_{G_i} - Q_{D_i} - V_i \sum_{j=1}^{n_b} V_j \left(G_{ij} \cos\theta_{ij} + B_{ij} \sin\theta_{ij}\right) = 0, \quad i = 1, 2, \ldots, n_b \\
& V_{G_i}^{\min} \le V_{G_i} \le V_{G_i}^{\max}, \quad i = 1, 2, \ldots, n_g \\
& Q_{G_i}^{\min} \le Q_{G_i} \le Q_{G_i}^{\max}, \quad i = 1, 2, \ldots, n_g \\
& V_{l_i}^{\min} \le V_{l_i} \le V_{l_i}^{\max}, \quad i = 1, 2, \ldots, n_{pq} \\
& Q_{C_i}^{\min} \le Q_{C_i} \le Q_{C_i}^{\max}, \quad i = 1, 2, \ldots, n_c \\
& T_{K_i}^{\min} \le T_{K_i} \le T_{K_i}^{\max}, \quad i = 1, 2, \ldots, n_t \\
& S_l \le S_l^{\max}, \quad i = 1, 2, \ldots, n_1 \\
& P_s^{\min} \le P_s \le P_s^{\max}
\end{aligned}
\tag{4.45}
$$

where P_s is the active power of slack bus, P_L represents the transmission loss of the power system, n_l is the number of transmission lines, g_k indicates the conductance of the line connecting i and j buses, V_i and V_j are the voltage value of the ith and jth buses, respectively, θ_{ij} represents the phase angle of the voltage value between the ith and jth buses, n_b is the total number of buses in the network, n_g is the number of generator buses, n_{pq} is the number of load buses, n_c is the number of existing VAR sources, n_t is the number of transformers, P_{G_i} is the power generated by the ith generator, P_{D_i} is the active power demand, G_{ij} and B_{ij} are the mutual conductance and susceptance between buses i and j, respectively, Q_{G_i} is the reactive power output of the ith generator, Q_{D_i} is the reactive power demand, V_{G_i}, V_{l_i}, Q_{C_i}, T_{K_i} are the voltage of generator i, voltage magnitudes of load bus i, reactive power injection of existing VAR source i, and tap setting of transformer i, respectively, S_l indicates the power flow in transmission line i, and other variables with min and max superscripts show the lower and upper bounds for the corresponding variables. Bhattacharya et al. (2012b) investigated the performance of a standard IEEE 14-Bus test system with nonlinear characteristics using hybrid CA.

4.5.10 Optimal Operation of Cascade Hydropower Station

Hydropower is one of the renewable energy resources for electricity generation. The management of water resources in hydropower generation is of greater importance. The optimal operation of the cascade hydropower station problem looks for an optimum strategy to reduce the abandoned water and maximise the resource utilisation and productivity of the power system (Lu et al. 2015).

Let us consider a hydropower system with N number of power stations for a study period T. Let us also consider variables $x_1, x_2, \ldots, x_N$ for the water levels at each power station. Then, the optimal operation of the cascade hydropower station problem can be formulated as an energy maximisation problem under the water level, power output, water discharge, reservoir, and hydraulic connection constraints as follows (Wang et al. 2009):

$$\begin{aligned}
&\text{Find: } X = [x_1, x_2, \ldots, x_N] \\
&\text{To maximise: } f(X) = \sum_{i=1}^{N} \sum_{t=1}^{T} A_i H_{i,t} Q_{i,t} M_t \\
&\text{Subject to:} \\
&Z_{i,t}^{\min} \le Z_{i,t} \le Z_{i,t}^{\max} \\
&N_{i,t}^{\min} \le N_{i,t} \le N_{i,t}^{\max} \\
&Q_{i,t}^{\min} \le Q_{i,t} \le Q_{i,t}^{\max} \\
&V_{i,t+1} = V_{i,t} + \left(I_{i,t} - Q_{i,t}\right) M_t \\
&I_{i,t} = Q_{i-1,t-\tau_{i-1}} + q_{i,t}
\end{aligned} \tag{4.46}$$

where X is the decision vector, in which each variable represents the water level at each power station, A_i represents the output power coefficient of the ith power station, $H_{i,t}$ is the head of power station i at time t, $Q_{i,t}$ is the discharge of station i at time t, M_t is the duration of the period, $Z_{i,t}$ is the water level of station i at time t, $N_{i,t}$ is the power output of power station i at time t, $V_{i,t}$ is the reservoir storage i at time t, $I_{i,t}$ is the inflow runoff of station i at time t, $q_{i,t}$ is the local inflow of station i at time t, and other variables with min and max superscripts show the lower and upper bounds for the corresponding variables, respectively.

Wang et al. (2009) employed Chaos Cultural PSO (CCPSO) for optimal operation of a hydropower system with two stations and compared the CCPSO against the classical approach. The results revealed the effectiveness of CCPSO. In another study, Xie et al. (2009) applied a PSO algorithm based on CA for the same problem and case study.

4.5.11 Optimal Design of Passive Power Filters

Harmonics refer to non-sinusoidal current and voltage waveforms generated by the equipment with nonlinear characteristics in the power systems (Masoum and Fuchs 2015a), which can significantly affect the performance of the power system. Today's electrical devices generate harmonics, such as computers, printers, fax machines, fluorescent, and Light-Emitting Diode (LED) lighting. (Masoum and Fuchs 2015b). To mitigate the effects of harmonics, hybrid power filters are used in electric devices, such as Passive Power Filters (PPFs) and Active Power Filters (APFs).

Guo et al. (2008) employed Multi-Objective CA (MCA) for the optimum design of PPFs to enhance their harmonics suppression capabilities, minimise the cost, and improve the reactive power compensation. This is a multi-objective optimisation problem that looks for the optimum parameters of PPFs and can be written as follows

(Guo et al. 2008):

$$\begin{aligned}
&\text{Find: } X = [x_1, x_2, \ldots, x_N] \\
&\text{To minimise: } \mathrm{THD}_i(X) = \sqrt{\sum\nolimits_{k=1}^{\infty} \left(\frac{I_k}{I_1}\right)^2} \times 100 \\
&\qquad C(X) = \sum_j \left(q_1 R_j + q_2 L_j + q_3 C_j\right) \\
&\text{To maximise : } Q(X) = \sum\nolimits_j Q_j \\
&\text{Subject to: } Q_{\min} \le Q_j \le Q_{\max}
\end{aligned} \tag{4.47}$$

where $THD_i(.)$ represents the total harmonics distortion of current, I_k is the rms value of kth harmonic current, I_1 is the rms value of fundamental current, $C(.)$ is the cost of PPF, q_1, q_2, and q_3 are the unit price factor for the resistance R_j, inductor L_j, and capacitor C_j of the PPF, respectively, $Q(.)$ indicates the reactive power compensation, and Q_j is the fundamental reactive power afforded by PPF which should be between $Q_{\min}$ and $Q_{\max}$.

Guo et al. (2008) compared the design results obtained by the multi-objective CA (MCA) for PPF under a three-phase full-wave controlled rectifier as the harmonic source. The results revealed that the application of MCA can result in better harmonics suppression effect and lower investment for equipment in comparison to the filter designs yielded by traditional filter design methods.

4.5.12 *Hydrothermal Scheduling*

Hydropower systems can be categorised as renewable energy resources, in which the energy of water moves from higher to lower elevations (Killingtveit 2020). The main aim of hydrothermal optimal scheduling is to reduce the energy generation cost in the hydro and thermal power plants. The problem formulates the optimum scheduling of the hydro and thermal power plants separately, which can be categorised as a multi-objective optimisation problem. In the hydrothermal optimal scheduling, the optimal amount of water discharge rates of hydroplants is obtained, while the problem also tries to find optimal power generation capacities of generators. The hydrothermal optimal scheduling problem can be seen as a mixed problem of economic/environmental dispatch and optimal operation of cascade hydropower stations, which were described earlier in Sects. 4.5.1 and 4.5.12, respectively. The problem can be mathematically expressed as follows (Lu et al. 2011):

$$\text{Find: } X = \left[p_1, p_2, \ldots, p_{N_T}, Q_1, Q_2, \ldots, Q_{N_S}\right]$$

To minimise:

$$
\begin{aligned}
& f_v(X) = \sum_{i=1}^{N_T} \left(a_i + b_i p_i - c_i p_i^2 + \left|d_i \sin\left[e_i\left(p_i^{\min} - p_i\right)\right]\right|\right) \\
& f_e(X) = \sum_{i=1}^{N_T} \left[10^{-2}\left(\alpha_i + \beta_i p_i + \gamma_i p_i^2\right) + \xi_i e^{\varphi_i p_i}\right] \\
& \text{Subject to:} \\
& \sum_{i=1}^{N_T} p_i + \sum_{i=1}^{N_H} p_{i,H} - p_D - p_L = 0 \\
& p_i^{\min} \le p_i \le p_i^{\max} \\
& p_{i,H}^{\min} \le p_{i,H} \le p_{i,H}^{\max}, \quad i = 1, 2, \ldots, N_H \\
& V_{i,\min} \le V_i \le V_{i,\max}, \quad i = 1, 2, \ldots, N_H \\
& Q_{i,\min} \le Q_i \le Q_{i,\max}, \quad i = 1, 2, \ldots, N_H
\end{aligned} \tag{4.48}
$$

in which:

$$
p_{i,H} = C_{i,1} V_i^2 + C_{i,2} Q_i^2 + C_{i,3} V_i Q_i + C_{i,4} V_j + C_{i,5} Q_i + C_{i,6} \tag{4.49}
$$

where $f_v(.)$ and $f_e(.)$ are the fuel cost and emissions of the thermal power plant, respectively, N_T is the number of thermal plants, N_H is the number of hydroplants, $p_{i,H}$ represents the power generated by the hydroplant i, $p_{i,H}^{\min}$ and $p_{i,H}^{\max}$ are the lower and upper bounds for the power generated by the hydroplant i, respectively, $C_{i,1}$, $C_{i,2}$, $C_{i,3}$, $C_{i,4}$, $C_{i,5}$ and $C_{i,6}$ are the power generation coefficients of the ith hydropower plant, V_i and Q_i are the reservoir storage volume and the water discharge rate of the ith hydropower plant, respectively, $V_{i,\min}$ is lower bound of reservoir storage of the ith hydropower plant, $V_{i,\max}$ is the upper bound of the reservoir storage in hydroplant i, and $Q_{i,\min}$ and $Q_{i,\max}$ are the upper and lower limits for the water discharge rate of the ith hydropower plant, respectively.

The definitions for the rest of the parameters in Eq. (4.48) are similar to those explained in Sect. 4.5.1. A comparison of Eq. (4.48) with the formulation presented for the economic/environmental dispatch problem in Eq. (4.31) in Sect. 4.5.1 reveals that they are similar. The problem formulated in Eq. (4.31) is extended in the hydrothermal optimal scheduling problem by adding the hydroplant parameters. The decision variables of the problem are the power obtained from each thermal generator $p_1, p_2, \ldots, p_{N_T}$ and the water discharge rate of the hydroplant $Q_1, Q_2, \ldots, Q_{N_S}$. Hence, the total number of decision variables in this problem is equal to $N_T + N_S$.

Yuan and Yuan (2006, 2009) employed standard CA to solve single-objective hydrothermal optimal scheduling problems by considering the fuel cost function $f_v(.)$ in Eq. (4.48). Lu et al. (2011) investigated the multi-objective hydrothermal optimal scheduling problem by considering both fuel cost $f_v(.)$ and emission $f_e(.)$ objective functions.

4.5.13 Hydroproduction Scheduling

Yuan et al. (2008) investigated the hydropower plant scheduling problem from a different perspective. For the hydroproduction scheduling problem, they have assumed the profit of the hydrocompany from selling energy in the day-ahead electric market as the objective function, in which different price scenarios are considered based on their probability values. The formulations adopted by the study assumes a set of practical constraints for hydropower plants, including minimum profit, power capacity, water discharge rate, and water storage volume constraints. In addition, the study considers the water transport delay between the reservoirs and water spillages in the hydropower network. By considering these assumptions, the optimum hydroproduction scheduling problem is formulated as follows (Yuan et al. 2008):

$$\text{Find: } X = \left[Q_1^0, Q_1^1, \ldots, Q_1^T, Q_2^0, Q_2^1, \ldots, Q_2^T, \ldots, Q_{N_H}^0, Q_{N_H}^1, \ldots, Q_{N_H}^T\right]$$

$$\text{To maximise: } f(X) = \sum_{k=1}^{n_s}\sum_{t=1}^{T}\left(\rho_k c_k^t \sum_{i=1}^{N_H} P_i^t\right)$$

$$\text{Subject to: } \sum_{t=1}^{T}\left(\sum_{i=1}^{N_H} P_i^t\right) \geq B_{\min}, \quad k = 1, 2, \ldots, n_s$$

$$p_i^{\min} \leq p_i^t \leq p_i^{\max}, \quad i = 1, 2, \ldots, N_H, \quad t = 1, 2, \ldots, T$$

$$Q_i^{\min} \leq Q_i^t \leq Q_i^{\max}, \quad i = 1, 2, \ldots, N_H, \quad t = 1, 2, \ldots, T$$

$$V_i^{\min} \leq V_i \leq V_i^{\max}, \quad i = 1, 2, \ldots, N_H, \quad t = 1, 2, \ldots, T$$

$$V_i^0 = V_i^B, V_i^T = V_i^E, \quad i = 1, 2, \ldots, N_H$$

$$V_i^t = V_i^{t-1} + I_i^t - Q_i^t - S_i^t + \left[\sum\nolimits_{m=1}^{N_u} Q_m^{t-\tau_{m,i}} + S_m^{t-\tau_{m,i}}\right] \tag{4.50}$$

In the above equation, ρ_k represents the kth price scenario, c_k^t is the price of the kth scenario at time t, N_H is the number of hydroplants, n_s indicates the total number of price scenarios, $B_{\min}$ is the assumed minimum profit value, p_i^t represents the power generated by the ith hydroplant at time t, V_i is the water storage volume of the ith hydroplant, Q_i^t is the water discharge rate from the ith hydroplant, V_i^B and V_i^E are initial and final water storage volume of the ith hydroplant, respectively, S_i^t is the water spillage from the ith hydroplant at time t, I_i^t is natural inflow into the ith reservoir at time t, N_u is the number of upstream hydroplants above the ith plant, $\tau_{m,i}$ indicates the water transport delay time between the reservoir m and i, and the parameters with min and max superscripts show the lower and upper bounds for the corresponding variables, respectively. As it can be seen from Eq. (4.50), the decision variables for this problem are the water discharge rates Q_i^t of hydroplants during the study period.

In a similar study, Gu and Wu (2010) used CA to solve the multi-objective water resources management problem, in which the power generation and water deficit are two conflicting objective functions.

4.5.14 Multi-robot Tasks Allocation

Nowadays, due to instantaneous changes in the market, there is serious competition between the companies and manufacturers to reduce the manufacturing time, enhance the quality of products and reduce production costs. Industrial robots play a key role in today's advanced manufacturing systems. Multi-robot systems facilitate the manufacturing process and reduce the manufacturing time (Czarnecki 1994). However, the mission planning of multi-robot systems and their coordination in a complex dynamic environment are challenging.

Yu et al. (2011) employed a heterogeneous hybrid CA for exploration mission planning of multi-robot systems. In a multi-robot system, each robot performs the allocated task by travelling from its position at a given time and completing the task. Hence, the travelling and mission execution times are two important parameters that affect the overall performance of the system.

Let us assume a multi-robot system with m number of robots represented by $R_1, R_2, \ldots, R_m$ and n number of missions denoted by $W_1, W_2, \ldots, W_n$. Let us also $X = [x_1, x_2, \ldots, x_m]$ be the mission allocation subsets for the robots. The multi-robot task allocation problem looks for the optimum set of allocated tasks S for different robots to minimise the total mission execution time and travelling time of the multi-robot system. The problem can be formulated as an optimisation problem as follows (Yu et al. 2011):

$$\text{Find: } X = [x_1, x_2, \ldots, x_m]$$

$$\text{To minimise: } f(X) = \sum_{i=1}^{m} (T_{\mathrm{m}} + T_{\mathrm{t}})^2 \tag{4.51}$$

where T_m represents the total mission time of the system and T_t is the total travel time of the robots between the different missions. As it can be seen from Eq. (4.51), the study converted the multi-objective problem into the equivalent single-objective one. It seems that the study has not considered any constraints among the tasks for the robots.

4.6 Industrial Engineering

Supply Chain Management and industrial automation are two problems in the area of industrial engineering that have been used to test the performance of CAs. In the following subsection, these problems are discussed.

4.6.1 Supply Chain Management

Supply Chain Management (SCM) is one of the most critical parts of different industries and plays a pivotal role in achieving global sustainable business strategies. The SCM can be defined as "*a complete set of approaches utilised to integrate all partners of a supply chain efficiently, including suppliers, manufacturers, stores and warehouses, so that products are manufactured and distributed to the right place at the right time with the right quantities, in order to minimise system-wide costs, meanwhile meeting service level requirements*" (Larson 2001).

With rising global environmental concerns, the companies are being criticised to reduce the Green House Gas (GHG) emissions resulting from the supply chain process. Hence, supply chain managers in different businesses have to develop a set of new strategies to not only minimise the costs and enhance profits but also reduce GHG and waste during the different stages of the supply chain. Rahmani et al. (2020) employed CA to solve a green dual-channel-based supply chain problem as shown in Fig. 4.12, in which the customers can purchase the products either from retailers or from plants directly. The main objective of this model is to achieve a green sustainable supply chain system through optimising the production technologies in the plants, prices, locations of plants, and flow of the products.

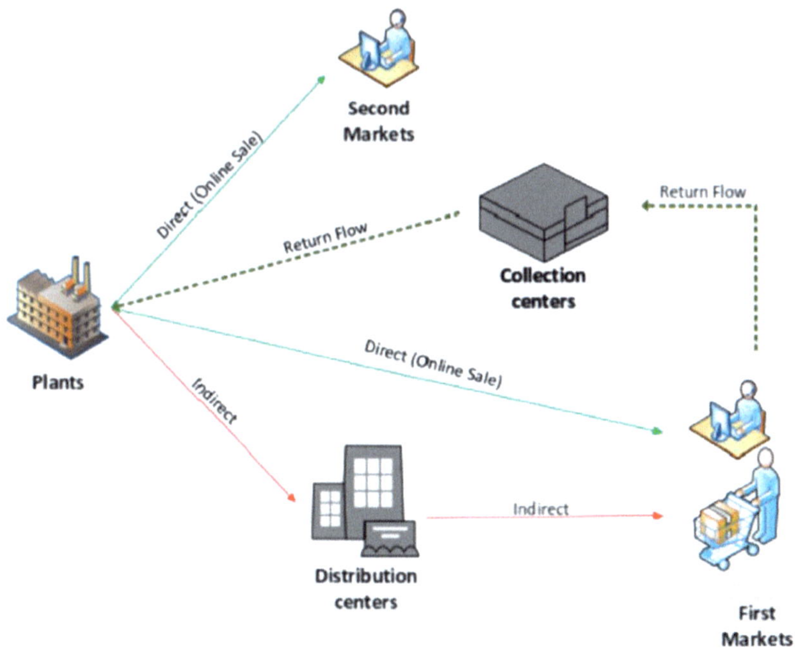

Fig. 4.12 Supply chain system with dual-channel investigated by Rahmani et al. (2020)

4.6.2 *Industrial Automation*

Integration in industrial automation is an important problem in industrial engineering. According to the definition, integration is "*the coordination of the operations of all company elements working together, to achieve an optimal compliance of the company mission*" (Aguilar et al. 2013). Terán et al. (2017) employed CA to optimise the industrial integration process based on a multi-agent system model, in which the coordination between the different agents and their interaction protocols within a company were optimised.

4.7 Summary

The chapter categorised the applications of CAs in civil engineering, mechanical engineering, chemical engineering, electrical engineering, and industrial engineering areas. The chapter briefly explained the different problems in these areas and provided detailed mathematical formulations for most of the problems. The formulations described the decisions variables, objective functions, and constraints assumed in different studies. The chapter showed that how CAs have attracted attention of research from a diverse range of research topics in different engineering fields. The simple framework of CAs and their efficient performance in dealing with optimisation problem made them one of the popular meta-heuristics in recent decades. CAs have also found a wide number of application studies in other areas, such as computer science, biology, physics, and economy, which will be discussed in the next chapter.

References

Aguilar J, Bolivar AR, Hidrobo F, Cerrada M (2013) Sistemas Multi Agentes y sus Aplicaciones en Automatizaci on industrial. Universidad de Los Andes, Merida, Venezuela

Aiyejina A, Chakrabarti DP, Pilgrim A, Sastry MKS (2011) Wax formation in oil pipelines: a critical review. Int J Multiph Flow 37:671–694. https://doi.org/10.1016/j.ijmultiphaseflow.2011.02.007

Alami J, El Imrani A (2008) Using cultural algorithm for the fixed-spectrum frequency assignment problem. J Mobile Commun 2(1):1–9

Alkayem NF, Cao M, Zhang Y, et al (2018) Structural damage detection using finite element model updating with evolutionary algorithms: a survey. Neural Comput Appl 30. https://doi.org/10.1007/s00521-017-3284-1

Ali MZ, Reynolds RG (2014) Cultural algorithms: a tabu search approach for the optimization of engineering design problems. Soft Comput 18(8):1631–1644

Ali MZ, Morghem A, Albadarneh J et al (2014b) Cultural algorithms applied to the evolution of robotic soccer team tactics: a novel perspective. In: 2014b IEEE congress on evolutionary computation (CEC). IEEE, pp 2180–2187

Ali MZ, Awad NH, Reynolds RG, Suganthan PN (2018) A balanced fuzzy cultural algorithm with a modified Levy flight search for real parameter optimization. Inf Sci 447. https://doi.org/10.1016/j.ins.2018.03.008

Amalo KA, Birninkudu SI, Bukata BB et al (2020) Cultured bat algorithm for optimized MPPT tracking under different shading conditions. In: 2020 international conference in mathematics, computer engineering and computer science (ICMCECS). IEEE, pp 1–8

Amiri-Ramsheh B, Safaei-Farouji M, Larestani A et al (2022) Modeling of wax disappearance temperature (WDT) using soft computing approaches: tree-based models and hybrid models. J Petrol Sci Eng 208:109774. https://doi.org/10.1016/j.petrol.2021.109774

Araghi S, Khosravi A, Creighton D (2015) A review on computational intelligence methods for controlling traffic signal timing. Expert Syst Appl 42. https://doi.org/10.1016/j.eswa.2014.09.003

Arora JS (2012) Optimum design problem formulation. In: Introduction to optimum design. Elsevier, pp 17–64

Arpaia P, Lucariello G, Zanesco A (2007) Automatic fault isolation by cultural algorithms with differential influence. IEEE Trans Instrum Meas 56:1573–1582. https://doi.org/10.1109/TIM.2007.903604

Assis FA, Leite da Silva AM, Resende LC et al (2022) Generation maintenance scheduling with renewable sources based on production and reliability costs. Int J Electr Power Energy Syst 134:107370. https://doi.org/10.1016/j.ijepes.2021.107370

Avci O, Abdeljaber O, Kiranyaz S et al (2021) A review of vibration-based damage detection in civil structures: from traditional methods to machine learning and deep learning applications. Mech Syst Signal Process 147. https://doi.org/10.1016/j.ymssp.2020.107077

Baig Mohammad G, Shitharth S (2021) Wireless sensor network and IoT based systems for healthcare application. Mater Today Proc. https://doi.org/10.1016/j.matpr.2020.11.801

Balaji G, Balamurugan R, Lakshminarasimman L (2016) Mathematical approach assisted differential evolution for generator maintenance scheduling. Int J Electr Power Energy Syst 82:508–518. https://doi.org/10.1016/j.ijepes.2016.04.033

Bathe K-J (2006) Finite element procedures. Klaus-Jurgen Bathe

Bhattacharya B, Mandal K, Chakraborty N (2012a) Knowledge based evolutionary programming: cultural algorithm approach for constrained optimization. In: Proceedings of the international conference on information systems design and intelligent applications 2012a (INDIA 2012a) held in Visakhapatnam, India, January 2012a, pp 93–101

Bhattacharya B, Mandal KK, Chakraborty N (2012b) Reactive power optimization using hybrid cultural algorithm. In: International conference on swarm, evolutionary, and memetic computing, pp 106–115

Brice R (2001) Let's stick together—recording consoles. In: Music engineering. Elsevier, pp 340–387

Cai Z, Zhang M, Liu Y (2018) Sea-surface weak target detection scheme using a cultural algorithm aided time-frequency fusion strategy. IET Radar Sonar Navig 12:711–720. https://doi.org/10.1049/iet-rsn.2018.0004

Chang K-C, Kim C-W (2016) Modal-parameter identification and vibration-based damage detection of a damaged steel truss bridge. Eng Struct 122. https://doi.org/10.1016/j.engstruct.2016.04.057

Chung DDL (2010) Composite materials: science and applications. Springer Science & Business Media

Coelho L dos S, Souza RCT, Mariani VC (2009) Improved differential evolution approach based on cultural algorithm and diversity measure applied to solve economic load dispatch problems. Math Comput Simul 79:3136–3147. https://doi.org/10.1016/j.matcom.2009.03.005

Coello CA, Becerra RL (2004) Efficient evolutionary optimization through the use of a cultural algorithm. Eng Optim 36:219–236. https://doi.org/10.1080/03052150410001647966

Cuthrell JE, Biegler LT (1989) Simultaneous optimization and solution methods for batch reactor control profiles. Comput Chem Eng 13:49–62. https://doi.org/10.1016/0098-1354(89)89006-4

Czarnecki CA (1994) Multi-robot systems: a perspective. IFAC Proc Vol 27:201–205. https://doi.org/10.1016/S1474-6670(17)46023-X

Digalakis JG, Margaritis KG (2002) A multipopulation cultural algorithm for the electrical generator scheduling problem. Math Comput Simul 60:293–301. https://doi.org/10.1016/S0378-4754(02)00021-6

Divoux T, Georges J-P, Harchi S (2015) A session protocol for wireless sensor networks. Application to oil spills monitoring. Comput Electr Eng 48:312–329. https://doi.org/10.1016/j.compeleceng.2015.06.015
dos Santos Coelho L, Mariani VC (2006) An efficient particle swarm optimization approach based on cultural algorithm applied to mechanical design. In: 2006 IEEE international conference on evolutionary computation. IEEE, pp 1099–1104
Dwivedi AK, Ghosh S, Londhe ND (2018) Review and analysis of evolutionary optimization-based techniques for FIR filter design. Circ Syst Signal Process 37:4409–4430. https://doi.org/10.1007/s00034-018-0772-1
Gao H, Diao M (2010) Differential cultural algorithm for digital filters design. In: 2010 second international conference on computer modeling and simulation. IEEE, pp 459–463
Gao H, Diao M (2011) Cultural firework algorithm and its application for digital filters design. Int J Model Ident Control 14(4):324–331
Gao H, Xu C (2013) Cultural quantum-inspired shuffled frog leaping algorithm for direction finding of non-circular signals. Int J Comput Sci Math 4(4):321–331
Gao HY, Chi PF, Du YN et al (2019) Digital watermarking based on wavelet transform and cultural invasive weed algorithm, pp 1620–1627
Gao M, Zhu Y, Cao C, Zhu Y (2020) A hybrid cultural harmony search algorithm for constrained optimization problem of diesel blending. IEEE Access 8:6673–6690. https://doi.org/10.1109/ACCESS.2019.2963244
Goudarzi A, Ahmadi A, Swanson AG, van Coller J (2016) Non-convex optimisation of combined environmental economic dispatch through cultural algorithm with the consideration of the physical constraints of generating units and price penalty factors. SAIEE Afr Res J 107:146–166. https://doi.org/10.23919/SAIEE.2016.8532239
Goudarzi A, Swanson AG, Tooryan F, Ahmadi A (2017) Non-convex optimization of combined environmental economic dispatch through the third version of the cultural algorithm (CA3). In: 2017 IEEE Texas power and energy conference (TPEC). IEEE, pp 1–6
Gu W, Wu Y (2010) Application of multi-objective cultural algorithm in water resources optimization. In: 2010 Asia-Pacific power and energy engineering conference. IEEE
Guo Y, Cheng J, Lin Y, Jiang X (2008) Optimal design of passive power filters based on multi-objective cultural algorithms. In: International conference on intelligent computing, pp 235–242
Guo Y, Chen M, Wang C (2013a) Multi-objective quantum cultural algorithm and its application in the wireless sensor networks' energy-efficient coverage optimization, pp 161–167
Guo Y, Liu D, Liu Y, Chen M (2013b) The coverage optimization for wireless sensor networks based on quantum-inspired cultural algorithm, pp 87–96
Guedria N ben, Hassine H (2019) Efficient cultural algorithm for structural damage detection problem based on modal data
Haldar V, Chakraborty N (2014) Minimisation of power loss by network reconfiguration and capacitor installation using modified cultural algorithm. Int J Swarm Intell 1(3):226–245
Haldar V, Chakraborty N (2015) Power loss minimization by optimal capacitor place-ment in radial distribution system using modified cultural algorithm. Int Trans Electr Energy Syst 25(1):54–71
Hamami L, Nassereddine B (2020) Application of wireless sensor networks in the field of irrigation: a review. Comput Electron Agric 179:105782. https://doi.org/10.1016/j.compag.2020.105782
Haikal A, El-Hosseni M (2011) Modified cultural-based genetic algorithm for process optimization. Ain Shams Eng J 2:173–182. https://doi.org/10.1016/j.asej.2011.09.002
Hosseinzadeh Y, Jalili S, Khani R (2020) Investigating the effects of flax fibers application on multi-objective optimization of laminated composite plates for simultaneous cost minimization and frequency gap maximization. J Build Eng 32:101477. https://doi.org/10.1016/j.jobe.2020.101477
Hu Z, Wang X, Taylor G (2010) Stochastic optimal reactive power dispatch: formulation and solution method. Int J Electr Power Energy Syst 32:615–621. https://doi.org/10.1016/j.ijepes.2009.11.018
Jalili S, Hosseinzadeh Y (2015) A cultural algorithm for optimal design of truss structures. Latin Am J Solids Struct 12. https://doi.org/10.1590/1679-78251547

Jalili S, Hosseinzadeh Y, Rabczuk T (2019) Simultaneous size and shape optimization of dome-shaped structures using improved cultural algorithm
Jalili S, Khani R, Hosseinzadeh Y (2021) On the performance of flax fibres in multi-objective design of laminated composite plates for buckling and cost. Structures 33:3094–3106. https://doi.org/10.1016/j.istruc.2021.06.026
Jalili S, Nallaperuma S, Keedwell E et al (2021b) Application of metaheuristics for signal optimisation in transportation networks: a comprehensive survey. Swarm Evol Comput 63. https://doi.org/10.1016/j.swevo.2021.100865
Jalili S, Khani R, Maheri A, Hosseinzadeh Y (2022) Performance assessment of meta-heuristics for composite layup optimisation. Neural Comput Appl 34:2031–2054. https://doi.org/10.1007/s00521-021-06519-0
Jafari-Asl J, Sami Kashkooli B, Bahrami M (2020) Using particle swarm optimization algorithm to optimally locating and controlling of pressure reducing valves for leakage minimization in water distribution systems. Sustain Water Resourc Manag 6:64. https://doi.org/10.1007/s40899-020-00426-3
Jafari M, Salajegheh E, Salajegheh J (2019) An efficient hybrid of elephant herding optimization and cultural algorithm for optimal design of trusses. Eng Comput 35. https://doi.org/10.1007/s00366-018-0631-5
Jayyousi TW, Reynolds RG (2014) Extracting urban occupational plans using cultural algorithms (Application Notes). IEEE Comput Intell Mag 9. https://doi.org/10.1109/MCI.2014.2326103
Joos H-D (2009) Flight control law clearance using optimization-based worst-case search. IFAC Proc Vol 42:331–336. https://doi.org/10.3182/20090616-3-IL-2002.00057
Kavousi-Fard A, Samet H (2013) Multi-objective performance management of the capacitor allocation problem in distributed system based on adaptive modified honey bee mating optimization evolutionary algorithm. Electr Power Compon Syst 41:1223–1247. https://doi.org/10.1080/15325008.2013.816981
Kessler GC, Hosmer C (2011) An overview of steganography, pp 51–107
Khan SU, Qureshi IM, Zaman F et al (2014) Correction of faulty sensors in phased array radars using symmetrical sensor failure technique and cultural algorithm with differential evolution. Sci World J:1–10. https://doi.org/10.1155/2014/852539
Khan SU, Qureshi IM, Zaman F, Khan W (2017) Detecting faulty sensors in an array using symmetrical structure and cultural algorithm hybridized with differential evolution. Front Inf Technol Electron Eng 18:235–245. https://doi.org/10.1631/FITEE.1500315
Khodabakhshian A, Hemmati R (2013) Multi-machine power system stabilizer design by using cultural algorithms. Int J Electr Power Energy Syst 44:571–580. https://doi.org/10.1016/j.ijepes.2012.07.049
Killingtveit Å (2020) Hydroelectric power. In: Future energy. Elsevier, pp 315–330
Kobti Z, Snowdon AW, Rahaman S et al (2006b) A cultural algorithm to guide driver learning in applying child vehicle safety restraint. In: 2006b IEEE international conference on evolutionary computation. IEEE, pp 1111–1118
Kopsaftopoulos FP, Fassois SD (2013) A functional model based statistical time series method for vibration based damage detection, localization, and magnitude estimation. Mech Syst Signal Process 39. https://doi.org/10.1016/j.ymssp.2012.08.023
Kulkarni VR, Desai V (2020) Sensor localization in wireless sensor networks using cultural algorithm. Int J Swarm Intell Res 11:106–122. https://doi.org/10.4018/IJSIR.2020010010
Lakshminarayanan S, Kaur D (2018) Optimal maintenance scheduling of generator units using discrete integer cuckoo search optimization algorithm. Swarm Evol Comput 42:89–98. https://doi.org/10.1016/j.swevo.2018.02.016
Larson PD (2001) Designing and managing the supply chain: concepts, strategies, and case studies. J Bus Logist 22:259
Lim HC, Tayeb YJ, Modak JM, Bonte P (1986) Computational algorithms for optimal feed rates for a class of fed-batch fermentation: numerical results for penicillin and cell mass production. Biotechnol Bioeng 28:1408–1420. https://doi.org/10.1002/bit.260280915

Liu Y, Yang Z, Han X (2012) The research on cultural algorithm of shortest path in the digital map, pp 843–850
Liu W-Y, Lin C-C (2015) Spatial forest resource planning using a cultural algorithm with problem-specific information. Environ Model Softw 71. https://doi.org/10.1016/j.envsoft.2015.06.002
Long J, Jiang S, He R, Zhao L (2021) Diesel blending under property uncertainty: a data-driven robust optimization approach. Fuel 306:121647. https://doi.org/10.1016/j.fuel.2021.121647
Lu Y, Zhou J, Qin H et al (2011) A hybrid multi-objective cultural algorithm for short-term environmental/economic hydrothermal scheduling. Energy Convers Manage 52:2121–2134. https://doi.org/10.1016/j.enconman.2010.12.003
Lu D, Wang B, Wang Y et al (2015) Optimal operation of cascade hydropower stations using hydrogen as storage medium. Appl Energy 137:56–63. https://doi.org/10.1016/j.apenergy.2014.09.092
Luus R (1993) Application of dynamic programming to differential-algebraic process systems. Comput Chem Eng 17:373–377. https://doi.org/10.1016/0098-1354(93)80029-M
Maldonado F, Ochoa A, Arreola J, et al (2013) Intelligent application to reduce transit accidents in a city using cultural algorithms
Masoum MAS, Fuchs EF (2015a) Introduction to power quality. In: Power quality in power systems and electrical machines. Elsevier, pp 1–104
Masoum MAS, Fuchs EF (2015b) The roles of filters in power systems and unified power quality conditioners. In: Power quality in power systems and electrical machines. Elsevier, pp 779–886
Meng W, Wei C, Tang J (2011) Cultural algorithm based on particle swarm optimization for clearance of flight control law. In: 2011 6th IEEE conference on industrial electronics and applications. IEEE, pp 418–423
Murugadass G, Sivakumar P (2020) A hybrid elephant herding optimization and cultural algorithm for energy-balanced cluster head selection scheme to extend the lifetime in WSNs. Int J Commun Syst e4538. https://doi.org/10.1002/dac.4538
Nabavi SR, Najafi M (2021) An optimal routing protocol using multi-objective cultural algorithm for wireless sensor networks (ORPMCA). Majlesi J Telecommun Devices 10:79–86. https://doi.org/10.52547/mjtd.10.2.79
Nallaperuma S, Jalili S, Keedwell E et al (2020) Optimisation of signal timings in a road network, pp 257–268
Neves AC, González I, Leander J, Karoumi R (2017) Structural health monitoring of bridges: a model-free ANN-based approach to damage detection. J Civil Struct Health Monit 7. https://doi.org/10.1007/s13349-017-0252-5
Nikbakt S, Kamarian S, Shakeri M (2018) A review on optimization of composite structures part I: laminated composites. Compos Struct 195:158–185. https://doi.org/10.1016/j.compstruct.2018.03.063
Nogueira FG, Barra Junior W, da Costa Junior CT, Lana JJ (2018) LPV-based power system stabilizer: identification, control and field tests. Control Eng Pract 72:53–67. https://doi.org/10.1016/j.conengprac.2017.11.004
Nuaekaew K, Artrit P, Pholdee N, Bureerat S (2017) Optimal reactive power dispatch problem using a two-archive multi-objective grey wolf optimizer. Expert Syst Appl 87:79–89. https://doi.org/10.1016/j.eswa.2017.06.009
Oliveira De Freitas CA, Limao de Oliveira RC, Azevedo Da Silva DJ et al (2018) Solution to economic—emission load dispatch by cultural algorithm combined with local search: case study. IEEE Access 6:64023–64040. https://doi.org/10.1109/ACCESS.2018.2877770
Ou L, Qin Z, Yin H, Li K (2016) Security and privacy in big data. In: Big data. Elsevier, pp 285–308
Ounahasaree Y, Suriyapraphadilok U, Bagajewicz M (2016) Global optimization of gasoline blending model using bound contraction technique, pp 1293–1298
Pan Z-L, Chen L, Zhang G-Z (2010) Cultural algorithm for minimization of binary decision diagram and its application in crosstalk fault detection. Int J Autom Comput 7:70–77. https://doi.org/10.1007/s11633-010-0070-2

Patil HK, Chen TM (2017) Wireless sensor network security. In: Computer and information security handbook. Elsevier, pp 317–337

Pedersen KS, Rønningsen HP (2003) Influence of wax inhibitors on wax appearance temperature, pour point, and viscosity of waxy crude oils. Energy Fuels 17:321–328

Pierezan J, Maidl G, Massashi Yamao E et al (2019) Cultural coyote optimization algorithm applied to a heavy duty gas turbine operation. Energy Convers Manage 199:111932. https://doi.org/10.1016/j.enconman.2019.111932

Qu BY, Zhu YS, Jiao YC et al (2018) A survey on multi-objective evolutionary algorithms for the solution of the environmental/economic dispatch problems. Swarm Evol Comput 38:1–11. https://doi.org/10.1016/j.swevo.2017.06.002

Rahmani D, Hasan Abadi MQ, Hosseininezhad SJ (2020) Joint decision on product greenness strategies and pricing in a dual-channel supply chain: a robust possibilistic approach. J Clean Prod 256:120437. https://doi.org/10.1016/j.jclepro.2020.120437

Rao SS (2019) Engineering optimization: theory and practice. Wiley

Reihani E, Sarikhani A, Davodi M, Davodi M (2012) Reliability based generator maintenance scheduling using hybrid evolutionary approach. Int J Electr Power Energy Syst 42:434–439. https://doi.org/10.1016/j.ijepes.2012.04.018

Reyes LC, Zezzatti CAOO, Santillán CG, et al (2010) A cultural algorithm for the urban public transportation pp 135–142

Reynolds RG, Chung C (1997) A cultural algorithm framework to evolve multiagent cooperation with evolutionary programming. In: international conference on evolutionary programming, Springer, pp 323–333

Reynolds RG, Peng B (2005) Cultural algorithms: computational modeling of how cultures learn to solve problems: an engineering example. Cybern Sys 36:753–771. https://doi.org/10.1080/01969720500306147

Reynolds R, Ali M (2008a) Embedding a social fabric component into cultural algorithms toolkit for an enhanced knowledge-driven engineering optimization. Int J Intell Comput Cybernet 1. https://doi.org/10.1108/17563780810919131

Reynolds RG, Ali MZ (2008b) Cultural algorithms: knowledge-driven engineering optimization via weaving a social fabric as an enhanced influence function. In: 2008b IEEE congress on evolutionary computation (IEEE world congress on computational intelligence). IEEE, pp 4192–4199

Saddique MS, Bhatti AR, Haroon SS et al (2020) Solution to optimal reactive power dispatch in transmission system using meta-heuristic techniques-status and technological review. Electr Power Syst Res 178:106031. https://doi.org/10.1016/j.epsr.2019.106031

Scott JF (2008) Ferroelectric nanostructures for device applications. In: Handbook of advanced dielectric, piezoelectric and ferroelectric materials. Elsevier, pp 541–569

Seyedpoor SM (2012) A two stage method for structural damage detection using a modal strain energy based index and particle swarm optimization. Int J Non-Linear Mech 47. https://doi.org/10.1016/j.ijnonlinmec.2011.07.011

Shaheen AM, El-Sehiemy RA, Farrag SM (2018) Integrated strategies of backtracking search optimizer for solving reactive power dispatch problem. IEEE Syst J 12:424–433. https://doi.org/10.1109/JSYST.2016.2573799

Soremekun G, Gürdal Z, Haftka RT, Watson LT (2001) Composite laminate design optimization by genetic algorithm with generalized elitist selection. Comput Struct 79:131–143. https://doi.org/10.1016/S0045-7949(00)00125-5

Talaq J (2012) Optimal power system stabilizers for multi machine systems. Int J Electr Power Energy Syst 43:793–803. https://doi.org/10.1016/j.ijepes.2012.06.030

Tam NT, Dat VT, Lan PN et al (2021) Multifactorial evolutionary optimization to maximize lifetime of wireless sensor network. Inf Sci 576:355–373. https://doi.org/10.1016/j.ins.2021.06.056

Tavakoli MR, Rasouli V, Nasajpour HR, Shaarbafchizadeh M (2014) A new simultaneous coordinated design of STATCOM controller and power system stabilizer for power systems using cultural algorithm. In: 2014 IEEE international energy conference (ENERGYCON). IEEE, pp 446–450

Tazay AF, Samy MM, Barakat S (2020) A techno-economic feasibility analysis of an autonomous hybrid renewable energy sources for university building at Saudi Arabia. J Electr Eng Technol 15:2519–2527. https://doi.org/10.1007/s42835-020-00539-x

Tefas A, Nikolaidis N, Pitas I (2009) Image watermarking. In: The essential guide to image processing. Elsevier, pp 597–648

Terán J, Aguilar J, Cerrada M (2017) Integration in industrial automation based on multi-agent systems using cultural algorithms for optimizing the coordination mechanisms. Comput Ind 91:11–23. https://doi.org/10.1016/j.compind.2017.05.002

Tremayne M, Chong SY, Bell D (2009) Optimisation of algorithm control parameters in cultural differential evolution applied to molecular crystallography. Front Comput Sci China 3:101–108. https://doi.org/10.1007/s11704-009-0009-3

Valera-Medina A, Banares-Alcantara R (2020) Techno-economic challenges of green ammonia as an energy vector. Academic Press

Vamvoudakis-Stefanou KJ, Sakellariou JS, Fassois SD (2018) Vibration-based damage detection for a population of nominally identical structures: unsupervised multiple model (MM) statistical time series type methods. Mech Syst Signal Process 111. https://doi.org/10.1016/j.ymssp.2018.03.054

Wang Y, Zhou J, Lu Y et al (2009) Chaos cultural particle swarm optimization and its application, pp 30–40

Wang W, Song Y, Xue Y et al (2015) An optimal vibration control strategy for a vehicle's active suspension based on improved cultural algorithm. Appl Soft Comput 28:167–174. https://doi.org/10.1016/j.asoc.2014.11.047

Xiao B, Xiao J, Dong X et al (2010) Research of PID parameter optimization based-on cultural based ant colony algorithm for superheated steam temperature. In: Proceedings of the 29th Chinese control conference, pp 5171–5176

Xie W, Ji C, Li X (2009) Particle swarm optimization based on cultural algorithm for short-term optimal operation of cascade hydropower stations. In: 2009 fifth international conference on natural computation. IEEE, pp 289–293

Xu W, Zhang L, Gu X (2010) A novel cultural algorithm and its application to the constrained optimization in ammonia synthesis. In: Life system modeling and intelligent computing. Springer, pp 52–58

Xu W, Wang R, Zhang L, Gu X (2012) A multi-population cultural algorithm with adaptive diversity preservation and its application in ammonia synthesis process. Neural Comput Appl 21:1129–1140. https://doi.org/10.1007/s00521-011-0749-5

Xue H (2020) Adaptive cultural algorithm-based cuckoo search for time-dependent vehicle routing problem with stochastic customers using adaptive fractional Kalman speed prediction. Math Probl Eng. https://doi.org/10.1155/2020/7258780

Yan X, Gong W, Wu Q (2017a) Contaminant source identification of water distribution networks using cultural algorithm. Concurrency Comput Pract Exp 29. https://doi.org/10.1002/cpe.4230

Ya-Li W, Li-qing X (2010) An improved cultural particle swarm optimization algorithm based on feedback mechanism for economic load dispatch problems. In: 2010 international conference on computational aspects of social networks. IEEE, pp 319–322

Yare Y, Venayagamoorthy GK (2010) Optimal maintenance scheduling of generators using multiple swarms-MDPSO framework. Eng Appl Artif Intell 23:895–910. https://doi.org/10.1016/j.engappai.2010.05.006

Yu L, Cai Z, Gao P, Liu X (2011) A spatial orthogonal allocation and heterogeneous cultural hybrid algorithm for multirobot exploration mission planning. J Control Theory Appl 9:171–176. https://doi.org/10.1007/s11768-011-9109-3

Yuan X, Yuan Y (2006) Application of cultural algorithm to generation scheduling of hydrothermal systems. Energy Convers Manag 47(15–16):2192–2201

Yuan X, Nie H, Yuan Y et al (2009) Hydrothermal systems generation scheduling using cultural algorithm. J Hydroinf 11:65–78. https://doi.org/10.2166/hydro.2009.056

Yuan X, Nie H, He L et al (2008) A cultural algorithm for scheduling of hydro producer in the power market. In: 2008 second international conference on genetic and evolutionary computing. IEEE, pp 364–367

Zechman EM, Ranjithan SR (2009) Evolutionary computation-based methods for characterizing contaminant sources in a water distribution system. J Water Resourc Plann Manag 135(5):334. https://doi.org/10.1061/(ASCE)0733-9496

Zhang R, Zhou J, Mo L et al (2013) Economic environmental dispatch using an enhanced multi-objective cultural algorithm. Electr Power Syst Res 99:18–29. https://doi.org/10.1016/j.epsr.2013.01.010

Chapter 5
Applications of Cultural Algorithms in Different Branches of Science

Abstract CAs have been employed to solve a variety of problems in other branches of science, such as computer, physical, social, and biological sciences. The main aim of this chapter is to review the problems in these areas and discuss related mathematical formulations.

Keywords Cultural algorithms · Computer science · Social science · Archaeology · Physics

5.1 Introduction

CAs have found interesting applications to real-world problems in areas that cannot be categorised into engineering problems. Providing an overall taxonomy for these problems is rather difficult due to their highly interdisciplinary nature. These problems belong to the different branches of science, such as computer, physical, social, and biological sciences. The application studies in computer science include a wide range of problems, such as job shop scheduling, travelling salesman, software testing, facility location, virtual machine replacement, online marketing, data mining, image matching, set covering, and timetabling problems. In social science, they have been used to discover the patterns of social networks among individuals in human societies. CAs have also been used to manage the cooperation between different business partners. In biology science, CAs have been used for the identification of the gene networks as well as the representation of the mitochondrial population in mammalian cells.

In this chapter, the applications of CAs in the aforementioned areas are briefly reviewed and the related mathematical formulations, including decision variables, objective functions, and constraints, are presented. The overall structure of this chapter is as follows. Section 5.2 focuses on the applications of CAs to a broad range of problems in computer science. Section 5.3 presents the application of CAs to problems in social science. Section 5.4 discusses how CAs can be applied for the interpretation of archaeological data to explain the ancient residential patterns. Section 5.5 presents the application of CAs in the field of electromagnetics in physics. There are also several other applications to other branches of science such as biology

S. Jalili, *Cultural Algorithms*, Engineering Optimization: Methods and Applications,
https://doi.org/10.1007/978-981-19-4633-2_5

and economics, which are presented in Sect. 5.6. A summary of contents of this chapter are also discussed in Sect. 5.7.

5.2 Computer Science

The applications of CAs in computer science include a broad range of problems, such as problems in function optimisation, scheduling, site location, cloud computing, fog computing, software development, and semantic networks. The following subsections provide overall details for these problems.

5.2.1 Function Optimisation

In the metaheuristics literature, it is common to evaluate the performance of different algorithms based on their capabilities in finding global optimums for a set of mathematical test functions. The literature shows that researchers investigated a set of benchmark mathematical functions to evaluate the performance of CA-based optimisation approaches. Table 5.1 presents the list of references in which CAs are applied to find the global optima in different kinds of mathematical benchmark functions. Table 5.2 also lists the different examples of unimodal and multimodal functions investigated by different references in the literature to assess the performance of CAs and their relevant versions.

These benchmark functions can be categorised from different perspectives based on their surface characteristic features in the search space, such as unimodal, multimodal, separable, and non-separable functions. The unimodal functions are referred to as functions with a single global optimum and one or no local optimum, whereas the multimodal functions have several local optima in the search space. The n-dimensional functions are called separable if they can be expressed as the sum of n functions with single variables. The non-separable multimodal functions are more

Table 5.1 List of references in which CAs are applied to mathematical function optimisation

Chung and Reynolds (1998), Raeesi et al. (2014), Raeesi et al. (2013), Becerra et al. (2004), Coello et al. (2002), Xue et al.(2011), Guo et al. (2011), Gao et al. (2006a), Nguyen et al. (2008), Gao et al. (2006b), Gao et al. (2007), Nguyen et al. (2006), Li et al. (2010), Ali et al. (2011a), Guo et al. (2018), Guo et al.(2012), Reynolds and Zhu (2001), He and Xu (2011), Sun et al. (2010), Sun et al. (2012), Dixit et al. (2019), Awad et al. (2017), Ali et al.(2016a), Ali et al. (2014), Ali et al. (2012), Xue and Guo (2007), Coello et al. (2002), Ali et al. (2016b), Alami et al. (2007), Huang et al. (2008), Wu et al. (2010), Stanley et al. (2019), Tang and Li (2008), Awad et al. (2013), Coello et al. (2003), Zhou et al. (2019), Reynolds et al. (2017a), Reynolds et al. (2017b), Ali et al. (2014a), Reynolds et al. (2012), Ali et al. (2011b), Ali et al. (2009), Reynolds et al. (2008c), Chung et al. (1997), Mao et al. (2020), Murugadass et al. (2020b), Singh et al. (2018), Muhamediyeva (2020), Chung et al. (1997)

Table 5.2 Different unimodal and multimodal benchmark functions

Function name	Formulation	Optimum point
Sphere	$f(X) = \sum_{i=1}^{n} x_i^2$	0
Schwefel 1.2	$f(X) = \sum_{i=1}^{n} \left(\sum_{j=1}^{i} x_j \right)^2$	0
Schwefel 2.22	$f(X) = \sum_{i=1}^{n} \lvert x_i \rvert + \prod_{i=1}^{n} \lvert x_i \rvert$	0
Schwefel 2.26	$f(X) = -\sum_{i=1}^{n} x_i \sin\left(\sqrt{\lvert x_i \rvert} \right)$	−12,569.48
Step	$f(X) = \sum_{i=1}^{n} \lfloor x_i + 0.5 \rfloor^2$	0
Rosenbrock	$f(X) = \sum_{i=1}^{n-1} \left(100\left(x_{i+1} - x_i^2\right)^2 + (x_i - 1)^2 \right)$	0
Rastrigin	$f(X) = \sum_{i=1}^{n} \left[x_i^2 - 10\cos(2\pi x_i) + 10 \right]$	0
Ackley	$f(X) = -20\exp\left(-0.2\sqrt{\frac{1}{n} \sum_{i=1}^{n} x_i^2} \right) - \exp\left(\frac{1}{n} \sum_{i=1}^{n} \cos 2\pi x_i \right) + 20 + e$	0
Griewank	$f(X) = \frac{1}{4000} \sum_{i=1}^{n} x_i^2 - \prod_{i=1}^{n} \cos\left(\frac{x_i}{\sqrt{i}} \right) + 1$	0

(continued)

Table 5.2 (continued)

Function name	Formulation	Optimum point
Penalised	$f(X) = \frac{\pi}{n}\left\{10\sin^2(\pi y_1) + \sum_{i=1}^{n-1}(y_i - 1)^2\left[1 + 10\sin^2(\pi y_{i+1})\right] + (y_n - 1)^2\right\} + \sum_{i=1}^{n} u(x_i, 10, 100, 4)$ $y_i = 1 + \frac{1}{4}(x_i + 1), u(x_i, a, k, m) = \begin{cases} k(x_i - a)^m & x_i > a \\ 0 & -a \leq x_i \leq a \\ k(-x_i - a)^m & x_i < -a \end{cases}$	0
Penalised- 2	$f(X) = 0.1\left\{\sin^2(3\pi x_1) + \sum_{i=1}^{n-1}(x_i - 1)^2\left[1 + \sin^2(3\pi x_{i+1})\right] + (x_n - 1)^2\left[1 + \sin^2(2\pi x_n)\right]\right\} + \sum_{i=1}^{n} u(x_i, 5, 100, 4)$	0
Hartman 3	$f(X) = -\sum_{i=1}^{4} c_i \exp\left[-\sum_{j=1}^{3} a_{ij}\left(x_j - p_{ij}\right)^2\right]$	−3.8627
Hartman 6	$f(X) = -\sum_{i=1}^{4} c_i \exp\left[-\sum_{j=1}^{6} a_{ij}\left(x_j - p_{ij}\right)^2\right]$	−3.3219
Colville	$f(X) = 100\left(x_1^2 - x_2\right)^2 + (x_1 - 1)^2 + (x_3 - 1)^2 + 90\left(x_3^2 - x_4\right)^2 + 10.1\left((x_2 - 1)^2 + (x_4 - 1)^2\right) + 19.8(x_2 - 1)(x_4 - 1)$	0

challenging from the solution finding viewpoint. In Table 5.2, the Sphere and Step functions are examples of unimodal separable functions, and Hartman 3, Hartman 6, Penalised, Penalised-2, Ackley, and Griewank functions can be categorised as multimodal non-separable functions.

Aforementioned function features can significantly affect the algorithm's performance in finding the global optima. For example, in some cases, the function surface consists of multiple local optimums with a very narrow global optimum area between them, which makes it difficult for the algorithm to escape from the local optimum areas. The dimensions of the functions can also affect the complexity of the search process for the global optimum. The interested readers are referred to Jamil et al. (2013) for more information on the test functions. Figure 5.1 presents the shapes of some test functions in two dimensions.

5.2.2 *Job-Shop Scheduling*

Job-shop Scheduling Problem (JSP) is an important optimisation problem in computer science and operational research. The JSP looks for the optimum schedule of a set of jobs that should be processed by a set of machines in a given sequence. Let us assume $j_1, j_2, \ldots, j_n$ jobs that should be processed by m machines represented by $m_1, m_2, \ldots, m_m$. Each job j_i is consisted of n_i sequence of operations indicated by $O_{i,1}, O_{i,2}, \ldots, O_{i,n_i}$, where $i = 1, 2, \ldots, n$. Each operation $O_{i,j}$ with the processing duration of $t_{i,j}$ can be processed on a given set of machines, and each machine can only perform one operation at a given time. The Flexible Job-shop Scheduling Problem (FJSP) is a specific case of JSP in which each operation $O_{i,j}$ can be processed on several available machines. In another case, which is called Hybrid Flow-shop Scheduling Problem (HFSP), each job in each stage can be processed on a set of parallel machines (Fernandez-Viagas 2022). These problems have been addressed in literature by considering different types of objective functions, constraints, and assumptions (Abdullah and Abdolrazzagh-Nezhad 2014; Çaliş and Bulkan 2015; Shao et al. 2020; Costa et al. 2020).

Cortés Rivera et al. (2007) adopted CA to solve the JSP. Becerra et al. (2005) employed standard CA to solve the JSP and compared its performance against Greedy Randomized Adaptive Search Procedure (GRASP) and Parallel GA (PGA) methods. The study reported the computational efficiency of CA. Researchers have also developed enhanced versions of CA to solve the mentioned problems. Ho et al. (2004) employed Cultural GA (CGA) to solve the FJSP in which the makespan minimisation is considered as the objective function. Niu et al. (2011) investigated the HFSP by using a hybrid CA and DE approach for fuzzy completion time (makespan) minimisation. Yang et al. (2014) utilised a Quadspace Cultural Genetic Tabu Algorithm (QSCGTA) to solve multi-objective JSP with the objective functions of makespan and mean flow time. Raeesi et al. (2012) solved the JSP using a multi-population version of CA. Zhang et al. (2012) proposed a hybrid Cultural PSO (CPSO) for the FJSP.

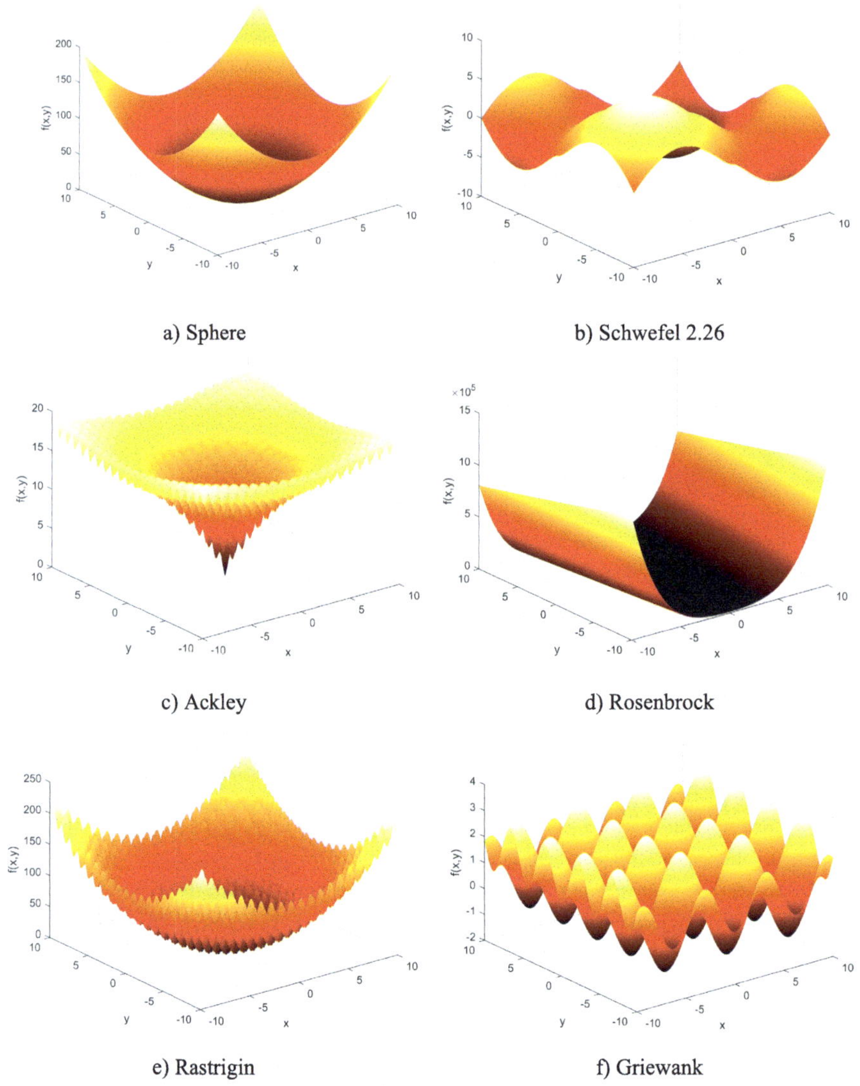

Fig. 5.1 Benchmark test functions: All in two dimensions with the interval $[-10, 10]$

5.2.3 *Facility Location*

Facility location is a problem in operational research that looks for optimum positioning and placement of different facilities to optimise the overall costs, transportation costs, service, waiting times, coverage, pollution, market share, energy costs, etc., while satisfying a variety of location constraints (Farahani et al. 2010).

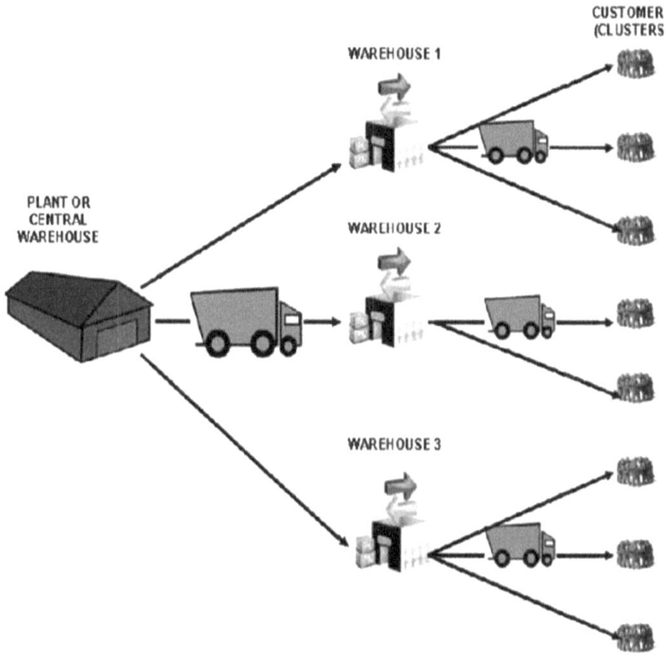

Fig. 5.2 Supply chain system considered in facility location problem (Cabrera et al. 2011)

Lagos et al. (2014) and Cabrera et al. (2011) applied CAs to solve the bi-objective un-capacitated facility location problem. In the facility location problem, a supply chain system of a company that distributes the products to the customers is considered as shown in Fig. 5.2. The primary aim of the problem is to find appropriate locations to build the plants, warehouses, and distribution centres from a set of available site alternatives, while the customers are assigned to serving facilities and the facilities are interconnected by flow assignment decisions (Lagos et al. 2014).

The overall formulation of the facility location problem can be described as follows. Let us consider a set of n potential facilities represented by $F_1, F_2, \ldots, F_n$ and m costumer clusters with demands $d_1, d_2, \ldots, d_m$. The opening cost of each facility is indicated by FC_i, and the cost of assigning the jth costumer cluster to the ith facility is represented by $c_{i,j}$. Each facility F_i has a covering area/distance indicated by $D_{\max}$ in which the customers receive the full service of the company without any interruption or delay. With these definitions, the faculty location problem can be mathematically formulated as a bi-objective optimisation problem as follows (Revelle and Laporte 1996):

$$
\begin{aligned}
&\text{Find: } X = [x_1, x_2, \ldots, x_n] \\
&\text{To minimise: } f_1(X) = \sum_{i=1}^{n} FC_i x_i + \sum_{i=1}^{n}\sum_{j=1}^{m} c_{i,j} y_{i,j} \\
&f_2(X) = -\sum_{j=1}^{m} d_j \sum_{i=1}^{n} y_{i,j} \\
&\text{Subject to:} \\
&\sum_{j=1}^{m} y_{i,j} = 1 \\
&y_{i,j} \le x_i, \quad i = 1, 2, \ldots, n, \quad j = 1, 2, \ldots, m. \\
&y_{i,j} \in \{0, 1\}, \quad i = 1, 2, \ldots, n, \quad j = 1, 2, \ldots, m. \\
&x_i \in \{0, 1\}, \quad i = 1, 2, \ldots, n
\end{aligned}
\tag{5.1}
$$

where $f_1(.)$ and $f_2(.)$ are the total operating cost and coverage objectives, respectively, x_i is a Boolean variable representing the openness of the facility which is equal to 1 if the ith facility is open and otherwise it is equal to 0, and $y_{i,j}$ is also a Boolean variable indicating the service performance of the ith facility for the jth costumer cluster which is equal to 1 if the demand of costumer j is fully supported by the ith facility. The first constraint ensures that only one facility should satisfy the demand of the jth customer cluster, while the second constraint indicates that at least one facility should be allocated for each costume cluster. The results reported by Lagos et al. (2014) and Cabrera et al. (2011) revealed that the CAs perform better than the multi-objective version of the GA, called NSGA-II, in finding optimal solutions for faculty location problems.

5.2.4 Set Covering

Set covering problem is a special case of facility location problem explained in Sect. 5.2.3 that considers only the opening cost of facilities as the objective function as follows (Farahani et al. 2012):

$$
\begin{aligned}
&\text{Find: } X = [x_1, x_2, \ldots, x_n] \\
&\text{To minimise: } f_1(X) = \sum_{i=1}^{n} FC_i x_i \\
&\text{Subject to:} \\
&\sum_{j=1}^{n} a_{ij} x_j \ge 1, \quad i = 1, 2, \ldots, m \\
&x_i \in \{0, 1\}, \quad i = 1, 2, \ldots, n
\end{aligned}
\tag{5.2}
$$

where a_{ij} is a binary parameter that is equal to 1 when the distance from the jth facility to customer cluster i is less than $D_{\max}$ and the definitions for all other parameters are similar to those explained in Eq. (5.1). The first constraint ensures that at least one facility should satisfy the demand of the jth costumer cluster. Crawford et al. (2007,

2013) applied CAs to solve the set covering problem presented in Eq. (5.2). The studies compared the results obtained from the CAs to those yielded by the standard GA and ACO methods. The comparison results verified the competent performance of CAs.

5.2.5 *Semantic Network*

Semantic networks can be described as a graph-based graphical knowledge representation in the form of interconnected nodes and arcs (He 2014). The arcs or links represent the semantic relations between the different concepts (Aronson 2003). Figure 5.3 presents an interesting example of the application of semantic networks in analysing the tweets on Twitter social media. The figure shows the relations between the term “financial risk” and other terms on Twitter.

In a semantic network, the network is categorised into several classes in which a set of nodes with similar attributes are represented. The membership of a given node to a specific class is determined by its attributes in comparison with other nodes in the network. The process of categorisation of the nodes into different classes and finding their most suitable locations in semantic networks is called classification. Within the classification process, a subsumption process is performed to obtain whether a given node can be linked to a potential parent node (Rychtyckyj and Reynolds 2000). The subsumption of semantic networks is a computationally expensive process, and the execution time can be dramatically increased for the semantic network with a huge number of nodes (Rychtyckyj and Reynolds 2005). In a set of studies reported by Rychtyckyj et al. (1998, 2000, 2001, 2002, 2005, 1999, 2003), the CAs were

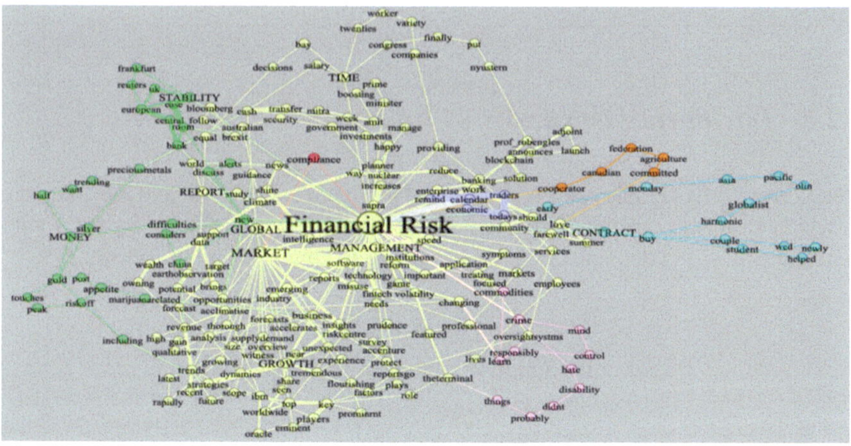

Fig. 5.3 An example semantic network for the tweets containing the term “financial risk” on Twitter in 2019 (Kwak and Grable 2021)

employed to reduce the computational complexity in the classification of semantic networks, in which the CAs were used as the guide operators for the subsumption process to identify a subset of useful attributes. The results suggested that the CAs can obtain more accurate results in comparison to manual inspections and significantly reduce the computation time.

5.2.6 *Program Understanding*

Program understanding or comprehension can be described as a set of activities that software engineers and programmers perform to understand how a given software works based on the available source code (Bennett et al. 2002). Reynolds et al. (1995) proposed a program understanding approach based on CAs to extract the functional knowledge for a program object.

5.2.7 *Timetabling*

University timetabling is a scheduling problem in operational research where the main aim is the allocation of classrooms in a given set of available time slots for the students under some resource constraints (Lindahl et al. 2018). Soza et al. (2007, 2011) employed CAs to solve the university course timetabling problem. The problem assumed a set of students and classrooms with a set of lectures that should be scheduled within a predefined set of timeslots during the week. The study considered a set of constraints on the availability of students and classrooms. The comparison results showed that the convergence rates of CAs are faster than other EAs.

5.2.8 *Intelligent Logistics*

Ochoa et al. (2010) employed CA to solve an intelligent logistic problem in a company in which a logistic system is investigated for the delivery of different products with different prices and demands. The study formulated the problem as follows:

$$\begin{aligned} &\text{Find: } X = [x_1, x_2, \ldots, x_n] \\ &\text{To minimise: } f(X) = -\sum_{i=1}^{n} r_i x_i \\ &\text{Subject to: } \sum_{i=1}^{n} v_i x_i < V_{\max} \end{aligned} \tag{5.3}$$

where X is the vector containing the quantity of each product, n is the number of products, $f(.)$ is the objective function of profit that should be maximised, the minus sign in $f(.)$ is introduced to convert it to the minimisation objective, r_i indicates the profit of the ith product, v_i is the volume of each product unit, and $V_{\max}$ is the maximum overall volume capacity of the logistic system.

5.2.9 *Artificial Neural Networks*

Nowadays, the Artificial Neural Networks (ANNs) have important applications in a wide range of research areas in science and technology. The meta-heuristic algorithms are necessary to train the ANNs based on the available data to improve their accuracy in the prediction of the outputs for any given input data in future. The researchers have been employed CAs and their different versions for training the ANNs and improving their accuracy.

Goli et al. (2019) developed an ANN model for market product demand prediction in a company. The study employed CA to optimise the weighting parameters of ANN and reduce the prediction errors. In a set of other studies, Chen et al. (2008, 2015 and Lin et al. 2008, 2009) developed ANNs with fuzzy membership functions, where the different versions of CAs have been employed to optimise the parameters of the fuzzy membership functions. Here, the main goal was to improve the performance of the ANN in predicting the outputs for a set of available training data. To this end, the studies considered the following objective function to improve the performance of ANNs with fuzzy rules:

$$f(X) = \frac{1}{1 + \sqrt{\frac{1}{N_t} \sum_{i=1}^{N_t} \left(y_i - \overline{y}_i\right)^2}} \tag{5.4}$$

where X is a vector containing parameters of fuzzy membership function that should be optimised, N_t indicates the number of training data, y_i is the output of the ANN obtained from the ith input data, and $\overline{y}_i$ is the desired output that is expected to yield from the ith input data.

In another study, Patil et al. (2021) employed CA to develop an ANN model for early diagnosis of type 2 diabetes based on the data set collected from different hospitals. The study used CA to minimise the Mean Square Error (MSE) of the ANN model for different data types.

5.2.10 *Multi-aircraft Target Allocation*

Li et al. (2012) employed CA to solve the multi-aircraft cooperative target allocation problem where the main objective is to select the best weapon-target allocation plan to achieve the most effective and precise attack performance of multiple aircrafts. Let us assume m friendly aircraft and n enemy aircrafts. Let $T_{i,j}$ be the threat from the ith friendly aircraft to enemy aircraft j and $\overline{T}_{j,i}$ be the threat from the jth enemy aircraft to friendly aircraft i. With these definitions, the study formulated the decision-making process in the multi-aircraft cooperative target allocation problem as the advantage maximisation problem of friendly aircrafts against the enemy aircrafts as follows (Li et al. 2012):

$$
\begin{aligned}
&\text{Find: } X = [x_1, x_2, \ldots, x_L] \\
&\text{To minimise: } f(X) = -\sum_{i=1}^{m}\sum_{j=1}^{n} v_{i,j} y_{i,j} u_{i,j} \\
&\text{Subject to:} \\
&\sum_{j=1}^{n} y_{i,j} \le Q_i, \quad i = 1, 2, \ldots, m \\
&\sum_{i=1}^{m} y_{i,j} \le P_j, \ j = 1, 2, \ldots, n \\
&y_{i,j} = 0, i \ne r, \ \text{if } v_{r,j} \ge v_f \\
&y_{i,j} = \begin{cases} 1, \text{ if } \lambda v_i^{\max} + (1-\lambda) v_i^{\min} \ge v_f \\ 0, \text{ otherwise} \end{cases} \\
&u_{i,j} = \begin{cases} 1, & v_{i,j} > 0 \\ -2, & v_{i,j} < 0 \end{cases}
\end{aligned}
\tag{5.5}
$$

in which:

$$
v_{i,j} = T_{i,j} - \overline{T}_{j,i} \tag{5.6}
$$

where Q_i indicates the number of enemy aircrafts that the ith friendly aircraft can attack, $L = \sum_{i=1}^{m} Q_i$ is the total number of enemy aircrafts that the friendly aircrafts can attack, $v_{i,j}$ represents the advantage of the ith friendly aircraft against the jth enemy, $y_{i,j}$ is a Boolean variable representing whether the ith friendly aircraft attacks the jth enemy aircraft, $u_{i,j}$ is a punishment parameter when the friendly aircraft has no advantage against the enemy aircraft, P_j represents the number of friendly aircrafts that can attack the enemy aircraft j, $v_i^{\min}$ and $v_i^{\max}$ are the lower and upper bounds for the ith friendly aircraft, respectively, v_f is a satisfaction factor, and λ is a Boolean variable representing the attack or evade modes.

The first constraint in Eq. (5.5) shows that each friendly aircraft can attack maximum Q_i number of enemy aircrafts. The second constraint limits the number of aircrafts that can attack each enemy aircraft. The third constraint is the performance satisfaction constraint. The fourth one is the pilot's balance constraint, and the last constraint is the penalty constraint to allocate penalty values for weak targeting plans.

5.2.11 Travelling Salesman Problem

TSP is one of the famous problems in computer science (Laporte 1992; Hoos and Stützle 2005; Öncan et al. 2009). The TSP considers a salesman that should visit different cities while satisfying some routing constraints, such as each city must be visited only once or the origin and destination should be the same. The objective of the problem is to find the shortest path or the route with the minimum cost for the salesman. Some researchers employed different versions of CAs to solve the TSP problem. Kim et al. (2009) applied CA with local search, Unold et al. (2016) employed Cultural ACO, and Unold et al. (2018) used multi-population CA to solve the TSPs.

5.2.12 Data Mining

Data mining is an important research field in computer science and statistics, which has wide applications in different areas of science and engineering (Mucherino et al. 2009; Yildirim et al. 2018; Alonso et al. 2018; Pan and Zhang 2021). The data mining techniques and algorithms are essential to discovering and extracting useful patterns from large sets of data which cannot be handled by the typical database software (Zaki and Meira Jr 2020; Nguyen et al. 2020; Gupta and Chandra 2020).

Feature selection in data mining is referred to the selection of useful subsets of features, information, or variables from the huge amount of data that represent the data in the best possible way. The feature selection methods play the role of a data processing scale compressor which keep the relevant information and remove the redundant and irrelevant features (Cai et al. 2018a). These techniques can reduce the computational cost and improve learning quality in the learning models (Cai et al. 2018a). Oloruntoba et al. (2019) applied a Clan-based CA (CCA) to solve the feature selection problem of different medical benchmark data sets.

One of the main challenges in data mining is the huge number of possible patterns that can be extracted from the big data sets which highlights the fact that the development of powerful algorithms is vital to efficiently extract useful patterns. The Decision Tree Induction (DTI) is a supervised learning method in data mining developed by Utgoff et al. (1997) in which the results of the data mining are represented in terms of the decision trees. The flexible framework of DTI makes its incremental extension for the new types of data without rebuilding the entire tree. However, building decision trees in the DTI can be computationally expensive. Reynolds et al. (1997, 1998) employed CA to optimise the decision trees within the DTI approach and enhance their accuracy, size, and interpretation. Jin and Reynolds (2000) applied CA to mine the knowledge in the large archaeological data sets. Similar studies have been done in the field of the data mining of large data sets based on CAs (Abdolrazzagh-Nezhad et al. 2020; Deng et al. 2020). In another study, the CA has been employed by Abitha and Vennila (2020) to improve the accuracy of association rule mining in a database of patients with an autism spectrum disorder.

5.2.13 Cloud Computing

The cloud computing paradigm can be viewed as a model of computing resource allocation and is one of the recent important computing model developments to support different activities on the Internet platform, ranging from data storage to running applications (Gurkok 2017). The US National Institute of Standards and Technology defines cloud computing as follows (Mell and Grance 2011):

> *"Cloud computing is identified as a model for enabling ubiquitous, convenient, on-demand network access to a shared pool of configurable computing resources (e.g., networks, servers, storage, applications, and services) that can be rapidly provisioned and released with minimal management effort or service provider interaction."*

In cloud computing, the users with different devices use the cloud services with given prices, usually obtained based on the pay-as-you-go pricing model, for different purposes on the Internet, such as access to computer programs, storage spaces, and application development platforms (Furht and Escalante 2010). Nowadays, the "Big Tech" companies, such as Google, Amazon, Meta, and Apple, offer a wide range of services based on cloud computing technology for their users. Cloud computing provides a range of advantages for both service providers and consumers, such as lower cost, flexibility, and ease of access (Avram 2014).

The CAs have been successfully employed to solve different kinds of optimisation problems in cloud computing. Mohammadhosseini et al. (2019) employed CA to solve the virtual machine replacement problem in the cloud systems, in which the optimum resource allocation in the virtualisation process was investigated. Mousavi Mojab et al. (2019) utilised CA to solve the big data workflow scheduling problem in the cloud systems, where the running cost on the cloud system is minimised under some work deadline constraints. Ma et al. (2008) employed CA for the routing problem of mobile agents, in which the node and route selections in the migration process of agents within the network are optimised to improve the network performance and reduce costs. In this section, the virtual machine replacement problem is selected to be discussed and related formulations are presented.

5.2.13.1 Virtual Machine Placement

Despite the abovementioned advantages, cloud computing systems are not efficient enough in terms of the energy usage and their environmental footprint has raised concerns (Lee and Zomaya 2012; Mastelic and Brandic 2015; Hameed et al. 2016). Recent research revealed that cloud computing systems are responsible for a significant part of global electric consumption (Mastelic and Brandic 2015). One of the efficient ways to reduce energy consumption, improve performance, and save costs in cloud computing is through virtualisation technology (Hassan and Hijazi 2017).

Virtualisation aims to share the computing resources among the different applications in order to provide a fast and low-cost service for customers. In virtualisation, the computing resources are divided into several execution environments, called Virtual Machines (VMs). Each VM is isolated and performs independently from other VMs (Masdari et al. 2016). The server is responsible to provide the required computing resources for each VM, such as CPU, memory, and storage (Vu and Hwang 2014). Figure 5.4 illustrates a simple example taken from Furht et al. (2010) for the virtualisation in a cloud with two servers. In this figure, the hypervisor is a software layer that controls the VMs inside each server. As it can be seen from Fig. 5.4, three applications running on non-cloud systems require three different servers, while the virtualisation technology makes it possible to run the applications on a cloud system with two servers. The advantages of virtualisation are much more significant in large could systems with numerous servers and customer applications.

Virtual Machine Placement (VMP) is one of the complex problems in cloud computing in which the main aim is to improve the performance of the cloud system by finding the optimum placements of different VMs within a set of available physical servers (Gahlawat and Sharma 2014). Mohammadhosseini et al. (2019) developed a CA-based VMP approach to improve the energy performance of cloud systems. In this subsection, the problem formulations proposed by Mohammadhosseini et al. (2019) are reviewed.

Let us consider q number of VMs. Let also $\vec{T}$ be the unit vector of total available computing resources in the physical server and $\vec{A}$ be the unit vector of the total allocated resources of the server. The vectors $\vec{T}$ and $\vec{A}$ can be written in terms of the

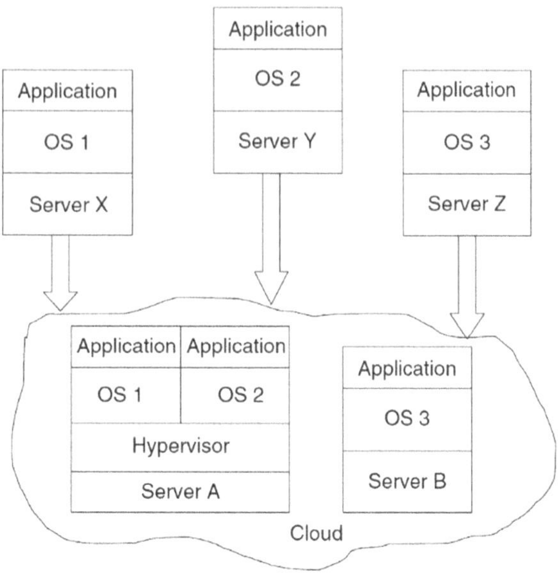

Fig. 5.4 A simple example of virtualisation in cloud computing (Furht et al. 2010) (adapted from Jones)

CPU, RAM, and bandwidth as follows (Mohammadhosseini et al. 2019):

$$\vec{T} = \frac{C_{\mathrm{T}}\vec{i} + R_{\mathrm{T}}\vec{j} + B_{\mathrm{T}}\vec{k}}{\sqrt{C_{\mathrm{T}}^2 + R_{\mathrm{T}}^2 + B_{\mathrm{T}}^2}} \tag{5.7}$$

$$\vec{A} = \frac{\left(\sum_{l=1}^{q} C_{\mathrm{A}_l}\right)\vec{i} + \left(\sum_{l=1}^{q} R_{\mathrm{A}_l}\right)\vec{j} + \left(\sum_{l=1}^{q} B_{\mathrm{A}_l}\right)\vec{k}}{\sqrt{\left(\sum_{l=1}^{q} C_{\mathrm{A}_l}\right)^2 + \left(\sum_{l=1}^{q} R_{\mathrm{A}_l}\right)^2 + \left(\sum_{l=1}^{q} B_{\mathrm{A}_l}\right)^2}} \tag{5.8}$$

where C_{T}, R_{T}, and B_{T} are the total available CPU, RAM, and bandwidth in the server, respectively, C_{A_l}, R_{A_l}, and B_{A_l} represent the allocated CPU, RAM, and bandwidth for the l th VM in the server, respectively.

In VM replacement, the balance in the allocation is defined as the equal distribution of computing resources between the VMs. Mohammadhosseini et al. (2019) defined the balance vector as a triangular area created by vectors $\vec{S}_{\mathrm{T}}$ and $\vec{S}_{\mathrm{A}}$ as follows:

$$\vec{\mathcal{B}} = \frac{1}{2}\vec{T} \times \vec{A} \tag{5.9}$$

where $\vec{\mathcal{B}}$ represents the unit balance vector. The unit balance vector can be reformulated by calculating the vector product in Eq. (5.9) as follows (Mohammadhosseini et al. 2019):

$$\vec{\mathcal{B}} = \frac{1}{2}\left(C_{\mathrm{b}}\vec{i} + R_{\mathrm{b}}\vec{j} + B_{\mathrm{b}}\vec{k}\right) \tag{5.10}$$

in which:

$$C_{\mathrm{b}} = \frac{1}{2}\frac{\sum_{l=1}^{q}\left(R_{\mathrm{T}}B_{\mathrm{A}_l} - B_{\mathrm{T}}R_{\mathrm{A}_l}\right)}{\sqrt{C_{\mathrm{T}}^2 + R_{\mathrm{T}}^2 + B_{\mathrm{T}}^2}\sqrt{\left(\sum_{l=1}^{q} C_{\mathrm{A}_l}\right)^2 + \left(\sum_{l=1}^{q} R_{\mathrm{A}_l}\right)^2 + \left(\sum_{l=1}^{q} B_{\mathrm{A}_l}\right)^2}} \tag{5.11}$$

$$R_{\mathrm{b}} = \frac{1}{2}\frac{-\sum_{l=1}^{q}\left(C_{\mathrm{T}}B_{\mathrm{A}_l} - B_{\mathrm{T}}C_{\mathrm{A}_l}\right)}{\sqrt{C_{\mathrm{T}}^2 + R_{\mathrm{T}}^2 + B_{\mathrm{T}}^2}\sqrt{\left(\sum_{l=1}^{q} C_{\mathrm{A}_l}\right)^2 + \left(\sum_{l=1}^{q} R_{\mathrm{A}_l}\right)^2 + \left(\sum_{l=1}^{q} B_{\mathrm{A}_l}\right)^2}} \tag{5.12}$$

$$B_{\mathrm{b}} = \frac{1}{2}\frac{\sum_{l=1}^{q}\left(C_{\mathrm{T}}R_{\mathrm{A}_l} - R_{\mathrm{T}}C_{\mathrm{A}_l}\right)}{\sqrt{C_{\mathrm{T}}^2 + R_{\mathrm{T}}^2 + B_{\mathrm{T}}^2}\sqrt{\left(\sum_{l=1}^{q} C_{\mathrm{A}_l}\right)^2 + \left(\sum_{l=1}^{q} R_{\mathrm{A}_l}\right)^2 + \left(\sum_{l=1}^{q} B_{\mathrm{A}_l}\right)^2}} \tag{5.13}$$

In the above equations, C_{b}, R_{b}, and B_{b} represent the balance values for the CPU, RAM, and bandwidth, respectively.

Now, let us consider a cloud system with n_{s} number of servers with given computing resources of C_{T_i}, R_{T_i}, and B_{T_i}, where $i = 1, 2, \ldots, n_{\mathrm{s}}$. With the above definitions, Mohammadhosseini et al. (2019) formulated the VMP as an optimisation problem as follows:

$$\text{Find: } X = \begin{bmatrix} x_{1,1} & x_{1,2} & \cdots & x_{1,q} \\ x_{2,1} & x_{2,2} & \cdots & x_{2,q} \\ \vdots & \vdots & \ddots & \vdots \\ x_{n_s,1} & x_{n_s,2} & \cdots & x_{n_s,q} \end{bmatrix}$$

$$\text{To minimise: } f(X) = \frac{\sum_{i=1}^{n_s} \left|\vec{\mathcal{B}}_i\right|}{\sum_{j=1}^{r} \left(\sum_{i=1}^{n_s} \left|\vec{\mathcal{B}}_i\right|\right)_j} + \frac{\sum_{i=1}^{n_s} E_i}{\sum_{j=1}^{r} \left(\sum_{i=1}^{n_s} E_i\right)_j}$$

Subject to:

$$g(X) = \begin{cases} 1 \text{ if } \left(\begin{array}{l} \left(\sum_{l=1}^{q} C_{\mathrm{A}_{l,i}} \leq C_{\mathrm{T}_i}\right) \cup \left(\sum_{l=1}^{q} R_{\mathrm{A}_{l,i}} \leq R_{\mathrm{T}_i}\right) \\ \cup \left(\sum_{l=1}^{q} B_{\mathrm{A}_{l,i}} \leq B_{\mathrm{T}_i}\right) \end{array} \right) \\ 0 \text{ otherwise} \end{cases}, i = 1, 2, \ldots, n_s \tag{5.14}$$

where X represents the placement matrix, in which $x_{i,j}$ is a Boolean variable representing whether the j th machine is placed in i th server, $\vec{\mathcal{B}}_i$ represents the balance vector for the i th server, E_i indicates the energy consumption of the i th server, $C_{\mathrm{A}_{l,i}}$, $R_{\mathrm{A}_{l,i}}$, and $B_{\mathrm{A}_{l,i}}$ indicate the allocated CPU, RAM, and bandwidth for the l th VM in the i th server, respectively, and r is the number of feasible solutions in each iteration of the optimisation algorithm. As it can be seen from Eq. (5.14), the objective function is to minimise the balance vector and energy usage within the cloud system. The elements of placement matrix X are decision variables of the problem which were optimised by using CA.

5.2.14 Fog Computing

Nowadays, the massive amounts of data produced by the Internet of Things (IoT) devices are transferred to the cloud applications in which the users store and process their data in third-party data centres of cloud technology (Huang and Wu 2018a; Yousefpour et al. 2019). However, due to the amount of data and the bandwidth limitations, the data transferring process from the IoT devices to the cloud-based systems is not efficient enough. On the other hand, geographically distant cloud technology can cause high latency for real-time applications. For example, some major cloud user companies have recently reported the availability and slow response of cloud services during critical periods of business transactions (Uhlir et al. 2016; Xu et al. 2021).

To remedy the aforementioned drawbacks of the centralised cloud-based systems, the fog computing technology performs the communication, networking, storage, and data management between the IoT devices and cloud systems by placing near-user edge devices (Huang and Wu 2018b; Yousefpour et al. 2019), which can significantly reduce the latency. Although fog computing is an efficient alternative to cloud computing, the optimum task scheduling in fog computing is a real challenge. To deal with the challenge, Hosseinioun et al. (2020) applied a hybrid CA and Invasive Weed Optimisation (IWO) algorithm for optimum task scheduling in the fog computing environment. The main objective of the problem is to find the optimum scheduling for the different tasks and processors to reduce the energy consumption and execution time in fog computing.

5.2.15 Online Marketing

In online marketing, recommendation systems play an important role in guiding the users to find the most suitable and attractive items based on their historical preferences. Nowadays, online retail businesses employ recommender systems to promote their deals and encourage users to buy a set of items. One of the most popular recommender system approaches is the Collaborative Filtering (CF), which predicts the future shopping decisions of each user based on their previous purchase history or based on the available information of other similar users (Felfernig et al. 2014).

Selvarajah et al. (2019a) applied CA to enhance the accuracy and performance of the CF-based recommender systems, in which the similarities between the users are considered. The study also considered the trust level of the users on the reviews and ratings for each product. The numerical results suggested that CA can exhibit better performance in comparison to the GA method.

5.2.16 Genetic Programming

Zannoni et al. (1997) developed a hybrid CA and Genetic Programming (GP) approach for symbolic regression problems. The individuals within the CA are defined as symbols related to the function and terminal sets. The number of decision variables is denoted by n where the n is in fact the number of nodes with the corresponding parse tree and the number of symbols in the list presentation of the GP. Cowan et al. (1999) used CA to assess the performance of programs generated by GP.

5.2.17 Games

CAs have been also applied to game development in the literature. Reynolds and Kinnaird-Heether (2003) integrated social fabric-based CA with The Open-source Racing Car Simulator (TORCS), which is a 3D racing game, to optimise the game controllers. The study optimised the car steering process through the optimisation of the angles between the car direction and the direction of the track axis. Vitale et al. (2011a, b) applied CA for learning group behaviour of caribou herds in the land bridge simulation. The study simulated the herding behaviour of caribou in a virtual environment.

5.2.18 *Software Testing*

The development of large software systems is a challenging process, in which the programmers should constantly test and find the errors as soon as they appear in every stage of the development process. In some cases, the delay in error identification may result in error propagation to other parts of the software which can make error identification a real challenge for the developers (Baresi and Pezzè 2006). Software testing is referred to a set of activities that should be performed during the software production process to assess the behaviour of the software from different programming perspectives and remedy possible errors and faults (Lonetti and Marchetti 2018).

With the rapid development of the software industry and its expansion speed, there is an urgent need for efficient software testing techniques with accurate and fast fault detection capabilities (Mayeda and Andrews 2021). Ostrowski and Reynolds (1999, 2005) developed an automatic software testing approach based on CA. The approach includes two black-box and white-box testing phases. The black-box testing aims to find several challenging test cases in which the software is likely to exhibit faults. The CA was employed in the black-box phase as an optimiser to find the most challenging test cases. In the white-box phase, the software is divided into a given number of slices and the CA is employed to search for the program segments and find the segments that the faults are resulting. In order to reduce the level of risk for fault detection, the study also provided a feedback loop between the white-box and black-box phases to generate additional test cases and enhance the reliability of the software testing process.

5.2.19 *Image Segmentation*

Image segmentation is one of the basic problems in computer vision (Minaee et al. 2021). As its name suggests, the main aim of image segmentation is to categorise the different objects in the images or videos (Haralick and Shapiro 1985; Zhang 1996). The regional and neighbourhood approaches are two basic approaches to image segmentation. Simply speaking, the region-based approach does not need initial information about features in the image environment. However, it is computationally expensive. On the other hand, the neighbourhood-based approach requires initial knowledge about the importance of different features in the image environment.

In order to remedy the computational drawback of the regional approach, Reynolds et al. (1995a, b) developed a new image segmentation approach by integrating CA and regional-based methods. The study proposed to perform the regional-based method guided by the CA as a learner mechanism and obtain the required knowledge about the image. The study investigated a sequence of three images produced by the archaeological excavation of Cueva Blanca, a cave located in the Valley of Oaxaca Mexico. The results revealed that the application of CA within the regional-based approach yields 95% results of the standard regional method, while the computational cost is significantly reduced.

5.2.20 Image Matching

Image matching is one of today's fundamental problems in artificial intelligence that has a wide range of real-world applications, such as medical science, security, remote sensing, and computer vision (Jiang et al. 2021). According to the definition in literature (Ma et al. 2021):

> *"image matching, also known as image registration or correspondence, aims to identify then correspond the same or similar structure/content from two or more images."*

Yan et al. (2017b) applied CA for a benchmark image matching problem. The comparison results between the CA and GA revealed that the CA is capable of establishing the correspondence between image patterns with a higher level of accuracy.

5.2.21 Sonar Systems

Sonar imaging is a long-distance detection technology that has important applications in the marine industry. The sonar imaging technology is employed in Remotely Operated Vehicles (ROVs) to scan the surrounding areas during underwater operations (Bai and Bai 2019). The ROVs have important applications in underwater excavation and inspections, cable laying, foundation installations, etc. (Shepherd 2001). The technology has also important applications in marine biodiversity research of marine species (Jones et al. 2021). Wang et al. (2017) applied CA to detect underwater sonar images.

5.2.22 Gesture Recognition

Gesture recognition is one of the sub-disciplines of computer vision in which the main goals are to recognise the human movements from the image sequences and interpret them for human–computer interaction (Turk and Athitsos 2020). Waris et al. (2015) studied the use of wearable control devices, such as smartwatches, for gesture-based control of the automotive human–machine interface. It is assumed that the driver has worn an android smartwatch with which she/he can use a set of gestures to control the automotive human–machine interface. The study considered a set of six gesture vocabularies for smartwatches and employed CA to enhance the gesture recognition process and optimise the human's best guess parameter values.

5.3 Social Science

CAs have found some applications in social science. They have been employed to discover the social networks between the individuals in the society as well as manage the social cooperation between different businesses partners. The following subsections describe these application studies.

5.3.1 *Anthropology*

According to the definition, "*anthropology is the study of humanity*" in which "*anthropologists generally observe the behavior of present-day people in order to explain and reconstruct their behavior patterns*" (Joukowsky 2015). The CA has been also applied to anthropology science.

Kobti et al. (2006a) employed CA to investigate the emergence of social networks among individuals in a real society. The study considered the resource exchange between the individuals to detect the social intelligence patterns within the population in the Mesa Verde Village region. The approach assumed that different knowledge about the agricultural opportunities and available resources obtained by the individuals are stored within knowledge components of the belief space of CA. The information about the distribution of agricultural land, wild plants, and animal resources was stored within the topographical knowledge component. The information about the rainfall distribution and water resources were assumed to be stored within the historical knowledge component. The domain knowledge was also considered to keep the agricultural planting and harvesting techniques. The study considered the kinship and economic networks, in which the social cooperation of individuals was modelled from the kinship and economic perspectives.

5.3.2 *Virtual Enterprise*

Virtual enterprise refers to "*a temporary alliance of enterprises that come together to share skills or core competencies and resources in order to better respond to business opportunities, and whose cooperation is supported by computer networks*" (Camarinha-Matos and Afsarmanesh 1999). Wei et al. (2012) applied a CA-based PSO algorithm to the virtual enterprise, in which the different businesses temporarily ally to cooperate. The study investigated a multi-objective virtual enterprise problem in which the total time, cost, and quality of the projects were considered as the objective function.

5.4 Archaeology

In archaeology, the interpretation of archaeological data to explain the residential patterns and social interactions between the inhabitants in archaic urban centres plays an important role in our understanding of the ancient people. The functionality of different locations in an archaic site can be processed to describe the morphology and social interactions at the site. This is one of the potential and most interesting application areas for the CA which can be efficiently used to build morphological models and reduce the complexity of archaeological raw data.

Reynolds and Ali (2008a, b) joined this research line and developed a new approach based on CA for the morphology of the Monte Albán archaic site in Mexico. The Monte Albán site includes more than 2000 terraces where the residents used to live and work. The study tried to detect the social identities and social interactions of the terrace occupants. In this approach, each individual in the population space of CA represents a potential terrace occupant with a set of social and economic backgrounds, such as where is she/he from, her/his occupation goal, etc. Each individual is represented by a Naive Bayesian decision tree structure. The social interactions between the individuals are modelled through the social fabric with a "Four Square" topology, which is the subject of discussion in Sect. 7.5 of Chap. 7. Reynolds and Ali (2008a, b) reported some interesting results on the residential patterns in Monte Albán archaic site. These patterns can be an opportunity for the archaeologist to understand the development of ancient urban centres as well as their cultural evolution process.

5.5 Physics

CA has found application in the field of electromagnetics in physics. The CA has been employed by dos Santos Coelho and Alotto (2009) for the optimum design problem of Loney's solenoid. The main aim of this problem is to generate a uniform magnetic flux density within a given interval on the axis of the solenoid based on a given current density distribution. Let B the magnetic flux density and J_0 be the constant current density in the coils. In this design problem, the goal is to keep the B uniform within the interval $[-z_0, z_0]$ on the axis of solenoid. The decision variables of the problem are the interval length s and the length of the correcting coils l. Thus, the optimum design problem can be formulated as follows (dos Santos Coelho and Alotto 2009):

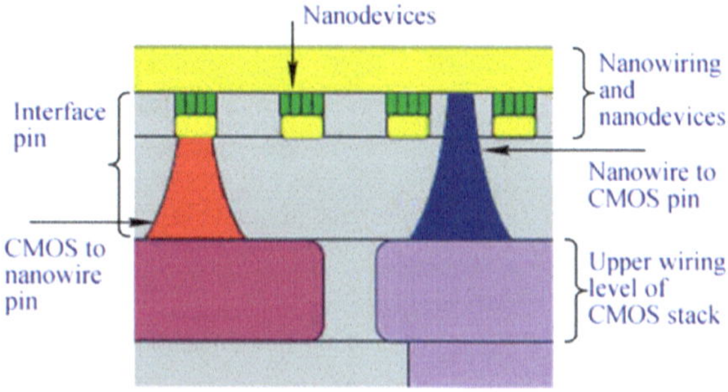

Fig. 5.5 Cross-section of the CMOS circuit with nanodevices (Pan et al. 2016)

$$\begin{aligned}&\text{Find: } X = [s, l]\\&\text{To minimise:} f(X) = \frac{B_{max} - B_{min}}{B_0}\end{aligned} \quad (5.15)$$

where B_{min} and B_{max} represent the minimum and maximum magnetic flux density within the interval $[-z_0, z_0]$, respectively, and B_0 indicates the flux density at point $z = 0$.

In another study, Pan et al. (2016) employed CA for cell allocation problems in Complementary Metal–Oxide–Semiconductor (CMOS) with nanodevices. The main aim of the problem is to find appropriate placements of nanodevices in CMOS and their optimum connections with nanowires. Figure 5.5 illustrates the cross-section of the CMOS circuit with nanodevices.

5.6 Others

In addition to the topics discussed in previous sections of this chapter, interesting applications of CAs in other areas of science have been also revealed in literature. Table 5.3 lists the studies in different areas ranging from economy, social science, Internet to biological science. In this section, these application areas are briefly explained and discussed.

There are application studies of CAs to some computer science-related problems. Ochoa et al. (2012) developed a mobile device based on CA to inform people about the security of the different parts of Ciudad Juarez city in Mexico. The device integrated the CA with the data mining technology. Prestwich et al. (2008) employed CA to solve the partially observable Markov decision process, which has important applications in different areas of machine learning. Sun et al. (2009) applied CA to solve the Quality of Service (QoS) problem, in which the computer network is represented by an undirected weighted graph consisting of several nodes and edges. The study assumed the energy consumption minimisation of the network as the objective function.

The fake news on social media appears in different forms, such as messages, videos, and images, which can make a complex challenge for governments, individuals, and society as a whole. Shah et al. (2020) investigated the detection of fake

Table 5.3 Applications of CAs in other areas

Area	Reference
Economical models	Ostrowski et al. (2004)
Biology	Alexiou et al. (2012), Judeh et al. (2013, 2014)
Social networks	Zadeh et al. (2015), Selvarajah et al. (2019b, 2020)
Security	Ochoa et al. (2012)
Fake news detection	Shah et al. (2020)
Markov decision process	Prestwich et al. (2008)
Quality of service	Sun et al. (2009)
Circuit Boolean problems	Reynolds et al. (1993)
Site location problem	Reynolds et al. (1997)
Web search	Zhang (2012)
Fraud detection in insurance companies	Sternberg and Reynolds (1997)

news between texts and images on social media. The results obtained based on the Weibo and Twitter data sets revealed that CA can efficiently detect fake content in comparison to ANN-based approaches.

Reynolds et al. (1993) developed an approach based on the CA to solve the circuit Boolean problems, which are basic problems in the complexity theory. Reynolds et al. (1997) investigated the decision-making problem for the site location based on the spatio-temporal databases. The study expressed the site knowledge in the form of the decision tree structures. Then, the CA is used to optimise the decision trees based on the influences of productivity, environmental location and distance from other sites. Zhang (2012) applied CA to enhance the web search on the internet.

The insurance firms are facing a serious amount of fraudulent claims which force them to consistently discover new fraud techniques and update their policies in order to reduce the budget losses. Sternberg and Reynolds (1997) employed CA for fraudulent claim detection in automotive insurance companies. The problem can be categorised as an optimisation problem with a dynamic environment. The approach categorises insurance claims into different categories, including fraud, non-fraud, false positive, and false negative claims. The false negative claim refers to the fraudulent claim predicted as a non-fraudulent claim and the false positive claim means that the non-fraudulent claim is predicted as a fraudulent claim. The numerical results reported by the study showed that the CA-based approach can accurately detect the claim categories under dynamically changing problem environments.

CAs have also been applied to different problems in social networks. Zadeh et al. (2015) employed CA to detect communities formed by individuals communicating with each other in the social network analysis. Selvarajah et al. (2019b, 2020) utilised CA for the identification of expert teams to maximise the chance of success of a given work project.

The applications of CAs have also appeared in biology science. Nowadays, big gene databases are available which can be processed and analysed for different purposes. Judeh et al. (2013, 2014) proposed the Gene Set CA to identify the networks that the genes are originally from. In other words, the CA is employed to discover the genes with similar features from a set of unordered gene databases. The results revealed that the CA-based gene identification is more accurate than the traditional gene set algorithms. In another study, Alexiou et al. (2012) employed CA for the representation of the mitochondrial population in mammalian cells, which shows how the CAs received good feedback from biological science as well.

5.7 Summary

The chapter discussed the application studies of CAs in different branches of science, such as computer, physical, social, and biological sciences. The investigated studies revealed that the diversity of applications in computer science is more significant than in other areas. Function minimisation, JSP, TSP, software testing, facility location, set covering, virtual machine replacement, gesture recognition, image matching, data mining, online marketing, and timetabling are examples of these problem classes in computer science. In social science, CAs have been employed to discover the social networks formed among individuals in a real human society. In archaeology, the residential patterns and social interactions between the inhabitants in archaic urban centres have been explained based on the computational intelligence of CAs. In biology science, CAs have been used for the identification of the gene networks as well as the representation of the mitochondrial population in mammalian cells. The chapter also provided a brief review of other applications of CAs in physics, security, economy, etc.

References

Abdolrazzagh-Nezhad M, Radgohar H, Salimian SN (2020) Enhanced cultural algorithm to solve multi-objective attribute reduction based on rough set theory. Math Comput Simul 170:332–350. https://doi.org/10.1016/j.matcom.2019.11.005

Abdullah S, Abdolrazzagh-Nezhad M (2014) Fuzzy job-shop scheduling problems: a review. Inf Sci 278:380–407. https://doi.org/10.1016/j.ins.2014.03.060

Abitha R, Vennila SM (2020) Discovering positive association of ASD attributes with class using multi objective cultural algorithm. ICTACT J Soft Comput 11:2222–2226

Alami J, el Imrani A, Bouroumi A (2007) A multipopulation cultural algorithm using fuzzy clustering. Appl Soft Comput 7:506–519. https://doi.org/10.1016/j.asoc.2006.10.010

Alexiou A, Vlamos P (2012) A cultural algorithm for the representation of mitochondrial population. Adv Artif Intell 2012:1–7. https://doi.org/10.1155/2012/457351

Ali MZ, Salhieh A, Snanieh RTA, Reynolds RG (2012) Boosting cultural algorithms with a heterogeneous layered social fabric influence function. Comput Math Organ Theory 18:193–210

Ali MZ, Awad NH, Suganthan PN et al (2016a) A novel hybrid cultural algorithms framework with trajectory-based search for global numerical optimization. Inf Sci 334–335:219–249. https://doi.org/10.1016/j.ins.2015.11.032

Ali MZ, Awad NH, Suganthan PN, Reynolds RG (2016b) A modified cultural algorithm with a balanced performance for the differential evolution frameworks. Knowl-Based Syst 111:73–86. https://doi.org/10.1016/j.knosys.2016.08.005

Ali MZ, Reynolds RG (2009) An intelligent social fabric influence component in cultural algorithms for knowledge learning in dynamic environments. In: 2009 IEEE/WIC/ACM International joint conference on web intelligence and intelligent agent technology. IEEE, pp 161–168

Ali M, Reynolds R, Ali R, Salhieh A (2011a) Knowledge-based constrained function optimization using cultural algorithms with an enhanced social influence metaphor. In: Computational intelligence. Springer, pp 103–119

Ali MZ, Salhieh A, Snanieh RTA, Reynolds RG (2011b) Boosting cultural algorithms with an incongruous layered social fabric influence function. In: 2011b IEEE Congress of evolutionary computation (CEC). IEEE, pp 1225–1232

Ali MZ, Awad N, Reynolds RG (2014a) Balancing search direction in cultural algorithm for enhanced global numerical optimization. In: 2014 IEEE Symposium on Swarm Intelligence. IEEE, pp 1–7

Alonso SG, de la Torre-Díez I, Hamrioui S et al (2018) Data mining algorithms and techniques in mental health: a systematic review. J Med Syst 42:161. https://doi.org/10.1007/s10916-018-1018-2

Aronson JE (2003) Expert systems. In: Encyclopedia of information systems. Elsevier, pp 277–289

Avram MG (2014) Advantages and challenges of adopting cloud computing from an enterprise perspective. Procedia Technol 12:529–534. https://doi.org/10.1016/j.protcy.2013.12.525

Awad NH, Ali MZ, Suganthan PN, Reynolds RG (2017) CADE: A hybridization of cultural algorithm and differential evolution for numerical optimization. Inf Sci 378:215–241. https://doi.org/10.1016/j.ins.2016.10.039

Awad NH, Ali MZ, Duwairi RM (2013) Cultural Algorithm with improved local search for optimization problems. In: 2013 IEEE Congress on evolutionary computation. IEEE, pp 284–291

Bai Y, Bai Q (2019) Subsea surveying, positioning, and foundation. In: Subsea engineering handbook. Elsevier, pp 81–121

Baresi L, Pezzè M (2006) An introduction to software testing. Electron Notes Theoret Comput Sci 148:89–111. https://doi.org/10.1016/j.entcs.2005.12.014

Becerra RL, Coello CAC (2004) A cultural algorithm with differential evolution to solve constrained optimization problems. In: Ibero-American conference on artificial intelligence, pp 881–890

Becerra RL, Coello CAC (2005) A cultural algorithm for solving the job shop scheduling problem. In: Knowledge incorporation in evolutionary computation. Springer, pp 37–55

Bennett KH, Rajlich VT, Wilde N (2002) Software evolution and the staged model of the software lifecycle, pp 1–54

Cabrera G, Rubio JM, Diaz D et al (2011) A cultural algorithm applied in a bi-objective uncapacitated facility location problem. In: International conference on evolutionary multi-criterion optimization, pp 477–491

Cai J, Luo J, Wang S, Yang S (2018) Feature selection in machine learning: a new perspective. Neurocomputing 300:70–79. https://doi.org/10.1016/j.neucom.2017.11.077

Çaliş B, Bulkan S (2015) A research survey: review of AI solution strategies of job shop scheduling problem. J Intell Manuf 26:961–973. https://doi.org/10.1007/s10845-013-0837-8

Camarinha-Matos LM, Afsarmanesh H (1999) The virtual enterprise concept, pp 3–14

Chen C-H, Liu Y-C, Lin C-J, Lin C-T (2008) A hybrid of cooperative particle swarm optimization and cultural algorithm for neural fuzzy networks. In: 2008 IEEE International conference on fuzzy systems (IEEE World congress on computational intelligence). IEEE, pp 238–245

Chen C-H, Yang S-Y (2015) Efficient DE-based symbiotic cultural algorithm for neuro-fuzzy system design. Appl Soft Comput 34:18–25. https://doi.org/10.1016/j.asoc.2015.04.026

Chung C, Reynolds RG (1997a) Function optimization using evolutionary programming with self-adaptive cultural algorithms, pp 17–26
Chung C-J, Reynolds RG (1998) CAEP: an evolution-based tool for real-valued function optimization using cultural algorithms. Int J Artif Intell Tools 7:239–291. https://doi.org/10.1142/S0218213098000135
Coello CAC, Becerra RL (2002) A cultural algorithm for constrained optimization, pp 98–107
Coello CAC, Becerra RL (2003) Evolutionary multiobjective optimization using a cultural algorithm. In: Proceedings of the 2003 IEEE swarm intelligence symposium. SIS'03 (Cat. No. 03EX706). IEEE, pp 6–13
Cortés Rivera D, Landa Becerra R, Coello Coello CA (2007) Cultural algorithms, an alternative heuristic to solve the job shop scheduling problem. Eng Optim 39:69–85. https://doi.org/10.1080/03052150600956811
Costa A, Fernandez-Viagas V, Framinan JM (2020) Solving the hybrid flow shop scheduling problem with limited human resource constraint. Comput Ind Eng 146:106545. https://doi.org/10.1016/j.cie.2020.106545
Cowan GS, Reynolds RG (1999) Learning to assess the quality of genetic programs using cultural algorithms. In: Proceedings of the 1999 Congress on evolutionary computation-CEC99 (Cat. No. 99TH8406). IEEE, pp 1679–1686
Crawford B, Lagos C, Castro C, Paredes F (2007) A cultural algorithm for solving the set covering problem. In: Analysis and design of intelligent systems using soft computing techniques. Springer, pp 408–415
Crawford B, Soto R, Monfroy E (2013) Cultural algorithms for the set covering problem. In: International conference in swarm intelligence, pp 27–34
Deng L, Yang P, Liu W (2020) Artificial immune network clustering based on a cultural algorithm. EURASIP J Wirel Commun Netw 2020:168. https://doi.org/10.1186/s13638-020-01779-1
Dixit A, Kumar S, Pant M, Bansal R (2019) CA-DE: Hybrid algorithm based on cultural algorithm and DE. In: Machine intelligence and signal analysis. Springer, pp 185–196
dos Santos CL, Alotto P (2009) Electromagnetic optimization using a cultural self-organizing migrating algorithm approach based on normative knowledge. IEEE Trans Magn 45:1446–1449. https://doi.org/10.1109/TMAG.2009.2012668
Farahani RZ, SteadieSeifi M, Asgari N (2010) Multiple criteria facility location problems: a survey. Appl Math Model 34:1689–1709. https://doi.org/10.1016/j.apm.2009.10.005
Farahani RZ, Asgari N, Heidari N et al (2012) Covering problems in facility location: a review. Comput Ind Eng 62:368–407. https://doi.org/10.1016/j.cie.2011.08.020
Felfernig A, Hotz L, Tiihonen J, Bagley C (2014) Configuration-related topics. In: Knowledge-based configuration. Elsevier, pp 21–27
Fernandez-Viagas V (2022) A speed-up procedure for the hybrid flow shop scheduling problem. Expert Syst Appl 187:115903. https://doi.org/10.1016/j.eswa.2021.115903
Furht B, Escalante A, others (2010) Handbook of cloud computing. Springer
Furht B, Escalante A (eds) (2010) Handbook of Cloud Computing. Springer US, Boston, MA
Gahlawat M, Sharma P (2014) Survey of virtual machine placement in federated clouds. In: 2014 IEEE International advance computing conference (IACC). IEEE, pp 735–738
Gao F, Cui G, Liu H (2006a) Integration of genetic algorithm and cultural algorithms for constrained optimization, pp 817–825
Gao F, Liu H, Zhao Q, Cui G (2006b) Hybrid model of genetic algorithm and cultural algorithms for optimization problem, pp 441–448
Gao F, Zhao Q, Liu H, Cui G (2007) Cultural particle swarm algorithms for constrained multi-objective optimization, pp 1021–1028
Goli A, Zare HK, Tavakkoli-Moghaddam R, Sadeghieh A (2019) An improved artificial intelligence based on gray wolf optimization and cultural algorithm to predict demand for dairy products: a case study. Int J Interact Multimedia Artif Intell 5:15. https://doi.org/10.9781/ijimai.2019.03.003
Guo Y, Cheng J, Cao Y, Lin Y (2011) A novel multi-population cultural algorithm adopting knowledge migration. Soft Comput 15:897–905. https://doi.org/10.1007/s00500-010-0556-4

Guo Y, Zhang P, Cheng J et al (2018) Interval multi-objective quantum-inspired cultural algorithms. Neural Comput Appl 30:709–722
Guo Y, Liu D, Cheng J (2012) Multi-population cooperative cultural algorithms, pp 199–206
Gupta MK, Chandra P (2020) A comprehensive survey of data mining. Int J Inf Technol 12:1243–1257. https://doi.org/10.1007/s41870-020-00427-7
Gurkok C (2017) Securing cloud computing systems. In: Computer and information security handbook. Elsevier, pp 897–922
Hameed A, Khoshkbarforoushha A, Ranjan R et al (2016) A survey and taxonomy on energy efficient resource allocation techniques for cloud computing systems. Computing 98:751–774. https://doi.org/10.1007/s00607-014-0407-8
Haralick RM, Shapiro LG (1985) Image segmentation techniques. Comput Vision Graph Image Process 29:100–132. https://doi.org/10.1016/S0734-189X(85)90153-7
Hassan NA, Hijazi R (2017) Future trends. In: Data hiding techniques in Windows OS. Elsevier, pp 291–298
He J, Xu F (2011) Chaotic-search-based cultural algorithm for solving unconstrained optimization problem. Modell Simul Eng 2011:1–6. https://doi.org/10.1155/2011/239743
He P (2014) Counter cyber attacks by semantic networks. In: Emerging trends in ICT security. Elsevier, pp 455–467
Ho NB, Tay JC (2004) GENACE: an efficient cultural algorithm for solving the flexible job-shop problem. In: Proceedings of the 2004 congress on evolutionary computation (IEEE Cat. No. 04TH8753). IEEE, pp 1759–1766
Hoos HH, Stützle T (2005) Travelling salesman problems. In: Stochastic local search. Elsevier, pp 357–416
Hosseinioun P, Kheirabadi M, Kamel Tabbakh SR, Ghaemi R (2020) A new energy-aware tasks scheduling approach in fog computing using hybrid meta-heuristic algorithm. J Parallel Distrib Comput 143:88–96. https://doi.org/10.1016/j.jpdc.2020.04.008
Huang D, Wu H (2018a) Introduction. In: Mobile cloud computing. Elsevier, pp 1–3
Huang D, Wu H (2018b) Edge clouds—pushing the boundary of mobile clouds. In: Mobile cloud computing. Elsevier, pp 153–176
Huang H, Liu M, Gu X (2008) A cultural algorithm based on multilayer belief spaces and its application in neural network fault classifier. In: 2008 7th World congress on intelligent control and automation. IEEE, pp 3990–3995
Jamil M, Yang X-S, Zepernick H-J (2013) Test functions for global optimization. In: Swarm intelligence and bio-inspired computation. Elsevier, pp 193–222
Jiang X, Ma J, Xiao G et al (2021) A review of multimodal image matching: methods and applications. Inf Fusion 73:22–71. https://doi.org/10.1016/j.inffus.2021.02.012
Jin X, Reynolds RG (2000) Mining knowledge in large scale databases using cultural algorithms with constraint handling mechanisms. In: Proceedings of the 2000 congress on evolutionary computation. CEC00 (Cat. No.00TH8512). IEEE, pp 1498–1506
Jones RE, Griffin RA, Unsworth RKF (2021) Adaptive resolution imaging sonar (ARIS) as a tool for marine fish identification. Fish Res 243:106092. https://doi.org/10.1016/j.fishres.2021.106092
Joukowsky MS (2015) Survey and excavation (field methods) in archaeology. In: International encyclopedia of the social & behavioral sciences. Elsevier, pp 742–746
Judeh T, Jayyousi T, Acharya L, et al (2013) Gene set cultural algorithm. In: Proceedings of the International conference on bioinformatics, computational biology and biomedical informatics. ACM, New York, NY, USA, pp 641–648
Judeh T, Jayyousi T, Acharya L, et al (2014) GSCA: Reconstructing biological pathway topologies using a cultural algorithms approach. In: 2014 IEEE Congress on evolutionary computation (CEC). IEEE, pp 2206–2213
Kim Y, Cho S-B (2009) A hybrid cultural algorithm with local search for traveling salesman problem. In: 2009 IEEE International symposium on computational intelligence in robotics and automation (CIRA). IEEE, pp 188–192

Kobti Z, Reynolds RG, Kohler TA (2006) THE emergence of social network hierarchy using cultural algorithms. Int J Artif Intell Tools 15:963–978. https://doi.org/10.1142/S0218213006003065
Kwak EJ, Grable JE (2021) Conceptualizing the use of the term financial risk by non-academics and academics using twitter messages and ScienceDirect paper abstracts. Soc Netw Anal Min 11:6. https://doi.org/10.1007/s13278-020-00709-9
Lagos C, Crawford B, Cabrera E et al (2014) Comparing evolutionary strategies on a biobjective cultural algorithm. Sci World J 2014:1–10. https://doi.org/10.1155/2014/745921
Laporte G (1992) The traveling salesman problem: an overview of exact and approximate algorithms. Eur J Oper Res 59:231–247. https://doi.org/10.1016/0377-2217(92)90138-Y
Lee YC, Zomaya AY (2012) Energy efficient utilization of resources in cloud computing systems. J Supercomput 60:268–280. https://doi.org/10.1007/s11227-010-0421-3
Li X, Liu K, Ma L, Li H (2010) A concurrent-hybrid evolutionary algorithms with multi-child differential evolution and Guotao algorithm based on cultural algorithm framework, pp 123–133
Li N, Yi W, Gong G (2012) Multi-aircraft cooperative target allocation in BVR air combat using cultural-genetic algorithm, pp 414–422
Lin C-J, Chen C-H, Lin C-T (2008) A hybrid of cooperative particle swarm optimization and cultural algorithm for neural fuzzy networks and its prediction applications. IEEE Trans Syst, Man, Cybern, Part C (Appl and Rev) 39(1):55–68
Lin C-J, Weng C-C, Lee C-L, Lee C-Y (2009) Using an efficient hybrid of cooperative particle swarm optimization and cultural algorithm for neural fuzzy network design. In: 2009 international conference on machine learning and cybernetics, 5, IEEE, pp 3076–3082
Lindahl M, Mason AJ, Stidsen T, Sørensen M (2018) A strategic view of university timetabling. Eur J Oper Res 266:35–45. https://doi.org/10.1016/j.ejor.2017.09.022
Lonetti F, Marchetti E (2018) Emerging software testing technologies, pp 91–143
Ma J, Zhang J, Yang J, Cheng L (2008) Research on cultural algorithm for solving routing problem of mobile agent. J China Univ Posts Telecommun 15:121–125. https://doi.org/10.1016/S1005-8885(08)60415-X
Ma J, Jiang X, Fan A et al (2021) Image matching from handcrafted to deep features: a survey. Int J Comput Vision 129:23–79. https://doi.org/10.1007/s11263-020-01359-2
Mao Z, Xiang Y, Zhang Y, Liu M (2020) A novel multi-objective cultural algorithm embedding five-element cycle optimization. In: 2020 IEEE congress on evolutionary computation (CEC). IEEE, pp 1–10
Masdari M, Nabavi SS, Ahmadi V (2016) An overview of virtual machine placement schemes in cloud computing. J Netw Comput Appl 66:106–127. https://doi.org/10.1016/j.jnca.2016.01.011
Mastelic T, Brandic I (2015) Recent trends in energy-efficient cloud computing. IEEE Cloud Comput 2:40–47. https://doi.org/10.1109/MCC.2015.15
Mayeda M, Andrews A (2021) Evaluating software testing techniques: a systematic mapping study, pp 41–114
Mell PM, Grance T (2011) The NIST definition of cloud computing. Gaithersburg, MD
Minaee S, Boykov YY, Porikli F, et al (2021) Image segmentation using deep learning: a survey. IEEE Trans Pattern Anal Mach Intell 1–1. https://doi.org/10.1109/TPAMI.2021.3059968
Mohammadhosseini M, Toroghi Haghighat A, Mahdipour E (2019) An efficient energy-aware method for virtual machine placement in cloud data centers using the cultural algorithm. J Supercomput 75:6904–6933. https://doi.org/10.1007/s11227-019-02909-3
Mousavi Mojab SZ, Ebrahimi M, Reynolds R, Lu S (2019) iCATS: scheduling big data workflows in the cloud using cultural algorithms. In: 2019 IEEE Fifth International conference on big data computing service and applications (BigDataService). IEEE, pp 99–106
Mucherino A, Papajorgji P, Pardalos PM (2009) A survey of data mining techniques applied to agriculture. Oper Res Int J 9:121–140. https://doi.org/10.1007/s12351-009-0054-6
Muhamediyeva DT (2020) Fuzzy cultural algorithm for solving optimization problems. J Phys: Conf Ser 1441:012152. https://doi.org/10.1088/1742-6596/1441/1/012152

Murugadass G, Sivakumar P (2020) A hybrid elephant herding optimization and cultural algorithm for energy-balanced cluster head selection scheme to extend the lifetime in WSNs. Int J Commun Syst e4538. https://doi.org/10.1002/dac.4538

Nguyen TT, Yao X (2006) Hybridizing cultural algorithms and local search, pp 586–594

Nguyen TT, Yao X (2008) An experimental study of hybridizing cultural algorithms and local search. Int J Neural Syst 18:1–17. https://doi.org/10.1142/S0129065708001397

Nguyen BH, Xue B, Zhang M (2020) A survey on swarm intelligence approaches to feature selection in data mining. Swarm Evol Comput 54:100663. https://doi.org/10.1016/j.swevo.2020.100663

Niu Q, Zeng T, Zhou Z (2011) A novel cultural algorithm based on differential evolution for hybrid flow shop scheduling problems with fuzzy processing time, pp 121–132

Ochoa A, García Y, Yañez J, Teymanoglu Y (2010) Using cultural algorithms to improve intelligent logistics, pp 127–134

Ochoa A, Trejo E, Azpeitia D et al (2012) Mobile device to measure ubiquitous danger in a great city based on cultural algorithms, pp 647–655

Oloruntoba O, Cosma G, Liotta A (2019) Clan-based cultural algorithm for feature selection. In: 2019 International conference on data mining workshops (ICDMW). IEEE, pp 465–472

Öncan T, Altınel İK, Laporte G (2009) A comparative analysis of several asymmetric traveling salesman problem formulations. Comput Oper Res 36:637–654. https://doi.org/10.1016/j.cor.2007.11.008

Ostrowski DA, Reynolds RG (1999) Knowledge-based software testing agent using evolutionary learning with cultural algorithms. In: Proceedings of the 1999 congress on evolutionary computation-CEC99 (Cat. No. 99TH8406). IEEE, pp 1657–1663

Ostrowski DA, Reynolds RG (2004) Using cultural algorithms to evolve strategies for recessionary markets. In: Proceedings of the 2004 congress on evolutionary computation (IEEE Cat. No. 04TH8753). IEEE, pp 1780–1785

Ostrowski DA, Reynolds RG (2005) Using cultural algorithms to evolve strategies in a complex agent-based system, pp 81–102

Pan Y, Zhang L (2021) A BIM-data mining integrated digital twin framework for advanced project management. Autom Constr 124:103564. https://doi.org/10.1016/j.autcon.2021.103564

Pan Z-L, Chen L, Zhang G-Z (2016) Efficient design method for cell allocation in hybrid CMOS/nanodevices using a cultural algorithm with chaotic behavior. Front Phys 11:116201. https://doi.org/10.1007/s11467-015-0531-8

Patil R, Tamane S, Patil K (2021) An experimental approach toward type 2 diabetes diagnosis using cultural algorithm, pp 405–415

Prestwich SD, Tarim SA, Rossi R, Hnich B (2008) A cultural algorithm for POMDPs from stochastic inventory control, pp 16–28

Raeesi N. MR, Kobti Z (2012) A multiagent system to solve JSSP using a multi-population cultural algorithm, pp 362–367

Raeesi N. MR, Kobti Z (2013) Heterogeneous multi-population cultural algorithm. In: 2013 IEEE congress on evolutionary computation. IEEE, pp 292–299

Raeesi N. MR, Kobti Z (2014) Heterogeneous multi-population cultural algorithm with a dynamic dimension decomposition strategy, pp 345–350

Revelle CS, Laporte G (1996) The plant location problem: new models and research prospects. Oper Res 44:864–874. https://doi.org/10.1287/opre.44.6.864

Reynolds R, Ali M (2008a) Embedding a social fabric component into cultural algorithms toolkit for an enhanced knowledge-driven engineering optimization. Int J Intell Comput Cybern 1. https://doi.org/10.1108/17563780810919131

Reynolds RG, Ali MZ (2008b) The social fabric approach as an approach to knowledge integration in cultural algorithms. In: 2008b IEEE congress on evolutionary computation (IEEE World congress on computational intelligence). IEEE, pp 4200–4207

Reynolds R, Al-Shehri H (1997) The use of cultural algorithms with evolutionary programming to control the data mining of large-scale spatio-temporal databases. In: 1997 IEEE International

conference on systems, man, and cybernetics. computational cybernetics and simulation. IEEE, pp 4098–4103
Reynolds R, Al-Shehri H (1998) The use of cultural algorithms with evolutionary programming to guide decision tree induction in large databases. In: 1998 IEEE International conference on evolutionary computation proceedings. IEEE World congress on computational intelligence (Cat. No. 98TH8360). IEEE, pp 541–546
Reynolds RG, Chung C (1997) A cultural algorithm framework to evolve multiagent cooperation with evolutionary programming, pp 323–333
Reynolds RG, Gawasmeh YA (2012) Evolving heterogeneous social fabrics for the solution of real valued optimization problems using cultural algorithms. In: 2012 IEEE congress on evolutionary computation. IEEE, pp 1–8
Reynolds RG, Kinnaird-Heether L (2017a) Population mechanics and cultural algorithms in the development of a cultural engine. In: 2017a IEEE symposium series on computational intelligence (SSCI). IEEE, pp 1–8
Reynolds RG, Kinnaird-Heether L (2017b) Problem solving using social networks in cultural algorithms with auctions. In: 2017b IEEE congress on evolutionary computation (CEC). IEEE, pp 2714–2721
Reynolds RG, Rolnick SR (1995a) Learning the parameters for a gradient-based approach to image segmentation using cultural algorithms. In: Proceedings first international symposium on intelligence in neural and biological systems. INBS'95. IEEE, pp 240–247
Reynolds RG, Rolnick SR (1995b) Learning the parameters for a gradient-based approach to image segmentation from the results of a region growing approach using cultural algorithms. In: Proceedings of 1995b IEEE international conference on evolutionary computation. IEEE, pp 819–824
Reynolds RG, Sverdlik W (1993) Learning the behavior of Boolean circuits from examples using cultural algorithms. In: Sadjadi FA (ed). pp 177–188
Reynolds RG, Shinin Zhu (2001) Knowledge-based function optimization using fuzzy cultural algorithms with evolutionary programming. IEEE Trans Syst Man Cybern Part B (Cybern) 31:1–18. https://doi.org/10.1109/3477.907561
Reynolds RG, Kinnaird-Heether L (2003) Survival of the fastest: Using cultural algorithms to optimize the design of a controller for a 3D racing game. Comput Math Organ Theory 9:227–254
Reynolds RG, Sverdlik W (1995) An evolution-based approach to program understanding using cultural algorithms. Int J Software Eng Knowl Eng 5:211–226. https://doi.org/10.1142/S0218194095000125
Rychtyckyj N, Reynolds RG (1998) Learning to re-engineer semantic networks using cultural algorithms, pp 179–190
Rychtyckyi N, Reynolds RG (1999) Using cultural algorithms to improve performance in semantic networks. In: Proceedings of the 1999 congress on evolutionary computation-CEC99 (Cat. No. 99TH8406). IEEE, pp 1651–1656
Rychtyckyj N, Reynolds RG (2000) Assessing the performance of cultural algorithms for semantic network re-engineering. In: Proceedings of the 2000 congress on evolutionary computation. CEC00 (Cat. No. 00TH8512). IEEE, pp 1482–1491
Rychtyckyj N, Reynolds RG (2001) Using cultural algorithms to improve knowledge base maintainability. In: Proceedings of the 2001 genetic and evolutionary computation conference GECCO-2001. pp 1405–1412
Rychtyckyj N, Reynolds RG (2002) Knowledge base maintenance using cultural algorithms: application to the DLMS manufacturing process planning system at Ford Motor Company. In: Proceedings of the 2002 congress on evolutionary computation. CEC'02 (Cat. No. 02TH8600). IEEE, pp 855–860
Rychtyckyj N, Ostrowski D, Schleis G, Reynolds RG (2003) Using cultural algorithms in industry. In: Proceedings of the 2003 IEEE swarm intelligence symposium. SIS'03 (Cat. No. 03EX706). IEEE, pp 187–192

Rychtyckyj N, Reynolds RG (2005) Using cultural algorithms to re-engineer large-scale semantic networks. Int J Software Eng Knowl Eng 15:665–693. https://doi.org/10.1142/S0218194005002506

Selvarajah K, Kobti Z, Kargar M (2019a) A cultural algorithm for determining similarity values between users in recommender systems, pp 270–283

Selvarajah K, Zadeh PM, Kargar M, Kobti Z (2019b) Identifying a team of experts in social networks using a cultural algorithm. Procedia Comput Sci 151:477–484. https://doi.org/10.1016/j.procs.2019.04.065

Selvarajah K, Kobti Z, Kargar M (2020) Cultural algorithms for cluster hires in social networks. Procedia Comput Sci 170:514–521. https://doi.org/10.1016/j.procs.2020.03.117

Shah P, Kobti Z (2020) Multimodal fake news detection using a cultural algorithm with situational and normative knowledge. In: 2020 IEEE congress on evolutionary computation (CEC). IEEE, pp 1–7

Shao W, Shao Z, Pi D (2020) Modeling and multi-neighborhood iterated greedy algorithm for distributed hybrid flow shop scheduling problem. Knowl-Based Syst 194:105527. https://doi.org/10.1016/j.knosys.2020.105527

Shepherd K (2001) Remotely operated vehicles (ROVs). In: Encyclopedia of ocean sciences, Elsevier, pp 742–747

Singh D, Zadeh PM, Kobti Z (2018) A multilevel cooperative multi-population cultural algorithm. In: 2018 Innovations in intelligent systems and applications (INISTA). IEEE, pp 1–5

Soza C, Landa R, Riff MC, Coello C (2007) A cultural algorithm with operator parameters control for solving timetabling problems. Foundations of fuzzy logic and soft computing. Springer, Berlin, Heidelberg, Berlin, Heidelberg, pp 810–819

Soza C, Becerra RL, Riff MC, Coello Coello CA (2011) Solving timetabling problems using a cultural algorithm. Appl Soft Comput 11:337–344. https://doi.org/10.1016/j.asoc.2009.11.024

Stanley SD, Kattan K, Reynolds R (2019) CAPSO: a parallelized multiobjective cultural algorithm particle swarm optimizer. In: 2019 IEEE congress on evolutionary computation (CEC). IEEE, pp 3060–3069

Sternberg M, Reynolds RG (1997) Using cultural algorithms to support re-engineering of rule-based expert systems in dynamic performance environments: a case study in fraud detection. IEEE Trans Evol Comput 1:225–243. https://doi.org/10.1109/4235.687883

Sun Y, Zhang L, Gu X (2012) A hybrid co-evolutionary cultural algorithm based on particle swarm optimization for solving global optimization problems. Neurocomputing 98:76–89. https://doi.org/10.1016/j.neucom.2011.08.043

Sun X-M, Lv X-Y, Duan X-M (2009) Novel QoS routing algorithm based on cultural-simulated annealing algorithm. In: 2009 Second international conference on intelligent networks and intelligent systems. IEEE, pp 209–212

Sun Y, Zhang L, Gu X (2010) Co-evolutionary cultural based particle swarm optimization algorithm, pp 1–7

Tang W, Li Y (2008) Constrained optimization using triple spaces cultured genetic algorithm. In: 2008 Fourth international conference on natural computation. IEEE, pp 589–593

Turk M, Athitsos V (2020) Gesture recognition. Computer vision: a reference guide 1–6

Uhlir V, Tomanek O, Kencl L (2016) Latency-based benchmarking of cloud service providers. In: Proceedings of the 9th International conference on utility and cloud computing, pp 263–268

Unold O, Tarnawski R (2016) Cultural ant colony optimization on gpus for travelling salesman problem, pp 317–329

Unold O, Tarnawski R (2018) A parallel framework for multi-population cultural algorithm and its applications in TSP, pp 470–482

Utgoff PE, Berkman NC, Clouse JA (1997) Decision tree induction based on efficient tree restructuring. Mach Learn 29:5–44. https://doi.org/10.1023/A:1007413323501

Vitale K, O'Shea J, Reynolds RG, Che X (2011a) Learning group behavior in games: using cultural algorithms: the land bridge game engine example. In: 2011b IEEE symposium on swarm intelligence. IEEE, pp 1–9

Vitale K, Reynolds R, O'Shea J, Meadows G (2011b) Exploring ancient landscapes under lake Huron using cultural algorithms. Procedia Comput Sci 6:303–310. https://doi.org/10.1016/j.procs.2011.08.056
Vu HT, Hwang S (2014) A track and power-aware algorithm for virtual machine placement in cloud data center. Int J Grid Distrib Comput 7:21–32
Wang X, Hao W, Li Q (2017) An adaptive cultural algorithm with improved quantum-behaved particle swarm optimization for sonar image detection. Sci Rep 7:17733. https://doi.org/10.1038/s41598-017-17945-3
Waris F, Reynolds RG (2015) Using cultural algorithms to improve wearable device gesture recognition performance. In: 2015 IEEE symposium series on computational intelligence. IEEE, pp 625–633
Wei Z, Yan-Ping B (2012) Cultural algorithm based on particle swarm optimization for partner selection of virtual enterprise. In: Proceedings of the 31st Chinese control conference, pp 2238–2241
Wu Y, Gao X-Z, Huang X-L, Zenger K (2010) A hybrid optimization method of particle swarm optimization and cultural algorithm. In: 2010 Sixth International conference on natural computation. IEEE, pp 2515–2519
Xu P, Goteng GL, He Y (2021) Modelling cloud service latency and availability using a deep learning strategy. Expert Syst Appl 182:115121. https://doi.org/10.1016/j.eswa.2021.115121
Xue Z, Guo Y (2007) Improved cultural algorithm based on genetic algorithm. In: 2007 IEEE International conference on integration technology. IEEE, pp 117–122
Xue X, Yao M, Cheng R (2011) A novel selection operator of cultural algorithm, pp 71–77
Yan X, Song T, Wu Q (2017) An improved cultural algorithm and its application in image matching. Multimedia Tools Appl 76:14951–14968. https://doi.org/10.1007/s11042-016-4313-2
Yang Y, Gu X (2014) Cultural-based genetic Tabu algorithm for multiobjective job shop scheduling. Math Probl Eng 2014:1–14. https://doi.org/10.1155/2014/230719
Yildirim P, Birant D, Alpyildiz T (2018) Data mining and machine learning in textile industry. Interdiscip Rev Data Min Knowl Disc 8:e1228. https://doi.org/10.1002/widm.1228
Yousefpour A, Fung C, Nguyen T et al (2019) All one needs to know about fog computing and related edge computing paradigms: a complete survey. J Syst Architect 98:289–330. https://doi.org/10.1016/j.sysarc.2019.02.009
Zadeh PM, Kobti Z (2015) A multi-population cultural algorithm for community detection in social networks. Procedia Comput Sci 52:342–349. https://doi.org/10.1016/j.procs.2015.05.105
Zaki MJ, Meira Jr W (2020) Data mining and machine learning: fundamental concepts and algorithms. Cambridge University Press
Zannoni E, Reynolds RG (1997) Learning to control the program evolution process with cultural algorithms. Evol Comput 5:181–211. https://doi.org/10.1162/evco.1997.5.2.181
Zhang YJ (1996) A survey on evaluation methods for image segmentation. Pattern Recogn 29:1335–1346. https://doi.org/10.1016/0031-3203(95)00169-7
Zhang Y (2012) Research of web search based on cultural algorithm new framework. Procedia Eng 29:3641–3645. https://doi.org/10.1016/j.proeng.2012.01.545
Zhang Y, Zhu X (2012) A cultural particle swarm optimization algorithm for flow shop scheduling problem, pp 1575–1583
Zhou Y, Zhang S, Luo Q, Abdel-Baset M (2019) CCEO: cultural cognitive evolution optimization algorithm. Soft Comput 23:12561–12583. https://doi.org/10.1007/s00500-019-03806-w

Part III
Variants of Cultural Algorithms

Chapter 6
Hybrid Schemes for Cultural Algorithms

Abstract The distinctive framework of CAs offers the flexibility of their components and operators to be hybridised with other algorithms. The belief space of CAs provides an efficient information transmission mechanism that can potentially guide the search process. This has motivated researchers to develop a variety of hybrid versions of CAs with other meta-heuristic algorithms. This chapter discusses the different hybridisation schemes developed in the literature for CAs.

Keywords Hybrid algorithm · Meta-heuristics · Evolutionary algorithms · Cultural algorithms

6.1 Introduction

As was discussed in Chap. 3, CAs have some characteristic features that make them unique in comparison with conventional EAs. The overall framework of CAs is designed to model both types of genetic and cultural evolutionary processes. In CAs, individuals within the population space can vertically and horizontally learn to produce more productive offspring. The horizontal learning capability is provided by the belief space consisting of a network of knowledge components, which can be viewed as a cultural information repository for the population space. The experiences and collective behaviours acquired by individuals over different generations are transmitted by belief space to improve the behaviours and decisions of the individuals within the population space.

The aforementioned features have attracted much attention from researchers to embed the belief space idea into the framework of other meta-heuristic algorithms. The studies have been mostly focused on the development of hybrid algorithms, in which the knowledge components of CAs are used to guide the search process. The main goal of this chapter is to provide an overview of such hybrid algorithms developed in the literature. In these algorithms, CAs have been hybridised with other meta-heuristic algorithms, such as PSO, HS, and DE. The chapter provides the key details of mentioned standard meta-heuristic algorithms to make the hybrid schemes easy to understand. The chapter also provides the general algorithmic ideas of hybrid

S. Jalili, *Cultural Algorithms*, Engineering Optimization: Methods and Applications,
https://doi.org/10.1007/978-981-19-4633-2_6

algorithms. In some cases, the hybridisation ideas are further explained by providing detailed algorithmic pseudo-codes.

The overall structure of this chapter is as follows. Sections 6.2 provides the details of hybridised versions of CAs with the PSO algorithm and discusses how the cultural knowledge components were used to guide the search direction in the PSO algorithm. Section 6.3 presents the hybridised CAs with the DE algorithm and show how new differential mutation operators were developed based on cultural information stored in belief space. Section 6.4 describes how the performance of the HS algorithm has been improved by using the normative intervals. Sections 6.5 and 6.6 explain the hybrid schemes of CAs with GAs and local search techniques. The hybridisation schemes of CAs with other meta-heuristic algorithms, such as FA, IWO, SFLA, FECO, and EHO, are presented in Sect. 6.7. The adoption of EP techniques within the framework of CAs are explained in Sect. 6.8. In the end, a summary of the chapter is presented in Sect. 6.9.

6.2 Hybrid CA and PSO

PSO developed by Kennedy and Eberhart (1995, 1997) in the 1990s belongs to the SI family. The algorithm was inspired by the collective intelligent behaviours exhibited by a group or swarm of species in searching for food sources in nature (Mavrovouniotis et al. 2017). Due to its efficiency and ease of implementation, PSO has attracted enormous attention from a wide variety of researchers in science and technology (Hajihassani et al. 2018; Jain et al. 2018; Elsheikh and Abd Elaziz 2019; Elbes et al. 2019).

The CAs have been hybridised with PSO to develop new algorithms with better performance properties. Table 6.1 lists the different versions of hybrid CAs with PSO proposed in the literature. In this section, the standard PSO are briefly explained first. Thereafter, the different hybridised versions of CAs with PSO will be discussed.

Table 6.1 Different hybrid versions of CAs with PSO

Reference	Algorithm	Comment
Gao et al. (2007)	Cultural PSO (CPSO)	• The population space consists of several sub-swarms
Zhang et al. (2012)	CPSO	• One global belief space is used, and the population space is divided into several sub-populations which are evolved by the PSO operators
Wei et al.(2012)	CPSO	• The population space is evolved by the PSO operators based on cultural knowledge components

6.2.1 Standard PSO

In PSO, the numerical optimisation is performed by a set of particles that can move within the search space with given velocities. These particles form a population space, in which their movements and velocities are obtained based on an information-sharing mechanism. To move within the search space, particles take the advantage of both the best global experience vector of the whole swarm, represented by $G^t = [g_1^t, g_2^t, \ldots, g_n^t]$, and their own best experience vector, denoted by $P_i^t = \left[p_1^t, p_2^t, \ldots, p_n^t\right]$. Let vector $X_i^t = \left[x_{i,1}^t, x_{i,2}^t, \ldots, x_{i,n}^t\right]$ and $V_i^t = \left[v_{i,1}^t, v_{i,2}^t, \ldots, v_{i,n}^t\right]$ be the position and velocity of the ith particle within the search space at time t, where $i = 1, 2, \ldots, N_P$ and N_P is the total number of particles in the population space. PSO spreads a set of particles represented by $X_1^0, X_2^0, \ldots, X_{N_P}^0$ randomly within search space and updates the local P_i^t and global G^t experience vectors based on the objective function values. Then, the algorithm updates the velocities and positions of particles by using the following formulas:

$$V_i^{t+1} = \omega V_i^t + c_1 \text{rand}\left(P_i^t - X_i^t\right) + c_2 \text{rand}\left(G^t - X_i^t\right) \tag{6.1}$$

$$X_i^{t+1} = X_i^t + V_i^{t+1} \tag{6.2}$$

where V_i^t and P_i^t are the velocity and best personal experience of the ith particle at time t, respectively, ω is inertia weight that controls the extent of influence of the particle's previous velocity on its current velocity, c_1 and c_2 are the acceleration parameters, G^t indicates the global best experience of the whole swarm until time t, and X_i^{t+1} is the new position of the ith particle at time $t + 1$. Algorithm 6.1 presents the overall pseudo-code of the PSO algorithm for a minimisation problem.

6.2.2 Multi-objective CA with Multiple PSO-Based Sub-Swarms

Gao et al. (2007) developed a hybrid Cultural PSO (CPSO) algorithm for multi-objective optimisation problems. In this algorithm, the population space is divided into several sub-swarms, in which each sub-swarm tries to solve a single objective function and evolves based on PSO formulations. Each sub-swarm updates the belief space based on its accepted elite individuals. In each generation, an upper-level search is also performed by the belief space based on the selection, cross-over, and mutation operators of GA.

Algorithm 6.1 The Pseudo-Code of PSO

Initialise PSO internal parameters, i.e., N_P, c_1, c_2, ω;

Initialise the velocity vectors V_i^0 of particles as n-dimensional zero vectors;

Initialise the N_P number of particles represented by vectors $X_1^0, X_2^0, \ldots, X_{N_P}^0$ randomly within the search space;

Evaluate the objective function values particles;

Initialise the personal experience vectors of particles as their current positions as follows: $P_i^0 = X_i^0, i = 1, 2, \ldots, N_P$;

Initialise the global experience as follows: $G^0 = \arg\min\{f(X_i^t)\}$;

Set $t = 0$;

while *stopping criteria are not met* **do**

  Set $t = t + 1$;

  for $i \leftarrow 1$ to N_P **do**

    $V_i^{t+1} = \omega V_i^t + c_1 rand(P_i^t - X_i^t) + c_2 rand(G^t - X_i^t)$;

    $X_i^{t+1} = X_i^t + V_i^{t+1}$;

    % Update local experience

    if $f(X_i^{t+1}) < f(P_i^t)$ **do**

      $P_i^{t+1} = X_i^{t+1}$;

    else

      $P_i^{t+1} = P_i^t$;

    end

    % Update global experience

    if $f(X_i^{t+1}) < f(G^t)$ **do**

      $G^t = X_i^{t+1}$;

    end

  end

  $G^{t+1} = G^t$;

end

6.2.3 *Hybrid Gaussian PSO and CA*

dos Santos Coelho and Mariani (2006) proposed a hybrid Gaussian PSO and CA (GPSO-CA) algorithm, in which the belief space is added to the framework of the Gaussian PSO (GPSO) algorithm. In GPSO-CA, the velocities of the particles are updated by using the cultural information and the social experience of the population. The authors proposed the following formula to update the velocities of the particles:

$$v_{i,j}^{t+1} = \begin{cases} \omega v_{i,j}^t + c_1 N(0,1)(p_{i,j}^t - x_{i,j}^t) + c_2 N(0,1)(g_j^t - x_{i,j}^t) \text{ if } x_{i,j}^t < x_{\min,j}^t \text{ or } x_{i,j}^t < g_j^t \\ \omega v_{i,j}^t + c_1 N(0,1)(p_{i,j}^t - x_{i,j}^t) + c_2 N(0,1)(g_j^t - x_{i,j}^t) \text{ if } x_{i,j}^t > x_{\max,j}^t \text{ or } x_{i,j}^t > g_j^t \\ \omega v_{i,j}^t + c_1 N(0,1)(p_{i,j}^t - x_{i,j}^t) + c_2 N(0,1)(g_j^t - x_{i,j}^t) \quad \text{otherwise} \end{cases} \tag{6.3}$$

where $N(0, 1)$ is a random number between 0 and 1 obtained from the Gaussian distribution, $p_{i,j}^t$ represents the jth variable of the best personal experience of particle i, and g_j^t is the jth variable of the best global experience. As can be seen from Eq. (6.3), GPSO-CA uses normative knowledge components to guide the search direction.

In a similar research, Xie et al. (2009) proposed a hybrid PSO and CA algorithm, in which the following influence function based on the PSO formula and normative knowledge components was proposed:

$$v_{i,j}^{t+1} = \begin{cases} \omega v_{i,j}^t + c_1 \text{rand}(p_{i,j}^t - x_{i,j}^t) + c_2 \text{rand}(g_j^t - x_{i,j}^t) \text{ if } x_{i,j}^t < x_{\min,j}^t \text{ or } x_{i,j}^t < p_{i,j}^t \\ \omega v_{i,j}^t + c_1 \text{rand}(p_{i,j}^t - x_{i,j}^t) + c_2 \text{rand}(g_j^t - x_{i,j}^t) \text{ if } x_{i,j}^t > x_{\max,j}^t \text{ or } x_{i,j}^t > p_{i,j}^t \\ \omega v_{i,j}^t \pm c_1 \text{rand}(p_{i,j}^t - x_{i,j}^t) + c_2 \text{rand}(g_j^t - x_{i,j}^t) \text{ otherwise} \end{cases} \tag{6.4}$$

where rand is a uniformly distributed random number between 0 and 1. The main differences between Eqs. (6.3) and (6.4) are the assumption of the uniform random variable instead of normal distribution and consideration of the personal experience instead of the global experience.

6.2.4 Modified PSO Update Formula Based on CA

Chen et al. (2008) developed a multi-population Cultural Cooperative PSO (CCPSO), in which the different sub-swarms are evolving with independent belief spaces. The study proposed the following formula to update the position of particles:

$$x_{i,j}^{t+1} = \begin{cases} x_{i,j}^t + v_{i,j}^t + \left|\text{rand}\left(x_{\max,j}^t - x_{\min,j}^t\right)\right| & \text{if } x_{i,j}^t \le x_{\min,j}^t \\ x_{i,j}^t + v_{i,j}^t - \left|\text{rand}\left(x_{\max,j}^t - x_{\min,j}^t\right)\right| & \text{if } x_{i,j}^t > x_{\max,j}^t \end{cases} \tag{6.5}$$

As it can be seen from the above equation, the particles are forced to move towards the normative interval represented by the normative bounds $x_{\min,j}^t$ and $x_{\max,j}^t$.

6.3 Hybrid CA and DE

In this section, the hybrid versions of CAs with the DE algorithm is presented. Table 6.2 lists the different versions of hybrid CA and DE algorithms in the literature.In the formulations presented in this section, it should be noted that parameters r_1, r_2, r_3, r_4 represent the random integers that are selected as $r_1, r_2, r_3, r_4 \in [1, 2, \ldots, N_I]$ and $r_1 \neq r_2 \neq r_3 \neq r_4$.

6.3.1 *Standard DE*

DE is an EA developed by Storn and Price (1997b; 2006), in which the evolutionary process is modelled based on the weighted differences between the different individual vectors in the population space. The algorithm consists of three main operators, mutation, cross-over, and selection operators. For a given individualX_i^t, the mutation operator picks a given number of other individuals in the population space and generates a mutant vector represented by $V_i^t = \left[v_{i,1}^t, v_{i,2}^t, \ldots, v_{i,n}^t\right]$. Various mutation formulas have been developed for DE, in which the types and number of selected individuals to generate the mutant vector are different. The most prominent mutation formulas proposed in the literature can be presented as follows (Qin et al. 2009; Das and Suganthan 2011):

- "DE\best\1":

$$V_i^t = X_{\text{Best}}^t + F\left(X_{r_1}^t - X_{r_2}^t\right) \tag{6.6}$$

- "DE\rand-to-best\1":

Table 6.2 Different versions of hybrid CAs with DE

Reference	Comment(s)
Yuan et al. (2008)	- Population space and situational knowledge are evolved based on DE operators
Becerra and Coello (2004)	- The population space is evolved by DE operators designed based on different knowledge components of belief space
Awad et al. (2017), Dixit et al. (2019), Niu et al. (2011)	- The evolution process is stochastically performed by DE and CA operators
Ali et al. (2016b), Wu et al. (2010)	- The population space of CA is evolved by DE-based influence functions

$$V_i^t = X_{r_1}^t + F\left(X_{\text{Best}}^t - X_{r_2}^t\right) + F\left(X_{r_3}^t - X_{r_4}^t\right) \quad (6.7)$$

- "DE\current-to-rand\1":

$$V_i^t = X_{r_1}^t + F\left(X_{r_2}^t - X_i^t\right) + F\left(X_{r_3}^t - X_{r_4}^t\right) \quad (6.8)$$

- "DE\current-to-best\1":

$$V_i^t = X_{r_1}^t + F\left(X_{\text{Best}}^t - X_i^t\right) + F\left(X_{r_2}^t - X_{r_3}^t\right) \quad (6.9)$$

In the above equations, X_{Best}^t represents the best solution vector obtained at time t, i.e., $X_{\text{Best}}^t = \text{argmin}\{f(X_i^t)\}$, F is a scaling factor, and the vectors $X_{r_1}^t$, $X_{r_2}^t$, $X_{r_3}^t$, and $X_{r_4}^t$ are the random individuals selected from the population space.

Compared to the parent individual X_i^t, the mutant vectors V_i^t obtained from Eqs. (6.6) up to (6.9) are different in all dimensions. However, producing a new generation of offspring that have purely different characteristics from their parents might not be an efficient strategy and can overwhelm the algorithm by imposing unnecessary exploration efforts. The cross-over operator of DE probabilistically produces a trial vector for each individual X_i^t, denoted by $U_i^t = \left[u_{i,1}^t, u_{i,2}^t, \ldots, u_{i,n}^t\right]$, as follows:

$$u_{i,j}^t = \begin{cases} v_{i,j}^t & \text{if rand} < CR \text{ or } j = \text{randi}(1, n) \\ x_{i,j}^t & \text{otherwise} \end{cases} \quad (6.10)$$

where $u_{i,j}^t$ and $v_{i,j}^t$ are the jth variable of the ith mutant and trial vectors, respectively, CR indicates the cross-over rate control parameter, and $\text{randi}(a, b)$ represents a random integer value within the interval $[a, b]$.

Once the trial vector U_i^t is generated by the cross-over operator in Eq. (6.10), DE investigates whether the ith individual in the population should be replaced by U_i^t. To this end, the algorithm uses the selection operator, in which a greedy selection approach is adopted. The selection operator of the DE algorithm can be written as:

$$X_i^{t+1} = \begin{cases} U_i^t & \text{if } f\left(U_i^t\right) < f\left(X_i^t\right) \\ X_i^t & \text{otherwise} \end{cases} \quad (6.11)$$

The pseudo-code of the standard DE algorithm can be presented as Algorithm 6.2 by putting together the mutation, cross-over, and selection operators discussed in this section.

6.3.2 CA with DE-Based Population Space

Ali et al. (2016b) developed a balanced hybrid CA with DE (b-hCA-DE) for function optimisation. In the b-hCA-DE, a set of new DE-based influence functions are developed to evolve the population space. These influence functions use different knowledge components to create the trial vectors U_i^t as follows (2016b):

- Situational knowledge-based influence:

$$u_{i,j}^t = s_j^t + F\left(x_{r_1,j}^t - x_{r_2,j}^t\right), i = 1, 2, \ldots, N_\mathrm{I}, j = 1, 2, \ldots, n \tag{6.12}$$

where s_j^t is the jth variable of the situational vector at time t.

Algorithm 6.2 Pseudo-Code of the Standard DE

```
Initialise the internal parameters N_I, F, and CR
Set t = 0;
Initialise the random population space with N_I number of random individuals, represented by
X_1^0, X_2^0, ..., X_{N_I}^0 , within the search space;
Evaluate the objective function values of individuals;
Obtain the best individual in the population as follows: X_Best^t = arg min{f(X_i^t)};
while stopping criteria are not met do
    for i ← 1 to N_I do
        Select the random integers r_1, r_2, r_3, r_4 ∈ [1, 2, ..., N_I] in such a way that r_1 ≠ r_2 ≠ r_3 ≠
        r_4 ≠ i;
        % Mutation operator
        Construct the mutant vector V_i^t based on one of the mutation formulas in equations (6.6)
        up to (6.9);
        for j ← 1 to n do
            % Cross-over operator
            if rand ≤ CR or j = randi(1, n) do
                u_{i,j}^t = v_{i,j}^t;
            else
                u_{i,j}^t = x_{i,j}^t;
            end
        end
        % Selection operator
        if f(U_i^t) < f(X_i^t) do
            X_i^{t+1} = U_i^t;
        else
            X_i^{t+1} = X_i^t;
        end
    end
    Set t = t + 1;
end
```

- Normative knowledge-based influence:

$$u_{i,j}^{t} = \begin{cases} x_{r_1,j}^{t} + F\left|x_{r_2,j}^{t} - x_{r_3,j}^{t}\right|, & \text{if } x_{r_1,j}^{t}\left\langle x_{\min,j}^{t} \text{ or } x_{r_1,j}^{t}\right\rangle x_{\max,j}^{t} \\ x_{r_1,j}^{t} + \left(\frac{x_{\max,j}^{t} - x_{\min,j}^{t}}{\max\left(x_{r_2,j}^{t-1} - x_{r_3,j}^{t-1}\right)}\right) FN(0,1)\left(x_{r_2,j}^{t} - x_{r_3,j}^{t}\right) & \text{otherwise} \end{cases} \quad (6.13)$$

where $\max\left(x_{r_2,j}^{t-1} - x_{r_3,j}^{t-1}\right)$ represents the maximum value of the difference vector found in the previous iteration and $N(0, 1)$ is the random number obtained from the normal distribution with zero mean and standard deviation of 1.

- Topographical knowledge-based influence:

The study assumes that the topographical knowledge divides the search space into different cells, in which each cell is represented by $c_j^t = \left[c_{\min,j}^t, c_{\max,j}^t\right]$ that stores the best solution intervals for each variable. The influence function uses topographical knowledge to create the trial vector as follows:

$$u_{i,j}^{t} = \begin{cases} x_{r_1,j}^{t} + F\frac{f(X_i^t)}{\sum_{k=1}^{N_1} f(X_k^t)}\left(x_{r_1,j}^{t} - x_{r_2,j}^{t}\right), \text{ if } x_{i,j}^{t} \in \left(c_{\min,j}^{t}, c_{\max,j}^{t}\right) \\ c_{\min,j}^{t} + F\left(c_{\max,j}^{t} - c_{\min,j}^{t}\right) \qquad \text{otherwise} \end{cases} \quad (6.14)$$

- History knowledge-based influence:

$$u_{i,j}^{t} = \begin{cases} e_{k,j}^{t} + FA_{\text{dist},j}\frac{\operatorname{sgn}\left(\sum_{z=1}^{k}\operatorname{sgn}\left(e_{z+1,j}^{t} - e_{z,j}^{t}\right)\right)}{\max\left(x_{r_1,j}^{t-1} - x_{r_2,j}^{t-1}\right)}\left|x_{r_1,j}^{t-1} - x_{r_2,j}^{t-1}\right| & \text{if rand} < \varepsilon \\ x_j^{\min} + \text{rand}\left(x_j^{\max} - x_j^{\min}\right) & \text{otherwise} \end{cases} \quad (6.15)$$

where $e_{k,j}^t$ represents the jth variable of the best current solution obtained before the last shift in the search space, and $x_j^{\min}$ and $x_j^{\max}$ are the lower and upper bounds of the problem in the jth dimension, respectively. $A_{\text{dist},j}$ indicates the average distance change in the jth dimension, which is calculated as follows:

$$A_{\text{dist},j} = \left(\sum_{z=1}^{k}\frac{\left|e_{z+1,j}^{t} - e_{z,j}^{t}\right|}{k}\right) \quad (6.16)$$

The b-hCA-DE algorithm employs a high-level teamwork hybrid model to stochastically decide which influence function should be applied for a given solution. The high-level teamwork hybrid model is explained in Sect. 6.3.4.

Arpaia et al. (2007) proposed a CA with DE-based influence functions for automatic fault isolation in the transport of dangerous goods, in which the situational, topographical, and historical knowledge components are used within the belief space. The study designed the influence functions based on the DE operators and information offered by the belief space. The algorithm defines the following situational knowledge-based influence function:

$$X_i^{t+1} = \mathcal{S}_i^t + F\left(X_{r_1}^t - X_{r_2}^t\right), i = 1, 2, \ldots, N_s \tag{6.17}$$

where X_i^{t+1} is the position of the ith individual in the population at time $t + 1$, $\mathcal{S}_i^t$ is the situational vector at time t, r_1 and r_2 are the random numbers selected as $r_1 \neq r_2 \neq i$, $X_{r_1}^t$ and $X_{r_2}^t$ represent the two random individuals selected from the population, F is the scaling factor, and N_s is the number of individuals stored in the situational knowledge component of the belief space.

The study by Arpaia et al. (2007) introduced another DE-based influence function using the topographical knowledge component C^t. The pseudo-code of the DE-based topographic influence function is shown in Algorithm 6.3.

Coelho et al. (2009) proposed an improved Cultural DE with a Measure of Diversity (CDEMD) for the economic dispatch problem. In the CDEMD, the mutation is performed based on the normative and situational knowledge components. Algorithm 6.4 presents the pseudo-code CDEMD algorithm. As it can be seen from Algorithm 6.4, the authors defined the control parameter CR of DE algorithm as a function of diversity measure $D(k, t)$. The diversity measure is calculated as follows:

$$D(k, t) = \frac{1}{N_I\left(x_k^{\max} - x_k^{\min}\right)} \sum_{i=1}^{N_I} \sqrt{\sum_{k=1}^{n} \left(x_{i,k}^t - \overline{x_k}\right)^2} \tag{6.18}$$

where $D(k, t)$ represents the diversity of the population in the kth dimension at time t, $x_k^{\max}$ and $x_k^{\min}$ are upper and lower bounds of the kth variable, respectively, $x_{i,k}^t$ is the kth variable of the ith individual, and $\overline{x_k}$ is the midpoint in the kth dimension of the search space.

Algorithm 6.3 Pseudo-Code of the DE-Based Influence Function Proposed by Arpaia et al. (2007)

---

for $i \leftarrow 1$ to N **do**

  For the ith individual, select the random numbers r_1, r_2, r_3 as follows:

  $r_1, r_2, r_3 \in [1, 2, \ldots, N]$ and $r_1 \neq r_2 \neq r_3 \neq i$

  Randomly select the i_rth row of the topographical knowledge component as follows:

  $i_r = U(1, K)$;

  for $k \leftarrow 1$ to n **do**

    Extract the normative interval stored for the kth variable in the i_rth cell of the topographical knowledge component as follows:

    $\varrho_{i_r,k}^t = [L_{i_r,k}^t, U_{i_r,k}^t]$;

    if $x_{r_3,k}^t \leq L_{i_r,k}^t$ **then**

      if rand$\leq CR$

        $x_{i,k}^{t+1} = x_{r_3,k}^t + F\left|x_{r_1,k}^t - x_{r_2,k}^t\right|$;

      else

        $x_{i,k}^{t+1} = x_{r_3,k}^t$;

      end

    elseif $L_{i_r,k}^t < x_{r_3,k}^t < U_{i_r,k}^t$ **then**

      if rand$\leq CR$

        $x_{i,k}^{t+1} = x_{r_3,k}^t + F\,(x_{r_1,k}^t - x_{r_2,k}^t)$;

      else

        $x_{i,k}^{t+1} = x_{r_3,k}^t$;

      end

    else

      if rand$\leq CR$

        $x_{i,k}^{t+1} = x_{r_3,k}^t + F\left|x_{r_1,k}^t - x_{r_2,k}^t\right|$;

      else

        $x_{i,k}^{t+1} = x_{r_3,k}^t$;

      end

    end

  end

end

Note: K denotes the number of the belief cells in the topographic knowledge component

Khan et al. (2014, 2017) proposed a hybrid CA and DE for sensor fault detection in phased array radars, in which the population space is evolved based on the normative knowledge and DE mutation operators.

Algorithm 6.4 Pseudo-Code of the Solution Generation in the CDEMD Algorithm (Coelho et al. 2009)

---

for $i \leftarrow 1$ to N_I **do**
  For the ith individual, select the random numbers r_1, r_2, r_3 as follows:
  $r_1, r_2, r_3 \in [1, 2, \ldots, N_I]$ and $r_1 \neq r_2 \neq r_3 \neq i$
  for $k \leftarrow 1$ to n **do**
    % Mutation:
    if $x^t_{r_1,k} < x^t_{\min,k}$ **then**
      $v^{t+1}_{i,k} = x^t_{r_1,k} + F\left|x^t_{r_2,k} - x^t_{r_3,k}\right|$;
    elseif $x^t_{r_3,k} > x^t_{\max,k}$ **then**
      $v^{t+1}_{i,k} = x^t_{r_1,k} - F\left|x^t_{r_2,k} - x^t_{r_3,k}\right|$;
    else
      $v^{t+1}_{i,k} = s^t_k - F\left|x^t_{r_2,k} - x^t_{r_3,k}\right|$;
    end
    $CR = \frac{D(k,t)}{D(1,t)}$;
    $CR = \min(0.9, \max(0.2, CR))$;
    % Crossover operator:
    if rand $\leq CR$ **then**
      $x^{t+1}_{i,k} = v^{t+1}_{i,k}$;
    else
      $x^{t+1}_{i,k} = x^t_{i,k}$;
    end
  end
end

Gao and Diao (2010) proposed a Differential CA (DCA) for digital filter design. The study proposes new DE-based normative and situational influence functions, in which the step sizes are calculated based on the random differential vectors and the search directions are obtained based on the normative and situational knowledge components as follows:

$$x^{t+1}_{i,j} = \begin{cases} x^t_{r_3,j} + k\left|\left(x^t_{r_2} - x^t_{r_1}\right)N(0,1)\right| & \text{if } x^t_{r_3,j} < x^t_{\min,j} \\ x^t_{r_3,j} - k\left|\left(x^t_{r_2} - x^t_{r_1}\right)N(0,1)\right| & \text{if } x^t_{r_3,j} > x^t_{\max,j} \\ x^t_{r_3,j} + \mathrm{ksize}\left(I^t_j\right)N(0,1) & \text{otherwise} \end{cases} \tag{6.19}$$

$$x_{i,j}^{t+1} = \begin{cases} x_{r_3,j}^t + \left|\left(x_{r_2}^t - x_{r_1}^t\right)N(0,1)\right| \text{ if } x_{r_3,j}^t < s_j^t \\ x_{r_3,j}^t - \left|\left(x_{r_2}^t - x_{r_1}^t\right)N(0,1)\right| \text{ if } x_{r_3,j}^t > s_j^t \\ x_{r_3,j}^t + \text{Ksize}\left(I_j^t\right)N(0,1) \quad \text{otherwise} \end{cases} \tag{6.20}$$

where $N(0, 1)$ is a random number obtained from the normal distribution with the mean value of zero and standard deviation of 1, k and K are the constant parameters, $r_1, r_2, r_3 \in [1, 2, \ldots, N_I]$, and $r_1 \neq r_2 \neq r_3 \neq i$.

Lu et al. (2011) developed a Hybrid Multi-Objective CA (HMOCA) for the short-term environmental/economic hydrothermal scheduling problem. The algorithm uses the DE operators to evolve the population space and uses a local search designed based on the situational knowledge components. The HMOCA stores a set of solution vectors in the situational knowledge component and employs the mutation, cross-over, and selection operators of DE to improve them in each iteration. Algorithm 6.5 presents the pseudo-code of HMOCA.

Algorithm 6.5 Pseudo-Code of the HMOCA

Initialise the random population space $\mathcal{P}^0$, evaluate the individuals, create the belief space $\mathcal{B}^0$, and initialise the internal parameters;

Set $t = 0$;

while *stopping criteria are not met* **do**

  Perform the mutation, cross-over, and selection operators for all individuals in the population space;

  For all solution vectors stored in the situational knowledge component of the belief space, perform the mutation, cross-over, and selection operators;

  Perform the local search for all solution vectors stored in the situational knowledge component of the belief space;

  Set $t = t + 1$;

end

6.3.3 *Hybrid CA and DE with Random Knowledge Influence*

Becerra and Coello (2004) employed DE-based mutation operators in CA to evolve the population space. The study designed several DE-based influence functions by using the cultural information provided by different knowledge components as follows (Becerra and Coello 2004):

- Influence function based on situational knowledge:

$$x_{i,j}^{t+1} = s_{i,j}^{t} + F\left(x_{r_1,j}^{t} - x_{r_2,j}^{t}\right),\ \ i = 1, 2, \ldots, N_{\mathrm{s}},\ \ j = 1, 2, \ldots, n \tag{6.21}$$

where $s_{i,j}^{t}$ represents the jth variables of ith individual in the situational knowledge component at time t, N_{s} is the number of individuals within the situational knowledge component, and F is the scaling factor.

- Influence function based on the normative knowledge:

$$x_{i,j}^{t+1} = \begin{cases} x_{r_3,j}^{t} + F\left|x_{r_1,j}^{t} - x_{r_2,j}^{t}\right| & \text{if } x_{r_3,j}^{t} < x_{\min,j}^{t} \\ x_{r_3,j}^{t} - F\left|x_{r_1,j}^{t} - x_{r_2,j}^{t}\right| & \text{if } x_{r_3,j}^{t} > x_{\max,j}^{t} \\ x_{r_3,j}^{t} + \frac{x_j^{\max} - x_j^{\min}}{d_j} F\left(x_{r_1,j}^{t} - x_{r_2,j}^{t}\right) & \text{otherwise} \end{cases},\ i = 1, 2, \ldots, N_{\mathrm{I}},\ \ j = 1, 2, \ldots, n \tag{6.22}$$

where d_j is a scaling factor which is defined as the greatest difference value of $\left|x_{r_1,j}^{t} - x_{r_2,j}^{t}\right|$ from the beginning of evolution.

- Influence function based on the topographical knowledge:

The study assumes that the topographical knowledge C^t component is consisted of a given number of best individuals found during the evolution process alongside their corresponding intervals for each dimension. Let us assume c_i^t be the topographical knowledge stored in the ith cell at iteration t. Then, the topographical knowledge-based influence function can be written as follows:

$$x_{i,j}^{t+1} = \begin{cases} x_{r_3,j}^{t} + F\left|x_{r_1,j}^{t} - x_{r_2,j}^{t}\right| & \text{if } x_{r_3,j}^{t} < l_{c,j}^{t} \\ x_{r_3,j}^{t} - F\left|x_{r_1,j}^{t} - x_{r_2,j}^{t}\right| & \text{if } x_{r_3,j}^{t} > u_{c,j}^{t} \\ x_{r_3,j}^{t} + F(x_{r_1,j}^{t} - x_{r_2,j}^{t}) & \text{otherwise} \end{cases},\ i = 1, 2, \ldots, N_{\mathrm{I}},\ \ j = 1, 2, \ldots, n \tag{6.23}$$

where $l_{c,j}^{t}$ and $u_{c,j}^{t}$ are the lower and upper bounds for the jth variable stored in the cth cell of the topographical knowledge component, respectively. The study does not provide the details for calculations of topographical bounds.

- Influence function based on the history knowledge:

The study assumes that the historic knowledge component $\mathcal{H}^t$ stores the best solutions obtained in each generation, and proposes the following influence function:

$$x_{i,j}^{t+1} = \begin{cases} e_{t,j} + A_{\text{dir,j}} F \left| x_{r1,j}^t - x_{r2,j}^t \right| & if\ rand < \alpha \\ e_{t,j} - \frac{1.5 A_{\text{dist,j}}}{d_j} F \left| x_{r1,j}^t - x_{r2,j}^t \right| & if\ rand < \beta\, i = 1, 2, \ldots, N_1\ j = 1, 2, \ldots, n \\ x_j^{\min} + rand\left(x_j^{\max} - x_j^{\min} \right) & \text{otherwise} \end{cases} \quad (6.24)$$

where $e_{t,j}$ represents the jth variable of the history knowledge stored at time t, $A_{\text{dir,j}}$ is the directional change in the jth dimension of the search space, $A_{\text{dist,j}}$ is the distance change in the jth dimension, α and β are the constant parameters. The parameters $A_{\text{dir,j}}$ and $A_{\text{dist,j}}$ are obtained by the following equations:

$$A_{\text{dir,j}} = \text{sgn}\left(\sum_{k=1}^{w-1} \text{sgn}\left(e_{k+1,j} - e_{k,j} \right) \right) \quad (6.25)$$

$$A_{\text{dist,j}} = \frac{\sum_{k=1}^{w-1} \left| e_{k+1,j} - e_{k,j} \right|}{w - 1} \quad (6.26)$$

where w is the size of the historic knowledge and sgn(.) represents the sign function.

Becerra and Coello (2004) suggest applying the influence functions in Eqs. (6.21) to (6.24) based on a given probability formula in the following equation:

$$p_{\text{ks}} = 0.1 + 0.6 \frac{\upsilon_{\text{ks}}}{\upsilon} \leq 0.1 \quad (6.27)$$

where p_{ks} represents the probability of applying a specific knowledge source for the influence function, υ_{ks} is the number of times that a given individual is updated based on a given knowledge source, and υ is the number of times that the individual produced better offspring using any type of knowledge source. It is clear that the probability value p_{ks} will be the same for all types of knowledge components at the beginning of the evolutionary process. Algorithm 6.6 presents the pseudo-code of hybrid CA and DE proposed by Becerra and Coello (2004).

Algorithm 6.6 Pseudo-Code of the Hybrid CA and DE Algorithm Proposed by Becerra and Coello (2004)

Initialise the internal parameters of the algorithm;
Initialise the belief space $\mathcal{B}^0$ with different knowledge components;
Initialise the population space $\mathcal{P}^0$ with random N_{I} individuals;
Evaluate the objective function values of individuals;
Assign equal probability value of $p_{\mathbf{ks}}$ to all knowledge components;
Set $t = 0$;
while *stopping criteria are not met* **do**
  Update the belief space and obtain $\mathcal{B}^{t+1}$;
  for $i \leftarrow 1$ to N_{I} **do**
    Select one of the influence functions in equations (6.21) to (6.24) based on the probability values $p_{\mathbf{ks}}$;
    Update the position of the ith individual using the selected influence function and obtain X_i^{t+1};
    Evaluate the objective function of X_i^{t+1};
    Update the probability values $p_{\mathbf{ks}}$;
  end
  Set $t = t + 1$;
end

Chen et al. (2015) proposed a similar approach, in which different DE-based mutation formulations were developed based on different knowledge sources of CA as follows:

- Influence function based on the normative knowledge:

$$v_{i,j}^{t+1} = \begin{cases} x_{r3,j}^t + F\left|x_{r1,j}^t - x_{r2,j}^t\right| & \text{if } x_{r3,j}^t < x_{\min,j}^t \\ x_{r3,j}^t - F\left|x_{r1,j}^t - x_{r2,j}^t\right| & \text{if } x_{r3,j}^t > x_{\max,j}^t \\ x_{r3,j}^t + F\left(x_{r1,j}^t - x_{r2,j}^t\right) & \text{otherwise} \end{cases} \quad i = 1, 2, \ldots, N_{\mathrm{I}}, \quad j = 1, 2, \ldots, n\,, \tag{6.28}$$

- Influence function based on the situational knowledge similar to Eq. (6.21)
- Influence function based on the domain knowledge:

  This influence function uses a trigonometric mutation strategy as follows:

$$v_{i,j}^{t+1} = \frac{\left(x_{r_1,j}^t + x_{r_2,j}^t + x_{r_3,j}^t\right)}{3} + (P_2 - P_1)\left(x_{r_1,j}^t - x_{r_2,j}^t\right) + (P_3 - P_2)\left(x_{r_2,j}^t - x_{r_3,j}^t\right) + (P_1 - P_3)\left(x_{r_3,j}^t - x_{r_1,j}^t\right) \tag{6.29}$$

where $P_1 = \frac{f\left(X_{r_1}^t\right)}{P'}$, $P_2 = \frac{f\left(X_{r_2}^t\right)}{P'}$, $P_3 = \frac{f\left(X_{r_3}^t\right)}{P'}$, and $P' = f\left(X_{r_1}^t\right) + f\left(X_{r_2}^t\right) + f\left(X_{r_3}^t\right)$.

- Influence function based on historical knowledge:

$$v_{i,j}^{t+1} = \begin{cases} x_{i,j}^t + F\left|r_3 - x_{i,j}^t\right| + F\left|x_{r_1,j}^t - x_{r_2,j}^t\right|, & \text{if } x_{i,j}^t < x_{\min,j}^t \\ x_{i,j}^t - F\left|e_{r_3} - x_{i,j}^t\right| - F\left|x_{r_1,j}^t - x_{r_2,j}^t\right|, & \text{if } x_{i,j}^t > x_{\max,j}^t, \\ x_{i,j}^t + F\left(e_{r_3} - x_{i,j}^t\right) - F\left(x_{r_1,j}^t - x_{r_2,j}^t\right), & \text{otherwise} \end{cases} \tag{6.30}$$

where e_{r_3} represents a random r_3th best individual stored within the historical knowledge component of the belief space.

To select which knowledge source should be used to mutate a given individual within the population space, Chen et al. (2015) suggested measuring the average fitness value of the individuals generated based on a given knowledge source and using the roulette wheel selection method to decide. Let us assume $X_1^t, X_2^t, \ldots, X_m^t$ be the m different individuals generated based on a given knowledge-based mutation. In this case, the average fitness values for each knowledge source will be calculated as follows:

$$\overline{a}_i^t = \frac{\sum_{j=1}^{m} f\left(X_j^t\right)}{m}, \quad i = 1, 2, \ldots, n_k \tag{6.31}$$

where $\overline{a}_i^t$ represents the average fitness value for the ih knowledge source by iteration t and n_k is the number of knowledge sources within the belief space. Then, the average fitness values are used to calculate the selection probability of a given knowledge-based mutation operator as follows:

$$p_i^t = \frac{\overline{a}_i^t}{\sum_{j=1}^{n_k} \overline{a}_j^t}, \quad i = 1, 2, \ldots, n_k \tag{6.32}$$

where p_i^t represents the selection probability of the ith knowledge source to mutate the individuals at time t. The probability values obtained from Eq. (6.32) can be used in the roulette wheel selection method to decide which knowledge source should be used in the mutation operator.

6.3.4 High-Level Teamwork Hybrid Model of CA and DE

Some research has been focused on the high-level teamwork hybrid model of CA and DE (Awad et al. 2017; Dixit et al. 2019). In this model, the operators of CA and DE are employed in a parallel way and the decision on which operator should be used in a specific stage of the solution-finding process is made by some probabilistic operators. It is similar to the definition of *hyper-heuristics* in literature (Burke et al. 2013b) if it is not the same. In this subsection, the approach is briefly explained based on Awad et al. 2017.

Let us assume a population of N_{I} individuals randomly generated in the search space represented by $X_1^0, X_2^0, \ldots, X_{N_{\mathrm{I}}}^0$. In the beginning, it is assumed that the population space is equally evolved by CA and DE operators. The role of each algorithm in the evolution process of the population is represented by a participation ratio as follows:

$$\Phi_i^0 = \frac{N_{\mathrm{I}}}{N_{\mathrm{s}}}, \quad i = 1, 2 \tag{6.33}$$

where Φ_i^0 represents the initial participation ratio of the ith search strategy and N_{s} is the number of available search strategies which is equal to 2 in this case. It should be noted that Φ_1^0 and Φ_2^0 represent the participation ratios for the DE and CA operators, respectively. Each search strategy will evolve a given portion of individuals represented by an auxiliary population APS_i^t, where APS_1^t and APS_2^t indicate the auxiliary populations for the DE and CA strategies, respectively. Based on the definition in Eq. (6.33), both DE and CA have equal opportunities to evolve the individuals at the beginning of the evolution process. In other words, the sizes of auxiliary population spaces, which are represented by n_1^t and n_2^t, will be initially same, i.e., $n_1^0 = n_2^0 = \Phi_i^0 = \frac{N_{\mathrm{I}}}{2}$. With this definition, the individuals in auxiliary populations of DE and CA can be represented by vectors $X_{1,1}^0, X_{1,2}^0, \ldots, X_{1,n_1^0}^0$ and $X_{2,1}^0, X_{2,2}^0, \ldots, X_{2,n_2^0}^0$, respectively.

After the evolution of initial auxiliary populations by each strategy, the participation ratios of each search strategy at each iteration, indicated by Φ_1^t and Φ_2^t, are calculated based on their performance in the previous iteration. Let us assume that the applications of DE and CA strategies have resulted in new trial vectors in the previous iteration represented by $U_{1,1}^{t-1}, U_{1,2}^{t-1}, \ldots, U_{1,n_1^{t-1}}^{t-1}$ and $U_{2,1}^{t-1}, U_{2,2}^{t-1}, \ldots, U_{2,n_2^{t-1}}^{t-1}$, respectively. The number of better trial vectors obtained by each strategy at the previous iteration η_i^{t-1} can be obtained as follows:

$$\eta_i^{t-1} = \sum_{j=1}^{n_i^{t-1}} b_{i,j}^{t-1}, \quad i = 1, 2 \tag{6.34}$$

which:

$$b_{i,j}^{t-1} = \begin{cases} 1 \text{ if } f\left(U_{i,j}^{t-1}\right) < f\left(X_{i,j}^{t-1}\right) \\ 0 \text{ otherwise} \end{cases}. \tag{6.35}$$

where $b_{i,j}^{t-1}$ determines whether the jth trial vector generated by the ith search strategy is better than the jth individual at previous iteration $t-1$ and $X_{i,j}^{t-1}$ is the jth individual in the auxiliary population belonging to the ith search strategy at previous iteration $t-1$.

To assess the performance of each search strategy, a quality function is calculated for each search strategy as follows:

$$\text{QF}_i^t = \frac{\Delta f_{\text{best}}^t - \Delta f_i^t}{\Delta f_{\text{best}}^t}, i = 1, 2 \tag{6.36}$$

where QF_i^t represents the quality function of the ith search strategy at iteration t, Δf_{best}^t indicates the cumulative fitness improvement of the search strategy with a better outcome in the previous iteration, and Δf_i^t is the cumulative fitness improvement of the ith search strategy in the previous iteration. The cumulative fitness improvement of the ith search strategy can be written as:

$$\Delta f_i^t = \sum_{j=1}^{\eta_i^{t-1}} a_{i,j}^{t-1} \tag{6.37}$$

where $a_{i,j}^{t-1}$ represents the fitness improvement resulting from the application of ith strategy and the jth successful trial vector at iteration $t-1$. Then, the participation of each search strategy Φ_i^t is calculated as follows:

$$\Phi_i^t = QF_i^t N_{\text{I}} \tag{6.38}$$

Algorithm 6.7 presents the overall pseudo-code of the high-level teamwork hybrid model of CA and DE. Without loss of generality, it is worth mentioning that the hybrid scheme can be expanded by considering additional search strategies based on different cultural knowledge components of CA, which will lead to the creation of multiple auxiliary populations. The b-hCA-DE algorithm proposed by Ali et al. (2016b) explained in Sect. 6.3.2 also employs the high-level teamwork technique to decide which knowledge component should be used to influence a given individual in the population.

6.4 Hybrid CA and HS

HS is a meta-heuristic algorithm inspired by the music improvisation process. The hybrid version of CAs with HS can also be found in the literature. In this section, the standard HS algorithm is reviewed first. Then, the hybrid CAs with HS will be explained.

Algorithm 6.7 Pseudo-Code of the High-Level Teamwork Hybrid Model of CA and DE Summarised Based on Awad et al. (2017)

Initialise the internal parameters of the algorithm;

Initialise N_{I} random individuals in the search space and evaluate their objective function values;

Set $n_1^0 = n_2^0 = \Phi_i^0 = \frac{N_{\text{I}}}{2}$;

Initialise the auxiliary populations $\boldsymbol{APS}_1^0$ and $\boldsymbol{APS}_2^0$ with n_1^0 and n_2^0 individuals, respectively;

Evolve the auxiliary populations $\boldsymbol{APS}_1^0$ and $\boldsymbol{APS}_2^0$ with DE and CA strategies, respectively, and create the trial vectors $U_{1,1}^0, U_{1,2}^0, \ldots, U_{1,n_1^0}^0$ and $U_{2,1}^0, U_{2,2}^0, \ldots, U_{2,n_2^0}^0$;

Set $t = 1$;

while *stopping criteria are not met* **do**

  Calculate the number of better trial vectors obtained by each strategy at the previous iteration η_i^{t-1} using equation (6.34);

  Calculate the quality function QF_i^t for each search strategy using equation (6.36);

  Obtain the participation ratios of each search strategy Φ_i^t using equation (6.38);

  From the new auxiliary populations $\boldsymbol{APS}_1^t$ and $\boldsymbol{APS}_2^t$ based on the participation ratios Φ_i^t;

  Evolve the auxiliary populations $\boldsymbol{APS}_1^t$ and $\boldsymbol{APS}_2^t$ with DE and CA strategies, respectively, and create the trial vectors $U_{1,1}^t, U_{1,2}^t, \ldots, U_{1,n_1^t}^t$ and $U_{2,1}^t, U_{2,2}^t, \ldots, U_{2,n_2^t}^t$;

  Set $t = t + 1$;

end

6.4.1 Standard HS

HS is one of the popular meta-heuristic algorithms which is originally proposed by Geem et al. (2001). The algorithm is inspired by the music improvisation process. In the real music improvisation process, the musicians try different combinations of music pitches to achieve the best possible harmony. HS imitates the music improvisation process performed by a set of musicians in an orchestra to create the most harmonious melody (Manjarres et al. 2013). In this algorithm, each decision variable represents a musician and each solution vector X_i^t is a musical harmony (Manjarres et al. 2013). The audience's aesthetic impression represents the performance of each harmony that can be metaphorically considered as the objective function in the optimisation problem.

Generally speaking, a musician has a set of options to improve the performance of a given harmony, such as (Wang et al. 2015b):

- Selecting one pitch from her/his memory
- Trying a new pitch generated by a minor modification of an already available pitch in her/his memory
- Generating a random possible pitch.

HS simulates the search for optimum solutions based on the aforementioned observations from the music improvisation process. The algorithm saves a set of harmonies (or solutions) in memory, called Harmony Memory (HM), which can be represented as a matrix as follows:

$$\mathrm{HM}^t = \begin{bmatrix} h_{1,2}^t & h_{1,2}^t & \cdots & h_{1,n}^t \\ h_{2,1}^t & h_{2,2}^t & \cdots & h_{2,n}^t \\ \vdots & \vdots & \cdots & \vdots \\ h_{H_s,1} & h_{H_s,2} & \cdots & h_{H_s,n} \end{bmatrix} \tag{6.39}$$

where HM^t represents the Harmony Memory at iteration t, H_s is the size of the HM, and $h_{i,j}^t$ indicates the jth variable of the ith harmony stored in HM.

To generate new harmonies/solutions, HS performs the harmony improvisation step, in which the new harmonies are generated either based on the harmonies stored in HM or based on a purely random behaviour. The algorithm defines the Harmony Memory Consideration Rate (*HMCR*) parameter which reflects the probability of considering the harmonies in HM by a musician in generating new harmonies. If the memory consideration condition is satisfied, the new harmony/solution, denoted by $H^{'} = \left[h_1^{'}, h_2^{'}, \ldots, h_n^{'}\right]$, will be generated with the probability of $HMCR$ as follows:

$$h_j^{'} = \begin{cases} h_{r,j}^t + \mathrm{rand} b_{\mathrm{w}}, & \text{if rand} \leq PAR \\ h_{r,j}^t, & \text{otherwise} \end{cases}, \quad j = 1, 2, \ldots, n \tag{6.40}$$

where h'_j represents the jth variable of the newly generated harmony, $h^t_{r,j}$ is the jth variable of selected harmony from HM where $r \in [1, 2, \ldots, H_s]$, b_w is a bandwidth parameter, and PAR is the pitch adjustment rate parameter. It should be noted that the selected harmony H^t_r from HM can be obtained by using a purely random method or other selection methods, such as the roulette wheel method.

If the memory consideration condition is not satisfied, the new harmony will be generated randomly in the search space. The newly generated harmony H' will be evaluated and used to update the HM. If the new harmony is better than the worst harmony in the memory, the worst harmony in HM will be replaced by the new harmony. Algorithm 6.8 presents the pseudo-code of the standard HS algorithm.

6.4.2 Hybrid CA and HS

Gao et al. (2020) developed a hybrid Cultural HS (CHS) algorithm for the diesel blending problem. In CHS, the positions of individuals are subsequently updated based on the HS scheme and the normative knowledge-based influence function. Algorithm 6.9 presents the pseudo-code of the CHS algorithm.

Algorithm 6.8 Pseudo-Code of the Standard HS

```
Initialise the internal parameters H_s, HMCR, b_w, and PAR
Set t = 0;
Initialise the HM^0 with H_s number of random harmonies generated within the search space;
Evaluate the objective function values of harmonies in HM^0 ;
while stopping criteria are not met do
    for j ← 1 to n do
        if rand ≤ HMCR do
            if rand ≤ PAR do
                h'_j = h^t_{r,j} + rand b_w;
            else
                h'_j = h^t_{r,j};
            end
        end
    end
    Update the HM with newly generated harmony and obtain HM^{t+1};
    Set t = t + 1;
end
```

6.5 Hybrid CAs with GAs

Literature shows some hybrid versions of CAs with GAs. Most of these hybrid algorithms embed the cultural information of belief space into GA operators to evolve the population space. To keep the chapter in a manageable size, interested readers are referred to studies reported by Gao et al. (2006a, b), Ho and Tay (2004), Li et al. (2012), Xue et al. (2007), and Tang and Li (2008).

Algorithm 6.9 Pseudo-Code of Solution Generation in CHS Algorithm

```
for i ← 1 to N do
    for k ← 1 to n do
        if rand ≤ HMCR
            b_w = N_bw(x_k^max − x_k^min);
            x_{i,k}^{t+1} = x_{i,k}^t + 2 rand b_w − b_w;
        else
            x_{i,k}^{t+1} = U(x_k^min, x_k^max);
        end
        if x_{i,k}^t < x_{min,j}^t
            x_{i,k}^{t+1} = x_{i,k}^t + |u rand (x_{max,j}^t − x_{min,j}^t)|;
        elseif x_{i,k}^t > x_{max,j}^t
            x_{i,k}^{t+1} = x_{i,k}^t − |u rand (x_{max,j}^t − x_{min,j}^t)|;
        else
            x_{i,k}^{t+1} = x_{i,k}^t;
        end
    end
end
```

Note: N_{bw}: adjusting parameter, $x_k^{\max}$ and $x_k^{\min}$: upper and lower bounds for the kth variable, respectively, b_w: bandwidth for the minor adjustment, $u \in (0,1)$, and $rand$ = a random number between 0 and 1

6.6 Hybrid CAs and Local Search Techniques

CAs can be hybridised with local search techniques to improve their exploitation capabilities. Hence, some studies have been focused on the hybridisation of CAs with local search methods. Table 6.3 presents the list of hybrid CAs with local search algorithms. Some of these hybridised versions are presented in the following subsections.

Table 6.3 Hybrid versions of CAs with local search techniques

Hybridisation schemes	References
CA + TS	Oliveira De Freitas et al. (2018a), Yang et al. (2014), Ali and Reynolds (2014)
CA + SA	Oliveira De Freitas et al. (2018a), Nguyen et al. (2006), Sun et al. (2009)
CA + Multiple Trajectory Search (MTS)	Ali et al.(2016a)
Niche CA + TS	Ali et al. (2014)
Others	Lu et al. (2011), Tang and Li (2008), Ali et al. (2014a), Kim et al. (2009)

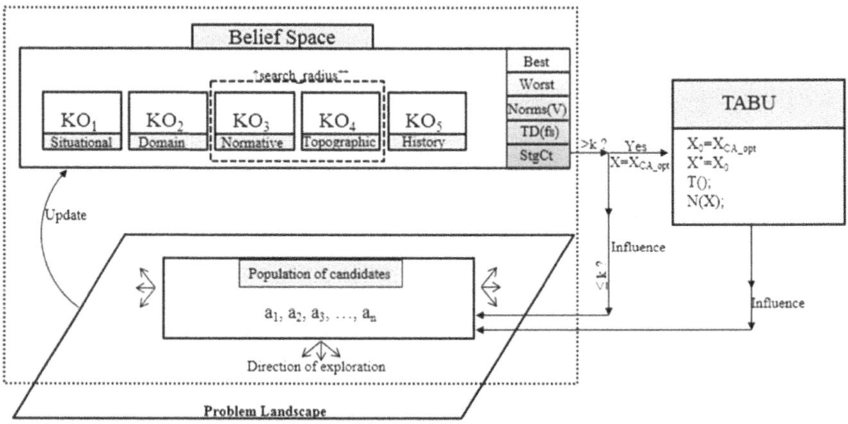

Fig. 6.1 The general framework of the CA-TS algorithm proposed by Ali and Reynolds (2014)

6.6.1 Hybrid CA and TS

Ali and Reynolds (2014) developed a hybrid CA and TS (CA-TS) for engineering design problems. In CA-TS, the best individuals generated by CA are fed into the TS process as shown in Fig. 6.1. CA-TS performs the local search around the best individuals based on TS to enhance the exploitation capability of CA.

6.6.2 Niche CA

Ali et al. (2014) proposed a Niche CA (NCA) for engineering optimisation. The algorithm uses niche techniques to form a set of sub-populations within the population space. The niche technique selects the higher-quality individuals and removes the individuals with the worst objective function values. The niche technique employed by Ali et al. (2014) uses two key parameters to create the sub-populations, including the sub-population size $\mathcal{K}$ and clearing radius σ.

6.6.3 *Balanced CA with Local Search*

As was discussed in Chap. 3, the belief space of the standard CA consists of different knowledge components, which can influence individuals in different ways. Some knowledge components provide an exploration mechanism for the algorithm, whereas others play the role of exploitation mechanism during the search process. Hence, an efficient balance is required between the application of different knowledge sources in the solution-finding process.

Ali et al. (2014a) proposed a balanced version of CA with local search, in which the performance of each knowledge source is considered during the search process. Let r_i^{t-1} be the ratio of individuals that use the ith knowledge source in their influence function at iteration $t-1$. The study updates the ratio r_i^t in each iteration, as follows:

$$r_i^t = r_i^{t-1} + 0.03 r_i^{t-1} \frac{\xi_{\text{best}}^t - \xi_i^t}{\xi_{\text{best}}^t} \tag{6.41}$$

where ξ_i^t represents the number of successful applications of ith knowledge component and ξ_{best}^t is the number of successful applications of the most successful knowledge component.

The balanced CA algorithm updates the follower ratios r_i^t for each knowledge source at each iteration. Then, if the global best individual X_{Best}^t does not improve for a set of subsequent iterations, say t_0, a local search mechanism will be employed. To this end, the knowledge components with the best and average follower rates will be selected to locally influence and perform the local search around each individual within the population space. Algorithm 6.10 presents the pseudo-code of the balanced CA approach proposed by Ali et al. (2014a).

6.7 Other Hybridisation Schemes

In this section, the hybrid versions of CAs with other meta-heuristic algorithms, including Fireworks Algorithm (FA), Invasive Wees Optimisation (IWO), Shuffled

Frog Leaping Algorithm (SFLA), Five-Element Cycle Optimisation (FECO), and Elephant Herding Optimisation (EHO), are presented.

Algorithm 6.10 Pseudo-Code of the Balanced CA with Local Search Proposed by Ali et al. (2014a)

```
Initialise the population space P^0 and belief space B^0;
Initialise the followers rate r_i^0 for each knowledge source in the belief space;
Evaluate the objective functions of individuals;
Set t = 1;
while stopping criteria are not met do
    Accept the elite individuals and update the belief space B^t;
    Influence the individuals;
    Evaluate the objective functions of individuals;
    Update the success rate of each knowledge source ξ_i;
    Update the follower rate r_i^t for each knowledge source using equation (6.41);
    % Local search
    if X_Best^t − X_Best^{t−t_0} < ε then
        Perform a simple local search around each individual based on knowledge sources with
        the best and average followers rates;
    end
    Set t = t + 1;
end
```

6.7.1 Hybrid CA and FA

FA is a population-based meta-heuristic that belongs to the family of SI algorithms. The main objective of this section is to describe the hybrid version of CAs with FA, called the Cultural Firework (CF) algorithm. To this end, the algorithmic details of FA are presented first. Then, the CF algorithm will be explained in detail.

6.7.1.1 Standard FA

FA is a SI algorithm developed by Tan and Zhu (2010) inspired by the firework explosion and spark formation process. In a real firework explosion, a set of sparks are formed around the firework explosion point in the sky. The spark distribution around the explosion point depends on the quality of the explosion. The distributions of sparks in desirable explosions are more condensed than those for low-quality explosions. The algorithm tries to imitate these phenomena to perform the exploitation and exploration during the search process for the optimum solution. The FA assumes each solution vector X_i^t as a firework. The algorithm starts with N_I number of random fireworks $X_1^0, X_2^0, \ldots, X_{N_\mathrm{I}}^0$ and considers them as the locations for the firework explosions. It is assumed that a set of sparks are formed around each firework. The number of sparks generated at each location X_i^t is different, and it is obtained as follows:

$$y_i = m\frac{f_{\max} - f\left(X_i^t\right) + \xi}{\sum_{j=1}^{N_\mathrm{I}}\left(f_{\max} - f(X_j^t)\right) + \xi}, i = 1, 2, \ldots, N_I \tag{6.42}$$

where y_i represents the number of sparks generated around firework i, m is a constant parameter that controls the number of sparks, $f_{\max}$ is the maximum objective function value in the population (i.e., $f_{\max} = \max_{i=1,2,\ldots,N_\mathrm{I}}\left(f(X_i^t)\right)$, and ξ is a constant small value introduced to prevent singularity. In FA, the values of y_i are limited to a given range to avoid the overwhelming effects of splendid fireworks as follows:

$$y_i = \begin{cases} \mathrm{round}(am) & \text{if } y_i < am \\ \mathrm{round}(bm) & \text{if } y_i > bm \\ \mathrm{round}(y_i) & \text{otherwise} \end{cases} \tag{6.43}$$

where a and b are the constant parameters satisfying the condition $a < b < 1$, and $round(.)$ is a function that rounds the input value to the nearest integer number.

The distances of the sparks from the firework location are obtained based on the amplitude of the explosion which is calculated by the following formula:

$$A_i = \widehat{A}\frac{f\left(X_i^t\right) - f_{\min} + \xi}{\sum_{j=1}^{N_\mathrm{I}}\left(f\left(X_j^t\right) - f_{\min}\right) + \xi}, i = 1, 2, \ldots, N_\mathrm{I} \tag{6.44}$$

where A_i represents the explosion amplitude for the ith firework, $\widehat{A}$ is the maximum explosion amplitude, and $f_{\min}$ is the minimum objective function value in the population (i.e., $f_{\min} = \min_{i=1,2,\ldots,N_1} \left(f(X_i^t)\right)$.

After calculation of the number of sparks generated around the ith firework X_i^t and amplitude of explosion A_i, the positions of the sparks can be obtained. The FA uses z parameter to define spark direction. It means that only $z$number of variables in firework location X_i^t will be changed to obtain the location of sparks. The value of z is randomly obtained as follows:

$$z = round(n\ rand) \tag{6.45}$$

where z represents the number of changes in the ith firework to obtain the sparks, n is the dimension of the problem, and $rand$ is a random number uniformly distributed between 0 and 1. The FA generates the sparks $U_l^{i,t}, l = 1, 2, \ldots, y_i$ around the ith firework X_i^t as follows:

$$u_{l,j}^{i,t} = \begin{cases} x_{i,j}^t + A_i rand(-1, 1) & \text{if} \quad j \in \{j_1, j_2, \ldots, j_z\} \\ x_{i,j}^t & \text{otherwise} \end{cases}, \; l = 1, 2, \ldots, y_i \tag{6.46}$$

where $u_{l,j}^{i,t}$ represents the jth variable of the lth spark generated from the ith firework at time t, $x_{i,j}^t$ is the jth variable of the ith firework at time t, rand$(-1, 1)$ represents the random number uniformly distributed between -1 and 1, and $j_1, j_2, \ldots, j_z$ represents the z number of variables that are selected to change.

The calculation of spark for each firework will result in a population space with a size of $N_1 + \sum_{i=1}^{N_1} y_i$, which is greater than the original population size N_1. To keep the size of the population space constant, FA uses a diversity measure. Let us assume the vectors $P_1^t, P_2^t, \ldots, P_{N_1+\sum_{i=1}^{N_1} y_i}^t$ represent the current solutions in memory, in which the firework and spark vectors are mixed and formed a population of solutions with the size of $N_1 + \sum_{i=1}^{N_1} y_i$. The FA calculates the accumulative distance of each solution vector from others as follows:

$$R_i = \sum_{j=1}^{N_1+\sum_{l=1}^{N_1} y_l} \|P_i^t - P_j^t\|, i = 1, 2, \ldots, N_1 + \sum_{l=1}^{N_1} y_l \tag{6.47}$$

where R_i represents the accumulative distance of the ith vector in the memory. Then, the algorithm calculates the selection probability of each vector based on the accumulative distance values as follows:

$$\mu_i = \frac{R_i}{\sum_{j=1}^{N_\mathrm{I}+\sum_{l=1}^{N_\mathrm{I}} y_l} R_j}, i = 1, 2, \ldots, N_\mathrm{I} + \sum_{l=1}^{N_\mathrm{I}} y_l \quad (6.48)$$

where μ_i represents the selection probability of ith vector in the memory for the next iteration. Based on the probability values of μ_i, N_I number of vectors will be selected from the new population space. Algorithm 6.11 shows the pseudo-code of FA. For more details about the FA, interested readers are referred to Tan and Zhu 2010.

Algorithm 6.11 The Pseudo-Code of FA

```
Initialise the population space P^0 with random N_I individuals;
Set t = 0;
while stopping criteria are not met do
    Evaluate the objective function values of fireworks;
    for i ← 1 to N_I do
        Calculate the number of sparks to be generated from ith firework y_i using equations
        (6.42) and (6.43);
        Calculate the explosion amplitude A_i using equation (6.44);
        for l ← 1 to y_i do
            Use equation (6.46) to generate the lth spark U_l^{i,t} from the ith firework;
        end
    end
    Calculate the accumulative distance R_i of each solution vector in the memory from the oth-
    ers;
    Calculate the selection probabilities μ_i;
    Keep N_I solution vector based on the selection probabilities μ_i, discard the rest solutions
      and form the new population space P^{t+1};
    Set t = t + 1;
end
```

6.7.1.2 Cultural Firework Algorithm

Gao et al. (2011) proposed a Cultural Firework (CF) algorithm, in which CA is combined with FA for digital filter design. The CF algorithm uses the spark generation framework in FA for the population space, while the knowledge components stored in the belief space are used for spark generation. The authors proposed the following spark generation formulations in the population space based on the situational and normative knowledge components:

$$u_{l,j}^{i,t} = \begin{cases} x_{i,j}^t + \left|\text{size}\left(I_j^t\right)N(0,1)\right| & \text{if } x_{i,j}^t < s_j^t \\ x_{i,j}^t - \left|\text{size}\left(I_j^t\right)N(0,1)\right| & \text{if } x_{i,j}^t > s_j^t \\ x_{i,j}^t + \eta size\left(I_j^t\right)N(0,1) & \text{otherwise} \end{cases}, \quad l = 1, 2, \ldots, y_i \tag{6.49}$$

$$u_{l,j}^{i,t} = \begin{cases} x_{i,j}^t + \left|\text{size}\left(I_j^t\right)N(0,1)\right| & \text{if } x_{i,j}^t < x_{\text{min},j}^t \\ x_{i,j}^t - \left|\text{size}\left(I_j^t\right)N(0,1)\right| & \text{if } x_{i,j}^t > x_{\text{max},j}^t \\ x_{i,j}^t + A_i \eta size\left(I_j^t\right)N(0,1) & \text{otherwise} \end{cases}, \quad l = 1, 2, \ldots, y_i \tag{6.50}$$

$$u_{l,j}^{i,t} = \begin{cases} x_{i,j}^t + \varepsilon\left|\text{size}\left(I_j^t\right)N(0,1)\right| & \text{if } x_{i,j}^t < x_{\text{min},j}^t \\ x_{i,j}^t - \varepsilon\left|\text{size}\left(I_j^t\right)N(0,1)\right| & \text{if } x_{i,j}^t > x_{\text{max},j}^t \\ x_{i,j}^t + A_i size\left(I_j^t\right)N(0,1) & \text{otherwise} \end{cases}, \quad l = 1, 2, \ldots, y_i \tag{6.51}$$

where $u_{l,j}^{i,t}$ is the jth variable of the lth spark generated from the ith firework at time t, η and ε are positive constant parameters, s_j^t is the jth variable of the situational knowledge component at time t, $x_{\text{min},j}^t$ and $x_{\text{max},j}^t$ are the lower and upper bounds for the normative knowledge at time t, respectively, and A_i is the amplitude of explosion of the ith firework obtained from Eq. (6.44). It should be noted that the abovementioned formulations will generate three sets of spark solutions that should be saved in the memory. All these spark solutions should be combined with the fireworks. Then, the N_{I} solution vectors should be selected to form the new population space for the next iteration based on the selection probability mechanism in FA. To clearly show the overall framework of CF, Algorithm 6.12 presents the pseudo-code of the algorithm.

Algorithm 6.12 The Pseudo-Code of the CF Algorithm

```
Initialise the population space P^0 with random N_I individuals;
Set t = 0;
while stopping criteria are not met do
    Evaluate the objective function values of fireworks;
    for i ← 1 to N_I do
        Calculates the number of sparks to be generated from ith firework y_i using equations
        (6.42) and (6.43);
        Calculate the explosion amplitude A_i using equation (6.44);
        for l ← 1 to y_i do
            Use equation (6.49) to generate the lth spark U_l^{i,t} from the ith firework;
        end
        Save spark solutions in the memory;
        for l ← 1 to y_i do
            Use equation (6.50) to generate the lth spark U_l^{i,t} from the ith firework;
        end
        Save spark solutions in the memory;
        for l ← 1 to y_i do
            Use equation (6.51) to generate the lth spark U_l^{i,t} from the ith firework;
        end
        Save spark solutions in the memory;
    end
    Calculate the accumulative distance R_i of each solution vector in the memory from the oth-
    ers;
    Calculate the selection probabilities μ_i;
    Keep N_I solution vector based on the selection probabilities μ_i, discard the rest solutions
      and form the new population space P^{t+1};
    Set t = t + 1;
end
```

6.7.2 *Hybrid CA and IWO*

IWO is another meta-heuristic algorithm that has been hybridised with CAs. The knowledge components in the belief space of CAs have been employed to improve the search performance of IWO. This section discusses the hybrid Cultural Invasive Weed Algorithm (CIWA) and sequential hybrid CA and IWO approaches. To provide a better explanation for the approaches, the standard IWO algorithm is briefly presented first.

6.7.2.1 Standard IWO

IWO originally developed by Mehrabian et al. (2006) is a meta-heuristic population-based algorithm inspired by weed colonisation. In IWO, each seed is represented by X_i^t, $i = 1, 2, \ldots, N_I$ which is a solution candidate for the problem. The algorithm consists of different phases, including initialisation, reproduction, and spatial dispersal. In the initialisation phase, the IWO spreads the seeds randomly in the search space. Depending on the objective function values, each seed produces other seeds in the reproduction phase. The spatial dispersal phase dispreads the newly produced seeds within the search space to grow new plants. The main algorithmic steps of the IWO can be summarised as follows.

Step 1: Initialisation

Similar to other meta-heuristics, IWO starts with a set of random seeds within the search space and their objective function values are evaluated.

Step 2: Reproduction

In this step, the number of new random seeds N_i is determined for each seed based on its objective function value as follows:

$$N_i = \text{round}\left(\frac{f(X_i^t) - f_{\min}}{f_{\max} - f_{\min}}(N_{\max} - N_{\min}) + N_{\min}\right) \tag{6.52}$$

where N_i represents the number of randomly generated seeds from the ith seed, $N_{\min}$ and $N_{\max}$ are the minimum and maximum numbers of random seeds generated from each seed, and $round(.)$ is a function that rounds the input number to the nearest integer value.

Step 3: Spatial dispersal

In this step, the seeds will be mutated within the search space based on the random function with normal distribution. The IWO uses normal distribution with zero mean and varying standard deviation. The standard deviation value is gradually reduced over time as follows:

$$\sigma^t = \frac{(t_{\max} - t)^c}{(t_{\max})^c}\left(\sigma^0 - \sigma^{t_{\max}}\right) + \sigma^{t_{\max}} \tag{6.53}$$

where σ^t represents the value of the standard deviation at iteration t, $t_{\max}$ is the maximum number of iterations, σ^0 and $\sigma^{t_{\max}}$ are the initial and final values of the standard deviation, respectively, and c is a nonlinear modulation index. By reducing the standard deviation value over time, the step size of the search process is reduced and the algorithm tries to gradually shift from the exploration phase to the exploitation phase.

Then, the new seeds are generated as follows:

$$y_{k,j}^{t} = x_{i,j}^{t} + N\left(0, \left(\sigma^{t}\right)^{2}\right), \quad k = 1, 2, \ldots, N_i, \quad j = 1, 2, \ldots, n \tag{6.54}$$

where $y_{k,j}^{t}$ represents the jth variable of the kth new seed generated from the ith seed and n is the dimension of the problem. The spatial dispersal step will result in $\sum_{i=1}^{N_{\mathrm{I}}} N_i$ number of new seeds represented by the vectors $Y_1^t, Y_2^t, \ldots, Y_{\sum_{i=1}^{N_{\mathrm{I}}} N_i}^t$. These new seeds will be combined with previous parent seeds represented by $X_1^t, X_2^t, \ldots, X_{N_{\mathrm{I}}}^t$ to form a population space with a size greater than N_{I}.

Step 4: Evaluation

The objective function values of the newly generated seeds $Y_1^t, Y_2^t, \ldots, Y_{\sum_{i=1}^{N_{\mathrm{I}}} N_i}^t$ are computed in this step. Then, N_{I} number of seeds with better objective function values will be selected to form the next generation and the rest of the seeds will be discarded.

Step 5: Termination or redo

Repeat steps 2–4 until the maximum number of iteration $t_{\max}$ is reached or the termination criterion is satisfied.

6.7.2.2 Hybrid CA and IWO

Gao et al. (2019) developed a hybrid CIWA approach for the robustness and imperceptibility optimisation of the watermarking systems. In CIWA, the belief space with situational and normative knowledge components is used to store the cultural information and guide the seeds within the search space. The main difference between the standard IWO and CIWA is in the spatial dispersal phase. In CIWA, the cultural information is used to generate new seeds in addition to the aforementioned simple normal distribution. Let us assume p_1 and p_2 be the positive constant parameters less than 1. The CIWA produces new seeds as follows:

- if rand $< p_1$:

$$y_{k,j}^{t} = x_{i,j}^{t} + N\left(0, \left(\sigma^{t}\right)^{2}\right) \tag{6.55}$$

  if $p_1 < \gamma \le p_2$:

$$y_{k,j}^{t+1} = \begin{cases} x_{i,j}^{t} + \left|\left(x_{\max}^{t} - x_{\min}^{t}\right) \times N(0,1)\right| & \text{if } x_{i,j}^{t} < x_{\min}^{t} \\ x_{i,j}^{t} - \left|\left(x_{\max}^{t} - x_{\min}^{t}\right) \times N(0,1)\right| & \text{if } x_{i,j}^{t} > x_{\max}^{t} \\ x_{i,j}^{t} + \eta\left(x_{\max}^{t} - x_{\min}^{t}\right) \times N(0,1) & \text{otherwise} \end{cases} \tag{6.56}$$

- if $p_2 < \gamma \le 1$:

$$y_{k,j}^{t+1} = x_{i,j}^{t} + rand\left(s_j^{t} - x_{,j}^{t}\right) + \left|b_j^{t} - x_{i,j}^{t}\right| N(0,1) \tag{6.57}$$

where σ^t is obtained from Eq. (6.53), $b_j^t = \frac{1}{N_I}\sum_{i=1}^{N_I} x_{i,j}^t$ is the average position of seeds in the jth dimension of the search space at time t, $x_{\min}^t$ and $x_{\max}^t$ are the upper and lower bounds for the normative knowledge, respectively, s_j^t is the jth variable of the situational knowledge at iteration t, and η is a constant parameter.

Algorithm 6.13 presents the pseudo-code of CIWA. The application results of CIWA reported by Gao et al. (2019) to watermarking images revealed that the CIWA performs better than the standard CA, PSO, and IWO methods.

Algorithm 6.13 Pseudo-Code of the CIWA Approach

Initialise the parameters of the algorithm: N_I, $N_{\min}$, $N_{\max}$, $t_{\max}$, σ^0, $\sigma^{t\max}$, c, p_1, p_2, η
Initialise the belief space $\mathcal{B}^0 = \{\mathcal{S}^0, \mathcal{N}^0\}$ with situational and normative knowledge components;
Initialise the population space P^0 with random N_I seeds represented $X_1^0, X_2^0, \ldots, X_{N_I}^0$;
Evaluate the objective function values of the seeds;
Set $t = 1$;
while *stopping criteria are not met* **do**
  Update the belief space and obtain $\mathcal{B}^{t+1}$;
  Set $p = 0$;
  % Spatial dispersal
  for $i \leftarrow 1$ to N_I **do**
    $N_i = round\left(\frac{f(X_i^t) - f_{\min}}{f_{\max} - f_{\min}}(N_{\max} - N_{\min}) + N_{\min}\right)$;
    for $k = p$ to N_i **do**
      for $j \leftarrow 1$ to n **do**
        if $rand < p_1$ **then**
          $y_{k,j}^{t+1} = x_{i,j}^t + N(0, (\sigma^t)^2)$;
        elseif $p_1 < \gamma \le p_2$ **then**
          if $x_{i,j}^t < x_{\min}^t$ **then**
            $y_{k,j}^{t+1} = x_{i,j}^t + |(x_{\max}^t - x_{\min}^t) \times N(0,1)|$;
          elseif $x_{i,j}^t > x_{\max}^t$ **then**
            $y_{k,j}^{t+1} = x_{i,j}^t - |(x_{\max}^t - x_{\min}^t) \times N(0,1)|$;
          else
            $y_{k,j}^{t+1} = x_{i,j}^t + \eta(x_{\max}^t - x_{\min}^t) \times N(0,1)$;
          end
        else
          $y_{k,j}^{t+1} = x_{i,j}^t + rnd(s_j^t - x_{i,j}^t) + |b_j^t - x_{i,j}^t| N(0,1)$;
        end
      end
    end
    Set $p = p + N_i$;
  end
  Evaluate the objective functions of vectors $Y_1^t, Y_2^t, \ldots, Y_{\sum_{i=1}^{N_I} N_i}^t$;
  Insert the seeds $Y_1^t, Y_2^t, \ldots, Y_{\sum_{i=1}^{N_I} N_i}^t$ into the population space P^t;
  Create new population space P^{t+1} by keeping N_I seeds with better objective function values in population space P^t and discard the rest of the seeds;
  Set $t = t + 1$;
end

6.7.2.3 Sequential Hybrid CA and IWO

Hosseinioun et al. (2020) proposed a hybrid IWO and CA algorithm for optimum task scheduling in fog computing. In this algorithm, the population space is evolved sequentially by the operators of standard IWO and CA.

6.7.3 *Hybrid CA and SFLA*

SFLA is a meta-heuristic algorithm that has been hybridised with CAs. The main aim of this section is to explain the hybrid Cultural Quantum Shuffled Frog Leaping (CQSFL) algorithm. The details of CQSFL will be presented after a brief review of the standard SFLA method.

6.7.3.1 Standard SFLA

The SFLA is a population-based evolutionary algorithm originally developed by Eusuff et al. (2006) based on the evolution process of memes carried by the interactive individuals and their global information exchange capability. The basics of the algorithm are inspired by the leaping behaviour of the frogs in a swamp, in which the frogs exchange information about the food resources with each other. The algorithm uses a sub-populations-based model for the population space. The overall framework of SFLA can be described as follows.

Let us assume n_{m} different frog communities with n_{F} number of frogs in each. In this case, the total number of frogs in the swamp will be equal to $N_{\mathrm{I}} = n_{\mathrm{m}} n_{\mathrm{F}}$. The algorithmic steps of the SFLA can be written as follows (Eusuff et al. 2006):

Step 1: Initialisation

By setting the iteration $t = 0$, the algorithm starts with the random N_{I} number of frogs in the search space represented by vectors $X_1^t, X_2^t, \ldots, X_{N_{\mathrm{I}}}^t$ with the performances of $f(X_1^t), f(X_2^t), \ldots, f(X_{N_{\mathrm{I}}}^t)$.

Step 2: Ranking

In this step, the frogs are ranked based on their objective function values. Then, the first frog with the best objective function value is recorded as follows: $X_{\mathrm{Best}}^t = X_1^t$

Step 3: Dividing the frogs

In the third step, the population of frogs are divided into n_{m} sub-populations, called memeplexes, with n_{F} number of frogs in each. Let us assume $Y_1^t, Y_2^t, \ldots, Y_{n_{\mathrm{m}}}^t$ represent the n_{m} number of memeplexes at iteration t, which can be formulated as follows:

$$Y_k^t = \left[X_k^t, X_{k+n_{\mathrm{m}}}^t, X_{k+2n_{\mathrm{m}}}^t, \ldots, X_{k+n_{\mathrm{m}}(n_{\mathrm{F}}-1)}^t\right], k = 1, 2, \ldots, n_{\mathrm{m}} \tag{6.58}$$

Step 4: Evolution of each memeplex

For each memeplex Y_k^t, the evolution process is performed for N_{E} times. To this end, a sub-memeplex is considered for each memeplex, in which a set of frogs are located. The SFLA uses a triangular stochastic approach for forming the sub-memeplex. In this approach, the frogs with better objective function values have a higher chance to be selected for sub-memeplex. The selection probability of each frog in each memeplex is calculated as follows:

$$p_j = \frac{2(n_{\mathrm{F}} + 1 - j)}{n_{\mathrm{F}}(n_{\mathrm{F}} + 1)}, \quad j = 1, 2, \ \ldots, \ n_{\mathrm{F}} \tag{6.59}$$

where p_j represents the selection probability of jth frog in memeplex to form the sub-memeplex. The sub-memeplex is created by the selection of q number of frogs ($q < n_{\mathrm{F}}$), in which the frogs are sorted based on their objective function values. Then, the best and worst frogs in the sub-memeplex of kth memeplex are selected, represented by $B_k^t = [b_1^t, b_2^t, \ \ldots, b_n^t]$ and $W_k^t = [w_1^t, w_2^t, \ \ldots, w_n^t]$, respectively.

At the beginning, the step tries to improve the performance of the worst frog in the sub-memeplex W_k^t by performing a local search as follows:

$$U_k^t = \begin{cases} B_k^t + \min\left\{\mathrm{int}\left[\mathrm{rand}\left(B_k^t - W_k^t\right)\right], S_{\max}\right\} & \text{for a positive step} \\ B_k^t + \max\left\{\mathrm{int}\left[\mathrm{rand}\left(B_k^t - W_k^t\right)\right], -S_{\max}\right\} & \text{for a negative step} \end{cases} \tag{6.60}$$

where U_k^t is a trial vector for the sub-memeplex of kth memeplex at iteration t, $S_{\max}$ represents the maximum allowable step size, int[.] is an integer function that rounds the input variable to the nearest integer value, and *rand* is a random number uniformly distributed between 0 and 1. If the quality of obtained vector U_k^t is better than the quality of W_k^t (i.e., $f\left(U_k^t\right) < f(P_{\mathrm{W}})$), then the W_k^t will be replaced by U_k^t. Otherwise, another vector U_k^t will be calculated based on a global search strategy as follows:

$$U_k^t = \begin{cases} W_k^t + \min\{\text{int}[\text{rand}(X_{\text{Best}}^t - W_k^t)], -S_{\max}\} & \text{for a positive step} \\ W_k^t + \max\{\text{int}[\text{rand}(X_{\text{Best}}^t - W_k^t)], -S_{\max}\} & \text{for a negative step} \end{cases} \quad (6.61)$$

where X_{Best}^t is the global best frog in the whole population space. If the evolution produces good results, i.e., $f(U_k^t) < f(W_k^t)$, the vector W_k^t will be replaced by U_k^t. Otherwise, U_k^t will be randomly generated in the search space and the worst frog W_k^t will be discarded. Then, the memeplex Y_k^t should be updated and sorted based on objective function values.

Step 5: Shuffle memeplexes

After performing the evolutionary steps explained above for each memeplex Y_k^t and for N times, the population space is updated and the frogs are ranked based on their objective function values. The global best solution X_{Best}^t needs to be updated as well.

Step 6: termination or redo

In this step, the process will be stopped if the convergence criteria are satisfied. Otherwise, steps 3–5 should be repeated.

The detailed aforementioned steps of the SFLA clearly show that the algorithm tries to keep a balance between exploitation and exploration by using Eqs. (6.60) and (6.61), respectively. Algorithm 6.14 presents the detailed pseudo-code of SFLA.

Algorithm 6.14 Pseudo-Code of SFLA Algorithm

```
Initialise the parameters of the algorithm: n_F, n_m, q, S_max, N_E
Initialise the population space P^0 with random N_I = n_F n_m frogs represented by X_1^t, X_2^t, ..., X_{N_I}^t;
Evaluate the objective function values of the frogs;
Rank the frogs based on their objective function values and find the global best X_Best^t;
Set t = 0;
while stopping criteria are not met do
    for k ← 1 to n_m do
        Set N = 0;
        while N ≤ N_E do
            % Formation of kth memeplex:
            Y_k^t = [X_k^t, X_{k+n_m}^t, X_{k+2n_m}^t, ..., X_{k+n_m(n_F-1)}^t];
            % Calculation of selection probabilities:
            for j ← 1 to n_F do
                p_j = 2(n_F+1-j) / (n_F(n_F+1));
            end
            Select q number of frogs randomly based on values of p_j to form the sub-memeplex
            for the kth memeplex;
            Sort selected q number of frogs and obtain the best frog B_k^t and worst frog W_k^t in the
            kth sub-memeplex;
            % Evolution of kth memeplex:
            if rand ≤ 0.5 then
                U_k^t = W_k^t + min{int[rand(B_k^t - W_k^t)], S_max};
            else
                U_k^t = W_k^t + max{int[rand(B_k^t - W_k^t)], -S_max};
            end
            if f(U_k^t) < f(W_k^t) then
                Replace W_k^t by U_k^t;
            else
                if rand ≤ 0.5 then
                    U_k^t = W_k^t + min{int[rand(X_Best^t - W_k^t)], S_max};
                else
                    U_k^t = W_k^t + max{int[rand(X_Best^t - W_k^t)], -S_max};
                end
                if f(U_k^t) < f(W_k^t) then
                    Replace W_k^t by U_k^t;
                else
                    Replace W_k^t by a random vector within the search space;
                end
            end
            N = N + 1;
        end
    end
    Rank the frogs based on their objective function values and update the global best X_Best^t;
    Set t = t + 1;
end
```

6.7.3.2 Cultural Quantum Shuffled Frog Leaping

Gao and Xu (2013) developed a hybrid Cultural Quantum Shuffled Frog Leaping (CQSFL) algorithm, in which the normative knowledge of belief space is used to evolve the frogs in each memeplex. The CQSFL mainly modified the fourth step of the SFLA. In CQSFL, the positions of the worst frogs in the kth memeplex are first modified as follows:

$$U_k^t = \left| W_k^t \cos\,\delta_{\mathrm{w}} - \sqrt{1 - \left(W_k^t\right)^2} \sin\,\delta_{\mathrm{w}} \right| \tag{6.62}$$

in which:

$$\delta_{\mathrm{w}} = \mathrm{rand}\left(W_k^t - X_{\mathrm{Best}}^t\right) + N(0, 1)\left(W_k^t - z_k\right) \tag{6.63}$$

where $N(0, 1)$ is a random number generated from the normal distribution with a mean and standard deviation of 0 and 1, respectively, and z_k represents the average position of quantum frogs in the kth memeplex, which is obtained as follows:

$$z_k = \frac{1}{n_{\mathrm{F}}} \sum_{j=1}^{n_{\mathrm{F}}} X_{k+n_{\mathrm{m}}(j-1)}^t \tag{6.64}$$

If the objective function value of U_k^t is improved, i.e., $f(U_k^t) < f(W_k^t)$, the W_k^t will be replaced by U_k^t. Otherwise, the vector U_k^t will be recreated based on the normative knowledge of the belief space as follows:

$$\delta'_{\mathrm{w},j} = \begin{cases} -\left|\mathrm{size}\left(I_j^t\right) N(0, 1)\right|, & \text{if } w_{k,j}^t < x_{\mathrm{min},j}^t \\ \left|\mathrm{size}\left(I_j^t\right) N(0, 1)\right|, & \text{if } w_{k,j}^t > x_{\mathrm{max},j}^t \\ \eta \cdot \mathrm{size}\left(I_j^t\right) N(0, 1) & \text{otherwise} \end{cases}, \quad j = 1, 2, \ldots, n \tag{6.65}$$

$$u_{k,j}^t = \left| w_{k,j}^t \cos\,\delta'_{\mathrm{w},j} - \sqrt{1 - \left(w_{k,j}^t\right)^2} \sin\,\delta'_{\mathrm{w},j} \right|, \quad j = 1, 2, \ldots, n \tag{6.66}$$

where $w_{k,j}^t$ represents the jth variable of the worst frog in the sub-memeplex of kth memeplex, $x_{\mathrm{min},j}^t$ and $x_{\mathrm{max},j}^t$ are the normative knowledge bounds for the jth variable, $u_{k,j}^t$ is the jth variable of the trial vector U_k^t for the k th memeplex, I_j^t is the normative interval for the jth variable, $size\left(I_j^t\right) = x_{\mathrm{max},j}^t - x_{\mathrm{min},j}^t$ gives the size of the normative interval, and η is a constant parameter.

Algorithm 6.15 shows the pseudo-code of the CQSFL algorithm. It should be noted that the belief space $\mathcal{B}^t$ needs to be updated at each iteration. The update mechanism of belief space is similar to that in the standard CA explained in Chap. 3.

6.7.4 Hybrid CA and FECO

FECO is another meta-heuristic that has been combined with CAs to develop a hybrid Multi-Objective CA and FECO (MOCAFECO) algorithm. The main idea of MOCAFECO is to use cultural information to further exploit the solutions generated by the standard FECO algorithm. The section briefly describes the standard FECO. Then, the MOCAFECO algorithm will be explained in detail.

6.7.4.1 Standard FECO

The FECO algorithm developed by Ye et al. (2019) is inspired by the generation and restriction mechanisms among five elements in Chinese traditional culture. In the FECO, each element is a solution candidate for the problem, represented by the vector X_i^t at iteration t. Each element has its own mass m_i^t which is subject to an external force F_i^t calculated by the following formula:

$$F_i^t = \omega_{\mathrm{gp}} \ln\left(\frac{m_{i-1}^t}{m_i^t}\right) - \omega_{\mathrm{rp}} \ln\left(\frac{m_{i-2}^t}{m_i^t}\right) - \omega_{\mathrm{ga}} \ln\left(\frac{m_i^t}{m_{i+1}^t}\right) - \omega_{\mathrm{ra}} \ln\left(\frac{m_i^t}{m_{i+2}^t}\right), \quad i = 1, 2, \ldots, L \tag{6.67}$$

where ω_{gp}, ω_{rp}, ω_{ga}, and ω_{ra} are the constant parameters within the range [0, 1] and L is the number of elements. Let q be the number of cycles. The FECO considers a population space with L elements in each cycle. With this definition, one can mathematically represent the population as a matrix:

Algorithm 6.15 Pseudo-Code of CQSFL Algorithm

Initialise the parameters of the algorithm: n_{F}, n_{m}, q, η, N_{E}
Initialise the population space $\mathcal{P}^0$ with random $N_{\mathrm{I}} = n_{\mathrm{F}} n_{\mathrm{m}}$ frogs represented by $X_1^t, X_2^t, \ldots, X_{N_{\mathrm{I}}}^t$;
Initialise the belief space $\mathcal{B}^0 = \mathcal{N}^0$ with normative knowledge component;
Evaluate the objective function values of the frogs;
Rank the frogs based on their objective function values and find the global best X_{Best}^t;
Set $t = 0$;
while *stopping criteria are not met* **do**
  for $k \leftarrow 1$ to n_{m} **do**
    Set $N = 0$;
    while $N \leq N_{\mathrm{E}}$ **do**
      % Formation of kth memeplex:
      $Y_k^t = \left[X_k^t, X_{k+n_{\mathrm{m}}}^t, X_{k+2n_{\mathrm{m}}}^t, \ldots, X_{k+n_{\mathrm{m}}(n_{\mathrm{F}}-1)}^t\right]$;
      % Calculation of selection probabilities:
      for $j \leftarrow 1$ to n_{F} **do**
        $p_j = \frac{2(n_{\mathrm{F}}+1-j)}{n_{\mathrm{F}}(n_{\mathrm{F}}+1)}$;
      end
      Select q number of frogs randomly based on values of p_j to form the sub-memeplex for the kth memeplex;
      Sort selected q number of frogs and obtain the best frog B_k^t and worst frog W_k^t in the kth sub-memeplex;
      % Evolution of kth memeplex:
      $z_k = \frac{1}{n_{\mathrm{F}}} \sum_{j=1}^{n_{\mathrm{F}}} X_{k+n_{\mathrm{m}}(j-1)}^t$;
      $\delta_{\mathrm{w}} = rand(W_k^t - X_{\mathrm{Best}}^t) + N(0,1)(W_k^t - z_k)$;
      $U_k^t = \left| W_k^t \cos \delta_{\mathrm{w}} - \sqrt{1 - (W_k^t)^2} \sin \delta_{\mathrm{w}} \right|$;
      if $f(U_k^t) < f(W_k^t)$ **then**
        Replace W_k^t by U_k^t;
      else
        for $j \leftarrow 1$ to n **do**
          if $w_{k,j}^t < x_{\min,j}^t$ **then**
            $\delta'_{\mathrm{w},j} = -\left| size(I_j^t) N(0,1) \right|$;
          elseif $w_{k,j}^t > x_{\max,j}^t$ **then**
            $\delta'_{\mathrm{w},j} = \left| size(I_j^t) N(0,1) \right|$;
          else
            $\delta'_{\mathrm{w},j} = \eta . size(I_j^t) N(0,1)$;
          end
          $u_{k,j}^t = \left| w_{k,j}^t \cos \delta'_{\mathrm{w},j} - \sqrt{1 - \left(w_{k,j}^t\right)^2} \sin \delta'_{\mathrm{w},j} \right|$;
        end
        Replace W_k^t by U_k^t;
      end
      $N = N + 1$;
    end
  end
  Rank the frogs based on their objective function values and update the global best X_{Best}^t;
  Update the belief space and obtain $\mathcal{B}^{t+1}$;
  Set $t = t + 1$;
end

$$P^t = \begin{bmatrix} X_1^{t,1} & X_2^{t,1} & \dots & X_L^{t,1} \\ X_1^{t,2} & X_2^{t,2} & \dots & X_L^{t,2} \\ \vdots & \vdots & \vdots & \vdots \\ X_1^{t,q} & X_2^{t,q} & \dots & X_L^{t,q} \end{bmatrix} \tag{6.68}$$

where $X_i^{t,k} = [x_{i,1}^{t,k}, x_{i,2}^{t,k}, \dots, x_{i,n}^{t,k}]$ represents the ith element of the kth cycle at iteration t and $x_{i,j}^{t,k}$ is the jth variable of the ith element in the kth cycle at iteration t.

For each of the elements in the population, the external force can be calculated by Eq. (6.67) and represented by $F_i^{t,k}$, which indicates the force exerted on ith individual in the kth cycle. The algorithm assumes a more general case of a multi-objective problem with M number of objective functions as follows: $f_1(X), f_2(X), \dots, f_M(X)$. In this case, the force $F_i^{t,k}$ can be calculated for each objective function. Let $F_{i_m}^{t,k}$ be the force exerted on ith individual within the kth cycle calculated based on the mth objective function, where $m = 1, 2, \dots, M$.

The FECO decides to update the position of ith element in the kth cycle based on the force values calculated for two random objective functions. Let r_1 and $r_2 \in [1, 2, \dots, M]$ be two random numbers. With this notation, $F_{i_{r_1}}^{t,k}$ and $F_{i_{r_2}}^{t,k}$ are the forces exerted on the ith element within the kth cycle calculated based on the r_1 th and r_2th objective function values, respectively. The algorithm decides to update the position of the ith element in the kth cycle based on the values of randomly selected forces as follows:

$$U_i^{t,k} = \begin{cases} X_i^{t,k} & \text{if } F_{i_{r_1}}^{t,k} \geq 0 \text{ and } F_{i_{r_2}}^{t,k} \geq 0 \\ X_i^{t,k} + V_i^{t,k}, & \text{otherwise} \end{cases} \tag{6.69}$$

in which:

$$V_i^{t,k} = \begin{cases} \omega V_i^{t-1,k} + c_1 \text{rand}\left(P_i^{t,k} - X_i^{t,k}\right) & \text{if } N(0,1) < P_u \\ \omega V_i^{t-1,k} + c_2 \text{rand}\left(G^t - X_i^{t,k}\right) & \text{if } N(0,1) \geq P_u \end{cases} \tag{6.70}$$

where $U_i^{t,k}$, $V_i^{t,k}$, and $P_i^{t,k}$ represent the trial vector, velocity, and best personal experience of the ith element within the kth cycle, respectively, ω, c_1, and c_2 are the constant coefficients, and G^t is the global best element obtained until iteration t.

As it can be seen from Eqs. (6.69) and (6.70), the position updating formulas of FECO are similar to those in the PSO algorithm. The FECO only updates the position of an element if at least one of the random forces is negative. The negativity of force value indicates the lower quality of an element. In Eq. (6.70), P_u is a parameter which controls the extrapolation and exploitation mechanisms in the algorithm.

In the next stage, the FECO mutates each element as follows:

$$Y_i^{t,k} = \begin{cases} U_i^{t,k} + U(-\sigma_1, \sigma_1), & \text{if } t \leq \frac{t_{\max}}{4} \\ U_i^{t,k} + C(0, \sigma_2), & \text{if } \frac{t_{\max}}{4} < t \leq \frac{3t_{\max}}{4} \\ U_i^{t,k} + N(0, \sigma_3), & \text{otherwise} \end{cases} \tag{6.71}$$

where $Y_i^{t,k}$ indicates the mutated ith element within the kth cycle at iteration t, $U(a, b)$, $C(a, b)$, and $N(a, b)$ are the random numbers obtained from the uniform, Cauchy, and normal distributions with parameters a and b, respectively, $t_{\max}$ represents the maximum number of iterations, and σ_1, σ_2, and σ_3 are the constant parameters.

6.7.4.2 Hybrid Multi-objective CA and FECO

Mao et al. (2020) proposed a hybrid Multi-Objective CA and FECO (MOCAFECO) algorithm for function optimisation. In the MOCAFECO, each element $Y_i^{t,k}$ evolved by the standard FECO algorithm is further updated by the influence function of CA. To this end, the study employed the following influence functions consisting of different knowledge components to calculate the final element vector for the next iteration $X_i^{t+1,k} = [x_{i,1}^{t+1,k}, x_{i,2}^{t+1,k}, \ldots, x_{i,n}^{t+1,k}]$:

- Normative knowledge-based influence:

$$x_{i,j}^{t+1,k} = y_{i,j}^{t,k} + \left|\alpha N(0, 1)\left(x_{\max,j}^t - x_{\min,j}^t\right)\right| \tag{6.72}$$

where $y_{i,j}^{t,k}$ is the jth variable of the ith element within cycle k calculated from Eq. (6.71), α is a constant parameter, $x_{\min,j}^t$ and $x_{\max,j}^t$ are the lower and upper normative bounds for the jth variable, respectively, and n is the dimension of the problem.

- Situational knowledge-based influence:

$$x_{i,j}^{t+1,k} = \begin{cases} y_{i,j}^{t,k} + \left|\gamma N(0, 1)\left(x_j^{\max} - x_j^{\min}\right)\right|, & \text{if } y_{i,j}^{t,k} < s_{r_1,j}^t \\ y_{i,j}^{t,k} - \left|\gamma N(0, 1)\left(x_j^{\max} - x_j^{\min}\right)\right|, & \text{if } y_{i,j}^{t,k} > s_{r_1,j}^t \\ y_{i,j}^{t,k} - \left|\gamma N(0, 1)\left(x_j^{\max} - x_j^{\min}\right)\right|, & \text{if } y_{i,j}^{t,k} = s_{r_1,j}^t \end{cases} \tag{6.73}$$

where $s^t_{r_1,j}$ is a randomly selected non-dominated situational vector, $x_j^{\min}$ and $x_j^{\max}$ are the lower and upper bounds in the jth dimension of the problem, and γ is a constant parameter. The study assumes that the non-dominated solutions are stored within the situational knowledge component.

- Mixed normative-situational knowledge-based influence:

$$x_{i,j}^{t+1,k} = \begin{cases} y_{i,j}^{t,k} + \left|\alpha N(0,1)\left(x^t_{\max,j} - x^t_{\min,j}\right)\right|, & \text{if } y_{i,j}^{t,k} < s^t_{r_1,j} \\ y_{i,j}^{t,k} - \left|\alpha N(0,1)\left(x^t_{\max,j} - x^t_{\min,j}\right)\right|, & \text{if } y_{i,j}^{t,k} > s^t_{r_1,j} \\ y_{i,j}^{t,k} - \left|\alpha N(0,1)\left(x^t_{\max,j} - x^t_{\min,j}\right)\right|, & \text{if } y_{i,j}^{t,k} = s^t_{r_1,j} \end{cases} \tag{6.74}$$

where the definitions of all parameters are similar to those in equations (6.72) and (6.73).

- Random influence:

$$x_{i,j}^{t+1,k} = y_{i,j}^{t,k} + \left|0.1N(0,1)\left(x_j^{\max} - x_j^{\min}\right)\right| \tag{6.75}$$

Algorithm 6.16 presents the pseudo-code of the MOCAFECO algorithm.

6.7.5 Hybrid CA and EHO

The EHO is a population-based stochastic algorithm inspired by the herding behaviour of elephant groups in nature, in which each elephant represents a potential candidate solution for the problem. The algorithm starts with random elephants divided into several clans that are gradually updated to find the optimum solution.

Algorithm 6.16 Pseudo-Code of the MOCAFECO Algorithm Developed by Mao et al. (2020)

Initialise the internal parameters and the population space $\mathcal{P}^0$ with q cycles and L elements in each;
Initialise the belief space;
Evaluate the fitness function of individuals;
Set $t = 0$;
while *stopping criteria are not met* **do**
  Accept the elite elements and update the belief space;
  for $k \leftarrow 1$ to q **do**
    for $i \leftarrow 1$ to L **do**
      Update the position of each element using equation (6.69) and obtain $U_i^{t,k}$;
      Obtain the mutated vector $Y_i^{t,k}$ using equation (6.71);
      Calculate the final updated element vector $X_i^{t+1,k}$ using one of randomly selected influence function from equations (6.72), (6.73), (6.74), and (6.75);
    end
  end
  Evaluate the objective function of updated elements in the population space;
  Set $t = t + 1$;
end

Jafari et al. (2019) proposed a hybrid CA and EHO for the optimal design of truss structures. During the optimisation process, EHO replaces the elephants with poor fitness values with some random elephants in the search space which can affect the performance of the algorithm negatively. To remedy this disadvantage, Jafari et al. (2019) employed the belief space within EHO to record the cultural knowledge of elephants within the population space. Then, the algorithm replaces the elephants with poor fitness values with the elephants randomly generated within the normative bounds as follows:

$$x_{i,j}^{t+1} = x_{\max,j}^{t} + rand\left(x_{\max,j}^{t} - x_{\min,j}^{t}\right) \tag{6.76}$$

In another study, Murugadass et al. (2020a) proposed a Hybrid Elephant Herding Optimisation and Cultural Algorithm (HEHO-CA), in which the individuals are simultaneously updated by the CA and EHO operators.

6.8 Hybrid CA and Evolutionary Programming (EP)

Reynolds et al. (2005) proposed a new configuration for the CAs, in which an EP-based configuration is used for the population space. Algorithm 6.17 shows the general configuration of CA based on the EP scheme. As can be seen from Algorithm 6.17, the new population space $\mathcal{P}^{t+1}$ is a combination of previous and mutated populations. This will help the algorithm to keep the previously obtained better individuals in the new population.

Algorithm 6.17 Pseudo-Code of the CA with EP-Based Population Update Proposed by Reynolds et al. (2005)

```
Initialise the population space P^0 with N_I individuals;
Evaluate the fitness function of individuals;
Set t = 0;
while stopping criteria are not met do
    Define the children population as a copy of the previous population as follows: C^{t+1} = P^t;
    Mutate the children population C^{t+1};
    Combine C^{t+1} and P^t, and randomly choose N_I fittest individuals to form the new population
    space P^{t+1};
    Eliminate the rest of the individuals from the memory;
end
```

6.9 Summary

The chapter presented the hybrid versions of CAs with other meta-heuristic algorithms developed in the literature, such as PSO, DE, HS, GAs, FA, IWO, SFLA, FECO, EHO, and local search methods. Explanation of hybridised algorithms is usually challenging, as they include several different concepts, parameters, and formulations. In order to deal with this challenge, the key details of other meta-heuristic algorithms were briefly explained to facilitate the understanding of the hybridised versions for the readers. The hybridised versions of CAs with other meta-heuristics were discussed, and their detailed algorithmic details were explained. In some cases, the hybridisation ideas were further explained by providing detailed algorithmic pseudo-codes, which can potentially help the readers to digest the overall frameworks of hybrid algorithms.

Generally speaking, the hybrid versions developed in literature try to take the advantage of the cultural information transmission mechanism of CAs in their frameworks. The hybridised versions of CAs with the PSO algorithm revealed that the cultural information stored in belief space can be used in PSO to guide the search directions. The studies employed normative knowledge intervals to choose the movement direction of particles in the search space. The combination idea of CAs with the DE algorithm has been mainly focused on the adoption of normative and situational knowledge components to design new differential mutation operators. The normative knowledge components have been also used in the hybrid version of CAs and HS to further modify the harmonies generated by the standard HS algorithm. Similar ideas can be observed in hybridised versions of CAs with other meta-heuristic algorithms.

References

Ali MZ, Reynolds RG (2014) Cultural algorithms: a Tabu search approach for the optimization of engineering design problems. Soft Comput 18. https://doi.org/10.1007/s00500-013-1169-5

Ali MZ, Awad N, Reynolds RG (2014a) Balancing search direction in cultural algorithm for enhanced global numerical optimization. In: 2014 IEEE symposium on Swarm intelligence. IEEE, pp 1–7

Ali MZ, Awad NH, Suganthan PN, Reynolds RG (2016a) A modified cultural algorithm with a balanced performance for the differential evolution frameworks. Knowl-Based Syst 111:73–86. https://doi.org/10.1016/j.knosys.2016.08.005

Ali MZ, Awad NH, Suganthan PN et al (2016b) A novel hybrid Cultural Algorithms framework with trajectory-based search for global numerical optimization. Inf Sci 334–335:219–249. https://doi.org/10.1016/j.ins.2015.11.032

Arpaia P, Lucariello G, Zanesco A (2007) Automatic fault isolation by cultural algorithms with differential influence. IEEE Trans Instrum Meas 56:1573–1582. https://doi.org/10.1109/TIM.2007.903604

Awad NH, Ali MZ, Suganthan PN, Reynolds RG (2017) CADE: a hybridization of cultural algorithm and differential evolution for numerical optimization. Inf Sci 378:215–241. https://doi.org/10.1016/j.ins.2016.10.039

Becerra RL, Coello CAC (2004) A cultural algorithm with differential evolution to solve constrained optimization problems. In: Ibero-American conference on artificial intelligence, pp 881–890

Burke EK, Gendreau M, Hyde M et al (2013) Hyper-heuristics: a survey of the state of the art. J Oper Res Soc 64:1695–1724. https://doi.org/10.1057/jors.2013.71

Chen C-H, Liu Y-C, Lin C-J, Lin C-T (2008) A hybrid of cooperative particle swarm optimization and cultural algorithm for neural fuzzy networks. In: 2008 IEEE international conference on fuzzy systems (IEEE world congress on computational intelligence). IEEE, pp 238–245

Chen C-H, Yang S-Y (2015) Efficient DE-based symbiotic cultural algorithm for neuro-fuzzy system design. Appl Soft Comput 34:18–25. https://doi.org/10.1016/j.asoc.2015.04.026

Coelho L dos S, Souza RCT, Mariani VC (2009) Improved differential evolution approach based on cultural algorithm and diversity measure applied to solve economic load dispatch problems. Math Comput Simul 79:3136–3147. https://doi.org/10.1016/j.matcom.2009.03.005

Das S, Suganthan PN (2011) Differential evolution: a survey of the state-of-the-art. IEEE Trans Evol Comput 15:4–31. https://doi.org/10.1109/TEVC.2010.2059031

Dixit A, Kumar S, Pant M, Bansal R (2019) CA-DE: hybrid algorithm based on cultural algorithm and DE. In: Machine Intelligence and Signal Analysis. Springer, Berlin, pp 185–196

dos Santos Coelho L, Mariani VC (2006) An efficient particle swarm optimization approach based on cultural algorithm applied to mechanical design. In: 2006 IEEE international conference on evolutionary computation. IEEE, pp 1099–1104

Elbes M, Alzubi S, Kanan T et al (2019) A survey on particle swarm optimization with emphasis on engineering and network applications. Evol Intel 12:113–129. https://doi.org/10.1007/s12065-019-00210-z

Elsheikh AH, Abd Elaziz M (2019) Review on applications of particle swarm optimization in solar energy systems. Int J Environ Sci Technol 16:1159–1170. https://doi.org/10.1007/s13762-018-1970-x

Eusuff M, Lansey K, Pasha F (2006) Shuffled frog-leaping algorithm: a memetic meta-heuristic for discrete optimization. Eng Optim 38:129–154. https://doi.org/10.1080/03052150500384759

Gao H, Diao M (2011) Cultural firework algorithm and its application for digital filters design. Int J Model Ident Control 14:324. https://doi.org/10.1504/IJMIC.2011.043157

Gao H, Xu C (2013) Cultural quantum-inspired shuffled frog leaping algorithm for direction finding of non-circular signals. Int J Comput Sci Math 4:321. https://doi.org/10.1504/IJCSM.2013.058066

Gao F, Cui G, Liu H (2006a) Integration of genetic algorithm and cultural algorithms for constrained optimization. Springer, Berlin, pp 817–825

Gao F, Liu H, Zhao Q, Cui G (2006b) Hybrid model of genetic algorithm and cultural algorithms for optimization problem. Springer, Berlin, pp 441–448

Gao F, Zhao Q, Liu H, Cui G (2007) Cultural particle swarm algorithms for constrained multi-objective optimization. Springer, Berlin, pp 1021–1028

Gao H, Diao M (2010) Differential cultural algorithm for digital filters design. In: 2010 second international conference on computer modeling and simulation. IEEE, pp 459–463

Gao HY, Chi PF, Du YN, et al (2019) Digital watermarking based on wavelet transform and cultural invasive weed algorithm. Springer, Berlin, pp 1620–1627

Gao M, Zhu Y, Cao C, Zhu Y (2020) A hybrid cultural harmony search algorithm for constrained optimization problem of diesel blending. IEEE Access 8:6673–6690. https://doi.org/10.1109/ACCESS.2019.2963244

Geem ZW, Kim JH, Loganathan GV (2001) A new heuristic optimization algorithm: harmony search. Simul 76:60–68. https://doi.org/10.1177/003754970107600201

Hajihassani M, Jahed Armaghani D, Kalatehjari R (2018) Applications of particle swarm optimization in geotechnical engineering: a comprehensive review. Geotech Geol Eng 36:705–722. https://doi.org/10.1007/s10706-017-0356-z

Ho NB, Tay JC (2004) GENACE: an efficient cultural algorithm for solving the flexible job-shop problem. In: proceedings of the 2004 congress on evolutionary computation (IEEE Cat. No.04TH8753). IEEE, pp 1759–1766

Hosseinioun P, Kheirabadi M, Kamel Tabbakh SR, Ghaemi R (2020) A new energy-aware tasks scheduling approach in fog computing using hybrid meta-heuristic algorithm. J Parallel Distrib Comput 143:88–96. https://doi.org/10.1016/j.jpdc.2020.04.008

Jafari M, Salajegheh E, Salajegheh J (2019) An efficient hybrid of elephant herding optimization and cultural algorithm for optimal design of trusses. Eng Comput 35. https://doi.org/10.1007/s00366-018-0631-5

Jain NK, Nangia U, Jain J (2018) A Review of Particle Swarm Optimization. J Inst Eng (India) Ser B 99:407–411. https://doi.org/10.1007/s40031-018-0323-y

Kennedy J, Eberhart R (1995) Particle swarm optimization. In: Proceedings of ICNN'95—international conference on neural networks. IEEE, pp 1942–1948

Kennedy J, Eberhart RC (1997) A discrete binary version of the particle swarm algorithm. In: 1997 IEEE international conference on systems, man, and cybernetics. Computational cybernetics and simulation. IEEE, pp 4104–4108

Khan SU, Qureshi IM, Zaman F et al (2014) Correction of faulty sensors in phased array radars using symmetrical sensor failure technique and cultural algorithm with differential evolution. Sci World J 2014:1–10. https://doi.org/10.1155/2014/852539

Khan SU, Qureshi IM, Zaman F, Khan W (2017) Detecting faulty sensors in an array using symmetrical structure and cultural algorithm hybridized with differential evolution. Front Inf Technol Electron Eng 18:235–245. https://doi.org/10.1631/FITEE.1500315

Kim Y, Cho S-B (2009) A hybrid cultural algorithm with local search for traveling salesman problem. In: 2009 IEEE international symposium on computational intelligence in robotics and automation—(CIRA). IEEE, pp 188–192

Li N, Yi W, Gong G (2012) Multi-aircraft cooperative target allocation in BVR air combat using cultural-genetic algorithm. Springer, Berlin, pp 414–422

Lu Y, Zhou J, Qin H et al (2011) A hybrid multi-objective cultural algorithm for short-term environmental/economic hydrothermal scheduling. Energy Convers Manage 52:2121–2134. https://doi.org/10.1016/j.enconman.2010.12.003

Manjarres D, Landa-Torres I, Gil-Lopez S et al (2013) A survey on applications of the harmony search algorithm. Eng Appl Artif Intell 26:1818–1831. https://doi.org/10.1016/j.engappai.2013.05.008

Mao Z, Xiang Y, Zhang Y, Liu M (2020) A novel multi-objective cultural algorithm embedding five-element cycle optimization. In: 2020 IEEE Congress on Evolutionary Computation (CEC). IEEE, pp 1–10

Mavrovouniotis M, Li C, Yang S (2017) A survey of swarm intelligence for dynamic optimization: algorithms and applications. Swarm Evol Comput 33:1–17. https://doi.org/10.1016/j.swevo.2016.12.005

Mehrabian AR, Lucas C (2006) A novel numerical optimization algorithm inspired from weed colonization. Eco Inform 1:355–366. https://doi.org/10.1016/j.ecoinf.2006.07.003

Murugadass G, Sivakumar P (2020a) A hybrid elephant herding optimization and cultural algorithm for energy-balanced cluster head selection scheme to extend the lifetime in WSNs. Int J Commun Syst e4538. https://doi.org/10.1002/dac.4538

Nguyen TT, Yao X (2006) Hybridizing cultural algorithms and local search. Springer, Berlin, pp 586–594

Niu Q, Zeng T, Zhou Z (2011) A novel cultural algorithm based on differential evolution for hybrid flow shop scheduling problems with fuzzy processing time. Springer, Berlin, pp 121–132

Price K, Storn RM, Lampinen JA (2006) Differential evolution: a practical approach to global optimization. Springer Science & Business Media

Oliveira De Freitas CA, Limao de Oliveira RC, Azevedo Da Silva DJ et al (2018) Solution to economic—emission load dispatch by cultural algorithm combined with local search: case study. IEEE Access 6:64023–64040. https://doi.org/10.1109/ACCESS.2018.2877770

Qin AK, Huang VL, Suganthan PN (2009) Differential evolution algorithm with strategy adaptation for global numerical optimization. IEEE Trans Evol Comput 13:398–417. https://doi.org/10.1109/TEVC.2008.927706

Reynolds RG, Peng B (2005) Cultural algorithms: computational modeling of how cultures learn to solve problems: an engineering example. Cybern Syst 36:753–771. https://doi.org/10.1080/01969720500306147

Storn R, Price K (1997) Differential evolution—a simple and efficient heuristic for global optimization over continuous spaces. J Global Optim 11:341–359. https://doi.org/10.1023/A:1008202821328

Sun X-M, Lv X-Y, Duan X-M (2009) Novel QoS routing algorithm based on cultural-simulated annealing algorithm. In: 2009 second international conference on intelligent networks and intelligent systems. IEEE, pp 209–212

Tan Y, Zhu Y (2010) Fireworks algorithm for optimization. Springer, Berlin, pp 355–364

Tang W, Li Y (2008) Constrained optimization using triple spaces cultured genetic algorithm. In: 2008 Fourth international conference on natural computation. IEEE, pp 589–593

Wang X, Gao X-Z, Zenger K (2015b) The overview of harmony search. Springer, Berlin, pp 5–11

Wei Z, Yan-Ping B (2012) Cultural algorithm based on particle swarm optimization for partner selection of virtual enterprise. In: Proceedings of the 31st Chinese control conference, pp 2238–2241

Wu Y, Gao X-Z, Huang X-L, Zenger K (2010) A hybrid optimization method of particle swarm optimization and cultural algorithm. In: 2010 sixth international conference on natural computation. IEEE, pp 2515–2519

Xie W, Ji C, Li X (2009) Particle swarm optimization based on cultural algorithm for short-term optimal operation of cascade hydropower stations. In: 2009 fifth international conference on natural computation. IEEE, pp 289–293

Xue Z, Guo Y (2007) Improved Cultural algorithm based on genetic algorithm. In: 2007 IEEE international conference on integration technology. IEEE, pp 117–122

Yang Y, Gu X (2014) Cultural-based genetic Tabu algorithm for multiobjective job shop scheduling. Math Probl Eng 2014:1–14. https://doi.org/10.1155/2014/230719

Ye C, Mao Z, Liu M (2019) A novel multi-objective five-elements cycle optimization algorithm. Algorithms 12:244. https://doi.org/10.3390/a12110244

Yuan X, Nie H, He L, et al (2008) A cultural algorithm for scheduling of hydro producer in the power market. In: 2008 second international conference on genetic and evolutionary computing. IEEE, pp 364–367

Zhang Y, Zhu X (2012) A cultural particle swarm optimization algorithm for flow shop scheduling problem. Springer, Berlin, pp 1575–1583

Chapter 7
Extended Cultural Algorithms

Abstract Depending on the type of the optimisation problem and the features of the search space, the efficiency of CAs could be sensitive to the update and influence functions as well as the type of knowledge sources. The success of CAs remarkably depends on the information exchange between the belief and the population spaces. In some cases, the information provided by the belief space is not productive enough to guide the individuals to efficiently explore the search space. For example, the excessively narrow normative intervals can reduce the step size in the influence functions and cause the algorithm to trap into local optimums. This chapter discusses the extended versions of CAs, in which new influence, acceptance, and update functions have been devised.

Keywords Influence function · Acceptance function · Communication topology · Fuzzy · Cultural algorithms

7.1 Introduction

Like other meta-heuristic algorithms, CAs do not guarantee the achievement of the global optimum solutions for all kinds of problems. The CAs with the standard operators described in Chap. 3 may exhibit poor performance in given types of problems. To improve the convergence properties, researchers have been focused on the improvement of different components of the standard CAs. This chapter discusses the recent extended versions of CAs in which the original operators have been modified to improve the performance of standard CAs in dealing with different problem types.

The overall structure of this chapter is as follows. In Sect. 7.2, the new influence functions developed in the literature are discussed. Section 7.3 presents the dynamic acceptance functions for CAs in which the number of accepted individuals changes over time. The extended domain knowledge is introduced in Sect. 7.4. The development of communication topologies, as well as the social fabric concept, are investigated in Sect. 7.5. Section 7.6 presents an extended version of CAs for multi-objective problems. The application of fuzzy theory for the development of

S. Jalili, *Cultural Algorithms*, Engineering Optimization: Methods and Applications,
https://doi.org/10.1007/978-981-19-4633-2_7

new operators for CAs is discussed in Sect. 7.7. In Sect. 7.8, the application of chaos theory in the development of chaotic CAs is investigated. Finally, a summary of the contents presented in this chapter is provided in Sect. 7.9.

7.2 New Influence Functions

The influence functions play a pivotal role in the information diffusion process during the overall evolution of individuals. They use the cultural information stored in the belief space to form a new generation of individuals. The influence functions need to be designed in an intelligent way to provide an efficient information diffusion mechanism within the population space.

The standard influence functions discussed in Sect. 3.5.2 of Chap. 3 may not exhibit satisfactory performance for some classes of problems. In some cases, they may fail to efficiently spread the cultural information which can lead to unwanted premature convergence. For example, the influence functions that use the size of normative intervals, as can be seen in Eqs. (3.8), (3.10), and (3.11) in Sect. 3.5.2 of Chap. 3, can easily fail to generate productive offspring. Despite the conservative approach of standard CAs in keeping the normative intervals wide enough, excessively narrow normative intervals may be observed during the evolutionary process. The excessively narrow normative intervals can potentially lead to a new generation of offspring without no significant different features compared to their parents. This situation is referred to as "premature convergence" in optimisation literature. To remedy this issue, researchers have been tried to design new influence functions with better information diffusion capabilities. This section focuses on these studies.

7.2.1 *Modified Influence Function*

Ali and Reynolds (2014) proposed a modified influence function based on the normative knowledge component. The main idea of their study is to randomly generate the offspring solution within the normative interval if the parent solution is outside the normative bounds. On the other hand, if the parent solution is already within the normative interval, the offspring solution will be obtained by considering the interval size and the fitness function values in the step size. This influence function can be expressed as follows:

$$x_{i,j}^{t+1} = \begin{cases} x_{i,j}^t + \text{rand}\left(x_{\max,j}^t - x_{\min,j}^t\right)\frac{f(X_i^t)}{\sum_{k=1}^{N_1} f(X_k^t)}, & \text{if } x_{\min,j}^t \le x_{i,j}^t \le x_{\max,j}^t \\ x_{\min,j}^t + \text{rand}\left(x_{\max,j}^t - x_{\min,j}^t\right), & \text{otherwise} \end{cases} \tag{7.1}$$

where N_1 is the number of individuals within the population space.

7.2.2 Gravity-Based Influence Function

Wang et al. (2017) developed a new influence function based on the idea of the centre of gravity of individuals in the population space. The study employed the idea of the centre of gravity from the Quantum Inspired PSO (IQPSO) algorithm. The new influence function uses the normative and situational knowledge components that can be written as:

$$x_{i,j}^{t+1} = \begin{cases} \frac{\left(r_1\, x_{i,j}^t + r_2\left(G_j^t - x_{i,j}^t\right)\right)}{r_1 + r_2}, & \text{if } f\left(X_i^t\right) < L_j^t \\ \frac{\left(r_1\, x_{i,j}^t + r_2\left(s_j^t - x_{i,j}^t\right)\right)}{r_1 + r_2}, & \text{if } L_j^t \le f\left(X_i^t\right) \le U_j^t \\ r_3\left(u_j - l_j\right), & \text{otherwise} \end{cases} \tag{7.2}$$

where r_1, r_2, and r_3 are three independent random numbers uniformly distributed between 0 and 1, L_j^t and U_j^t are the lower and upper normative objective function bounds for the jth variable at iteration t, respectively, u_j and l_j are the upper and lower bounds in the jth dimension of the problem, and G_j^t is the centre of gravity coefficient of individuals in the jth dimension of the problem, which is calculated as follows:

$$G_j^t = \frac{1}{N_I} \sum_{i=1}^{N_I} x_{i,j}^t, \quad j = 1, 2, \ldots, n \tag{7.3}$$

where n is the number of dimensions of the problem.

7.2.3 The Truncated Geometric Distribution-Based Influence Function

Jalili et al. (2019) proposed a new influence function based on the truncated geometric distribution to improve the performance of the standard CA, in which the normative and situational knowledge components are used. The influence function tries not to change all dimensions of the parent individual X_i^t. To this end, it creates a trial vector U_i^{t+1} for the ith individual as follows:

$$U_{i,j}^{t+1} = \begin{cases} x_{\min,j}^t + \text{rand}\left(x_{\min,j}^t - x_{i,j}^t\right) + \text{rand}\left(s_j^t - x_{i,j}^t\right) & \text{if } x_{i,j}^t < x_{\min,j}^t \\ x_{\max,j}^t + \text{rand}\left(x_{\max,j}^t - x_{i,j}^t\right) + \text{rand}\left(s_j^t - x_{i,j}^t\right) & \text{if } x_{i,j}^t > x_{\max,j}^t \\ s_j^t + \text{rand}\left(x_{i,j}^t - s_j^t\right) & \text{otherwise} \end{cases} \tag{7.4}$$

where $U_{i,j}^{t+1}$ indicates the jth variable of the ith trial vector, rand is a random number between 0 and 1, s_j^t represents the jth variable of the situation knowledge at iteration

t, and other parameters were introduced in Chap. 3. After calculation of the trial vector, the influence function uses a truncated geometric distribution to calculate the number of dimension changes in the parent individual X_i^t as follows:

$$q^t = \left\lceil \frac{\ln\left(1 - \left(1 - \left(1 - p_c^t\right)^n\right)\text{rand}\right)}{\ln\left(1 - p_c^t\right)} \right\rceil + q_0 - 1; q^t \in [1, 2, \ldots, n] \tag{7.5}$$

where q^t represents the number of changes at iteration t, p_c^t is the control parameter at iteration t $\left(p_c^t < 1, p_c^t \neq 0\right)$, and q_0 is the minimum number of changes. The greater values for the control parameter p_c^t will result in a smaller number of changes in parent individuals X_i^t, and vice versa. In the beginning of the solution-finding process, the diversity of the population is high and fewer changes can make significant changes in the fitness values. However, the diversity of the population is reduced in the final stages, which means larger changes will be required to exploit the area within the global optima region. To this end, the influence function uses the following formula for updating the control parameter p_c^t in each iteration:

$$p_c^t = 1 - 2\frac{t}{t_{\max}}, t = 1, 2, \ldots, t_{\max} \tag{7.6}$$

where t is the iteration number and $t_{\max}$ is the maximum number of iterations. After calculating the number of changes q^t, the q^t number of variables are randomly selected from the trial vector U_i^{t+1} to build the new individual as follows:

$$x_{i,j}^{t+1} = \begin{cases} U_{i,j}^{t+1} & \text{if } j \in \left[r_1, r_2, \ldots, r_{q^t}\right] \\ x_{i,j}^t & \text{otherwise} \end{cases} \tag{7.7}$$

where $r_1, r_2, \ldots, r_{q^t}$ are q^t number of random integers between 1 and n. The authors reported that the new influence function performs better than the original influence functions of the standard CAs.

7.2.4 *Cuckoo Search-Based Influence*

Xue (2020) developed an adaptive CA-based Cuckoo Search (CACS) algorithm for the time-dependent vehicle routing problem with stochastic customers. In CACS, the following equation is employed as an influence function within the CA (Xue 2020):

$$X_i^{t+1} = X_i^t + \sigma \oplus \text{levy}(\beta) \tag{7.8}$$

where X_i^{t+1} represents the new position of the individual i at iteration $t + 1$, X_i^t is the previous position of individual i at iteration t, σ is a vector that controls the search direction, $\oplus$ is the entry-wise multiplication operator, levy(.) is the levy flight function that controls the step size, and β is the index coefficient. In Eq. (7.8), the step size σ is calculated as follows:

$$\sigma = \sigma_0\left(X_j^t - X_i^t\right) \tag{7.9}$$

where σ_0 is a constant parameter and X_j^t is a randomly selected individual from the population space. The fly flight function is also calculated by the following equation to control the step size of the search:

$$\text{levy}(\beta) = t^{-1-\beta}, 0 < \beta \leq 2 \tag{7.10}$$

7.3 Dynamic Acceptance

The acceptance function in CAs has a direct influence on the quality of cultural information stored within the belief space. As was discussed in Sect. 3.5.1 of Chap. 3, the standard CAs use a given percentage of individuals with better qualities to adjust the belief space. However, it might not be a good idea to use a constant number of individuals to update the belief space in certain classes of problems. The quality of cultural information stored in belief space is directly related to the number as well as features of selected individuals. Therefore, alternative dynamic acceptance functions may be needed for some application case studies. In this section, two similarity-based and fitness-based dynamic acceptance functions proposed in the literature are described.

7.3.1 *Similarity-Based Acceptance Function*

Wang et al. (2015a) employed a dynamic acceptance function in CA, in which the similarities between the individuals in the population space are considered during the acceptance process. The authors suggested evaluating the similarity between the individuals based on the Euclidean distance as follows:

$$\left\|X_i - X_j\right\| = \sqrt{\sum_{k=1}^{n}\left(x_{i,k} - x_{j,k}\right)^2} \tag{7.11}$$

where $x_{i,k}$ and $x_{j,k}$ are the kth variable of the individuals i and j, respectively, and X_i and X_j are the positions of two different individuals in the population space (i.e., $i \neq j$). The authors defined a Euclidean distance parameter L to decide whether any pair of individuals are similar or not. When $\left\|X_i - X_j\right\| < L$ $\left(\left\|X_i - X_j\right\| > L\right)$, it means that the similarities between the individuals are high (low) and the number of accepted individuals needs to be reduced (increased). Based on these assumptions, the number of accepted individuals is dynamically calculated by the following equation:

$$n_{\text{accept}} = \begin{cases} N_{\text{I}}\beta\left(1 - \dfrac{1}{c_{\text{f}}}\right) & \text{if } \sum \|X_i - X_j\| < L \\ N_{\text{I}}\beta + 2 & \text{otherwise} \end{cases} \tag{7.12}$$

where c_{f} is crowding factor and β is the acceptance rate.

7.3.2 Fitness-Based Acceptance Function

The acceptance functions can be designed by considering the variations in the fitness values of individuals. Wang et al. (2017) proposed the following acceptance function based on the average and worst objective function values of individuals in the population space:

$$n_{\text{accept}} = \frac{\overline{f}^t - f_{\text{w}}^t}{\sum_{i=1}^{N_{\text{I}}} \left| f(X_i^t) - \overline{f}^t \right|} N_{\text{I}} \tag{7.13}$$

where $\overline{f}^t$ represents the average objective function values of the individuals in the population space at iteration t, f_{w}^t is the worst objective function value at iteration t in the population space, and N_{I} is the number of individuals in the population. Based on Eq. (7.13), the number of accepted individuals will be larger (lower) when there are significant (minor) differences between the fitness function values of individuals. Hence, it can be stated that the approach works based on the diversity of the population in a sense.

7.4 Extended Domain Knowledge

As was discussed in Sect. 3.4.5 of Chap. 3, the domain knowledge component stores the individuals with the best fitness function values obtained over different generations to form an archive memory to guide the evolution process. Due to its growing size over time, the update process of the domain knowledge component can be computationally expensive for large-scale problems. This highlights the fact that the computational framework of domain knowledge needs to be improved to facilitate an easy update process of belief space in large-scale problems.

Coello Coello and Becerra (2002) joined this line of research by proposing a 2^n-trees model to partition the search space into different cells and obtain the feasibility of each cell within the domain knowledge component of CA. The study tried to reduce the computational cost of information stored in the domain knowledge of the belief space by using 2^n-trees model. However, the efficiency of their approach is affected by the number of decision variables and the computational complexity can be increased in high-dimensional problems.

7.5 Communication Topologies for CAs

One of the important factors that can affect the convergence properties of CAs is the efficiency of the knowledge distribution between the individuals within the population space. The population structures with neighbourhood or topological properties can significantly enhance the diversity preservation capacity of the population-based meta-heuristics (Lynn et al. 2018). From this perspective, the knowledge propagation process in CAs can be controlled by devising adequate communication topologies between the individuals in the population space. Researchers have been tried to improve the quality of knowledge diffusion between individuals in CAs based on the social fabric concept, in which different population structures with neighbourhood or topological properties were employed as shown in Fig. 7.1 (Reynolds and Kinnaird-Heether 2003, 2017a, b; Reynolds et al. 2008a; Reynolds and Ali 2008a, c,; Ali and Reynolds 2009; Ali et al. 2011a, b, 2012; Reynolds and Gawasmeh 2012). This section describes the social fabric concept in CAs, in which the connections between social groups are modelled.

7.5.1 *Social Fabric*

Many real-world systems can be described as complex networks with numerous highly interconnected components, such as the Internet, communication, transportation, electric supply, and water supply systems. Due to the dynamic behaviour of different components in complex networks, the properties of a given single component do not reflect the overall properties of the network (Estrada 2012). G*raph theory* is a branch of discrete mathematics that can be applied to analyse the emerged complex networks in different branches of science and engineering.

Social networks between individuals in human societies are also examples of such complex networks. A real human society consists of different forms of social networks with numerous individuals who are interacting with each other based on complex interrelation behaviours. To simulate these interactions in the social networks within CA, a *social fabric* concept was developed by Reynolds et al. According to the definition stated by Reynolds et al. (2008a):

> *"The social fabric can be viewed metaphorically as a living skin created out of the engineered emergence of agents. The tissue fibers are formed by the connectivity of each agent and the generator that determines the topology type of interaction and controls the newly created kinetic sculpture."*

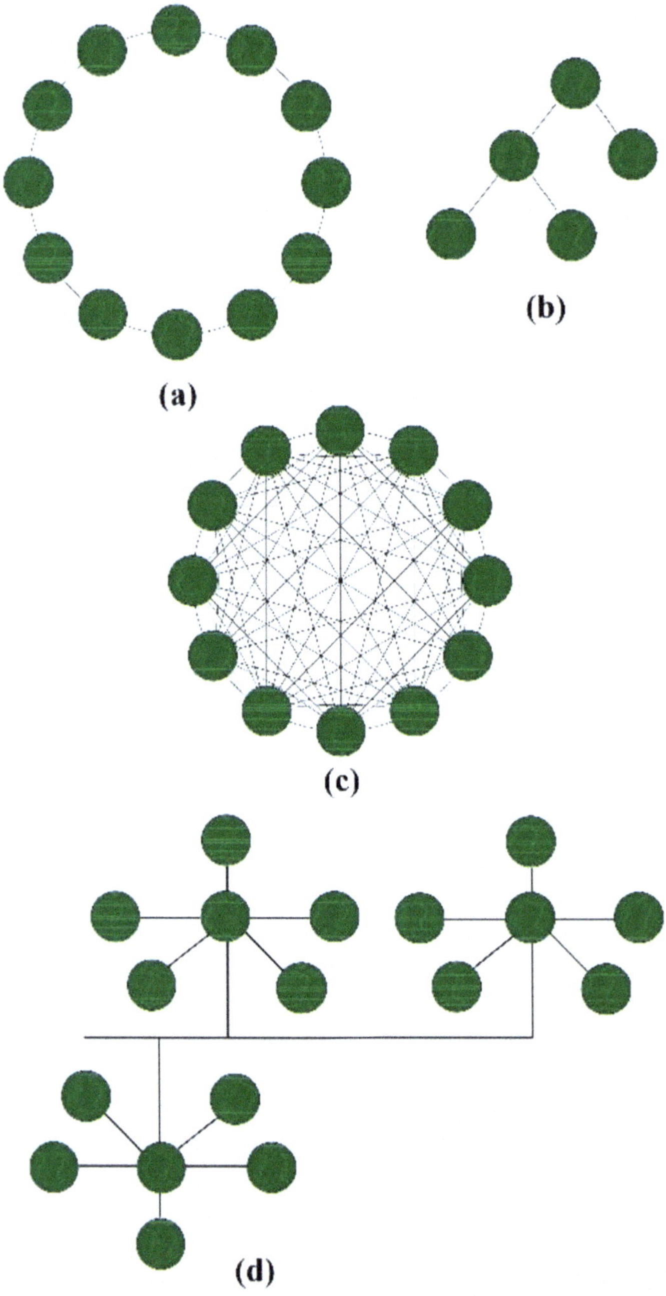

Fig. 7.1 Different communication topologies defined within the CA in literature **a** ring topology (lBest), **b** tree topology, **c** global topology (gBest), and **d** hybrid tree topology (Ali et al. 2012)

Theoretically, the cultural system of a real society consists of different network layers which can dynamically change over time. Social networks can emerge in various forms, such as political, kinship, religious, and scientific networks. The topologies of these networks can be static or dynamic which are controlled by the knowledge distribution and the type of interactions between the individuals. In the standard version of CAs, cultural knowledge is used to evolve the population space without considering the position or features of individuals. The main goal of the social fabric concept is to control the knowledge flow between the individuals in the population space.

Figure 7.2 illustrates the social fabric model developed by Reynolds et al. (2008a) for CAs. In the social fabric model, the individuals are socially interacted with each other based on a given topology, which represents the social network within the population space. The employed topology can be any topology type applied to the meta-heuristic algorithms in the literature.

At each iteration, the influence function of CA will apply a given knowledge component to each individual. Then, each individual will share the type of influenced knowledge component with other individuals based on the social network topology. Other individuals will score the knowledge components of belief space based on the information received from others in the communication network. This

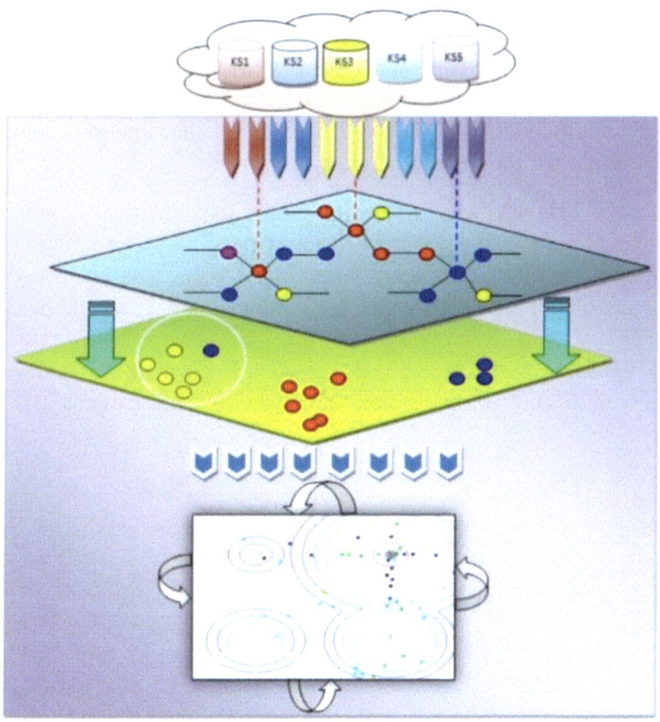

Fig. 7.2 Social fabric model in CAs (Ali et al. 2011a)

information-sharing mechanism will provide the possibility for each individual in the population space to choose the knowledge component with the highest score. It is worth mentioning that if the scores of all knowledge components are the same, a set of other scenarios can be modelled, such as the application of more frequently applied or rarely used knowledge components.

There are a series of studies in the literature on the impacts of social fabric on the performance of CAs (Reynolds and Kinnaird-Heether 2003, 2017a, b; Reynolds et al. 2008a; Reynolds and Ali 2008a, c; Ali and Reynolds 2009; Ali et al. 2011a, b, 2012; Reynolds and Gawasmeh 2012). The results revealed that embedding social fabric into CAs can significantly improve their performance.

7.6 Multi-objective CAs

Standard CAs presented in Chap. 3 have been originally developed for single-objective optimisation problems. Literature reveals that there have not been significant efforts to extend the applicability of CAs for multi-objective problems. In this section, an extended version of CAs for multi-objective problems is discussed.

Zhang et al. (2013a) proposed an Enhanced Multi-Objective CA (EMOCA) to solve the environmental/economic dispatch problem. In EMOCA, the population space is evolved by the PSO algorithm and the information stored in the situational knowledge component is used to obtain the Pareto front solutions. Let us assume the situational knowledge $\mathcal{S}^t$ has the capacity of storing N_{sc} non-dominated individuals. For each accepted individual X_i^t, $i = 1, 2, \ldots, n_{accept}$, the situational knowledge is updated by considering the following rules:

- if the situational knowledge component is empty, the individual X_i^t will be stored.
- if the situational knowledge contains a set of previously stored individuals and X_i^t is not dominated by any already stored individual, X_i^t will be added to $\mathcal{S}^t$. In addition, any other individuals which are dominated by X_i^t should be eliminated from the $\mathcal{S}^t$.
- To keep the size of the situational knowledge constant, the non-dominated solutions with less crowding distance values should be discarded.

To build a well-distributed Pareto front, the study employed a crowding distance criterion, originally proposed by Deb et al. (2002). Let us consider a set of n_s non-dominated solutions in the objective space represented by $\mathcal{I} = \left[X_1^t, X_2^t, \ldots, X_{n_s}^t\right]$, which are sorted based on the mth objective function value. In this criterion, for each individual, two nearest individuals in the objective space are selected and the crowding distance is calculated as follows (Deb et al. 2002):

$$\Gamma_i^{\mathrm{D}} = \sum_{j=1}^{m} \frac{f_m\left(X_{i+1}^t\right) - f_m\left(X_{i-1}^t\right)}{f_m^{\max} - f_m^{\min}}, i = 2, 3, \ldots, n_s - 1 \quad (7.14)$$

In this equation, Γ_i^{D} represents the crowding distance for the ith particle, $f_m\left(X_{i+1}^t\right)$ and $f_m\left(X_{i-1}^t\right)$ are the mth objective function values for the particles $i + 1$ and $i - 1$ at iteration t, respectively, $f_m^{\max}$ and $f_m^{\min}$ are the maximum and minimum values of the mth objective function in the objective space. Algorithm 7.1 shows the pseudo-code of the crowding distance calculations.

Algorithm 7.2 shows the overall pseudo-code of EMOCA. To keep the code as simple as possible, the equations of the PSO algorithm were not mentioned in Algorithm 7.2. However, it should be noted that the study used standard PSO formulations. EMOCA uses the crowding distance criterion to keep the diversity of solutions within the situational knowledge at a high level that can be useful to obtain well-distributed Pareto fronts.

Algorithm 7.1 Crowding distance calculation for a set of non-dominated $\mathcal{I}$ (Deb et al. 2002)

```
Set the crowding distance for all individuals equal to zero, i.e., Γ_i^D = 0, i = 1, 2, ..., n_s
for m = 1: N_O do
    Sort the non-dominated solution set I with respect to mth objective function values;
    Set the crowding distance values of the boundary points as follows:
    Γ_1^D = Γ_{n_s}^D = ∞;
    for i = 2: n_s − 1 do
        Γ_i^D = (f_m(X_{i+1}^t) − f_m(X_{i−1}^t)) / (f_m^max − f_m^min);
    end
end
```

Note N_O =Number of the objective functions in the problem.

Algorithm 7.2 Pseudo-code of the EMOCA

```
Initialise the population space P^0 with N_I individuals;
Initialise the belief space B^0;
Evaluate the fitness function of individuals;
Set t = 0;
while stopping criteria are not met do
    for i = 1: N_I do
        Generate new trial individuals U_i^{t+1} using PSO operators;
        if X_i^t and U_i^{t+1} are both feasible and U_i^{t+1} dominates X_i^t then
            Add the solution U_i^{t+1} to the population space P^{t+1};
        elseif X_i^t and U_i^{t+1} are both feasible and X_i^t dominates U_i^{t+1}, then
            Add the solution X_i^t to the population space P^{t+1};
        elseif X_i^t and U_i^{t+1} are both infeasible then
            Add the solution vector with a lower constraint violation to the population P^{t+1};
        elseif X_i^t and U_i^{t+1} are both feasible and U_i^{t+1} dominates X_i^t then
            Add the solution vector U_i^{t+1} to the population P^{t+1};
        elseif X_i^t and U_i^{t+1} are both feasible and X_i^t dominates U_i^{t+1} then
            Add the vector X_i^t to the population space P^{t+1} and discard U_i^{t+1};
        elseif X_i^t and U_i^{t+1} are non-dominated with each other then
            Add both X_i^t and U_i^{t+1} to the population space P^{t+1};
        end
    end
    If the size of the population is greater than N_I, sort the individuals and select the individuals
      with longer crowding distance using Algorithm 7.1;
    Update the belief space B^t;
    Set t = t + 1;
end
```

7.7 Fuzzy CAs

In recent years, the different components of CAs have been redefined based on the fuzzy sets theory to accelerate their convergence speed. In this section, the application of fuzzy sets theory in CAs is discussed and the different fuzzy-based operators, such as fuzzy acceptance, update, and influence functions, are presented. For the basics of fuzzy set theory, the readers are referred to Dubois (1980) and Bede (2013).

7.7.1 Fuzzy Acceptance

Reynolds and Zhu (2001) and Ali et al. (2018) suggested a fuzzy acceptance function based on a fuzzy similarity matrix, which is used to perform the fuzzy classification of the population space. The studies suggested dividing the individuals into two categories, the elite and inferior individuals. Algorithm 7.3 presents the pseudo-code of the fuzzy acceptance function proposed by Reynolds and Zhu (2001) and Ali et al. (2018). In this subsection, the fuzzy acceptance function is explained in detail.

The fuzzy similarity matrix M determines the similarities between each pair of individuals in the population space. The similarity M is a symmetric matrix with the size of $(N_\mathrm{I} \times N_\mathrm{I})$, where N_I is the number of individuals in the population space. This matrix can be written as follows:

$$M = \begin{bmatrix} m_{1,1} & m_{1,2} & \dots & m_{1,N_\mathrm{I}} \\ m_{2,1} & m_{2,2} & \dots & m_{2,N_\mathrm{I}} \\ \vdots & \vdots & \vdots & \vdots \\ m_{N_\mathrm{I},1} & m_{N_\mathrm{I},2} & \dots & m_{N_\mathrm{I},N_\mathrm{I}} \end{bmatrix}, M \in \mathbb{C}^{N_\mathrm{I} \times N_\mathrm{I}} \tag{7.15}$$

in which each element of the similarity matrix $m_{i,j}$ is calculated as follows:

$$m_{i,j} = 1 - \frac{\left| f(X_i^t) - f\left(X_j^t\right) \right|}{\sum_{k=1}^{N_\mathrm{I}} f(X_k^t)}; \tag{7.16}$$

Algorithm 7.3 Pseudo-code of the fuzzy acceptance function proposed by Reynolds and Zhu (2001) and Ali et al. (2018)

```
Set ε = 0.4t_max;
Set φ = 2.79;
Set μ_β = 0.3;
Set σ_β = 0.1;
if t < ε then
    β^t = 1 − φ(t−ε)/t_max;
else
    β^t = N(μ_β, σ_β);
end
for i = 1: N_I do
    for j = 1: N_I do
        m_{i,j} = 1 − |f(x_i^t) − f(x_j^t)| / Σ_{k=1}^{N_I} f(x_k^t);
        if m_{i,j} > β^t then
            β^t_{m_{i,j}} = 1;
        else
            β^t_{m_{i,j}} = 0;
        end
    end
end
% Build the matrix the β-Cuts matrix:
M_β = [β_{m_{1,1}} β_{m_{1,2}} ... β_{m_{1,N_I}};
       β_{m_{2,1}} β_{m_{2,2}} ... β_{m_{2,N_I}};
       ⋮ ⋮ ⋮ ⋮;
       β_{m_{N_I,1}} β_{m_{N_I,2}} ... β_{m_{N_I,N_I}}];
Calculate the number of accepted individuals γ using M_β
```

To facilitate the fuzzy classification process and reduce the computational time, Reynolds and Zhu (2001) and Ali et al. (2018) employed the β-Cuts matrix M_β to classify the individuals, which is a Boolean matrix defined as:

$$M_\beta = \begin{bmatrix} \beta_{m_{1,1}} & \beta_{m_{1,2}} & \cdots & \beta_{m_{1,N_I}} \\ \beta_{m_{2,1}} & \beta_{m_{2,2}} & \cdots & \beta_{m_{2,N_I}} \\ \vdots & \vdots & \vdots & \vdots \\ \beta_{m_{N_I,1}} & \beta_{m_{N_I,2}} & \cdots & \beta_{m_{N_I,N_I}} \end{bmatrix}, \ M_\beta \in \mathbb{C}^{N_I \times N_I} \tag{7.17}$$

in which:

$$\beta^t_{m_{i,j}} = \begin{cases} 1 & \text{if } m_{i,j} > \beta^t \\ 0 & \text{if } m_{i,j} \le \beta^t \end{cases} \tag{7.18}$$

where $\beta^t \in [0, 1]$ is a refinement level at iteration t. The unit value of an arbitrary entry $m_{i,j}$ of the matrix the β-Cuts matrix M_β shows that the individuals X_i^t and X_j^t are similar from the fitness viewpoint, and they can be potential individuals for adjusting the belief space. By using the matrix M_β, the number of individuals to be accepted, represented by γ, is obtained. The number of accepted individuals γ is the number of unit entries in the first row of the matrix M_β, i.e., unit values in the vector $m_{1,j}$.

Ali et al. (2018) suggested to assume β is equal to 1. The authors also used the following formula to control and reduce the refinement level β^t during the solution-finding process:

$$\beta^t = \begin{cases} N(\mu_\beta, \sigma_\beta) & \text{if } t > \varepsilon \\ 1 - \dfrac{\varphi(t - \varepsilon)}{t_{\max}} & \text{if } 0 \le t \le \varepsilon \end{cases} \tag{7.19}$$

where $N(\mu_\beta, \sigma_\beta)$ is the number obtained from a normal distribution with the mean value of μ_β and standard deviation of σ_β, φ and ε are the constant values, and t is the iteration number. The study suggests setting the mentioned parameters as follows: $\mu_\beta = 0.3$, $\sigma_\beta = 0.1$, $\varphi = 2.79$, and $\varepsilon = 0.4t_{\max}$.

7.7.2 *Fuzzy Influence*

Reynolds and Zhu (2001) and Ali et al. (2018) proposed a new fuzzy influence function based on the upper-bound fuzzy function as follows:

$$x_{i,j}^{t+1} = \begin{cases} x_{i,j}^t + \lambda(x_{i,j}^t)\left|\text{size}(I_j^t).N(0,1)\right| & \text{if } x_{i,j}^t < x_{\min,j}^t \\ x_{i,j}^t - \lambda(x_{i,j}^t)\left|\text{size}(I_j^t).N(0,1)\right| & \text{if } x_{i,j}^t > x_{\max,j}^t \\ x_{i,j}^t + \lambda(x_{i,j}^t)\text{size}(I_j^t).N(0,1) & \text{otherwise} \end{cases} \tag{7.20}$$

in which:

$$\lambda(x_{i,j}^t) = \begin{cases} 1 - t_{\max}\left(x_{i,j}^t\right)^p, & \text{if } x_{i,j}^t < \frac{1}{\sqrt{t_{\max}}} \\ 0, & \text{if } x_{i,j}^t \ge \frac{1}{\sqrt{t_{\max}}} \end{cases} \tag{7.21}$$

where $\lambda(t)$ represents the upper-bound fuzzy function at iteration t, p is a constant parameter, and $t_{\max}$ is the maximum number of iterations. The mentioned studies considered the value of parameter p as equal to 0.1.

7.7.3 Fuzzy Update

Reynolds and Zhu (2001) and Ali et al. (2018) presented a fuzzy function to update the knowledge components of the belief space in CAs. Their proposed approach considers the closed interval $\left[0.9c^t_j, 1.10c^t_j\right]$ as the kernel of the normative knowledge in the jth dimension of the problem as shown in Fig. 7.3, where $c^t_j = \frac{x^t_{\max,j} - x^t_{\min,j}}{2}$ is the centre of the normative knowledge in the jth dimension at iteration t and $j = 1, 2, \ldots, n$. This kernel c^t_j depends on the normative bounds, and it changes whenever the belief space is updated.

In the fuzzy update function proposed by Reynolds and Zhu (2001) and Ali et al. (2018), the normative knowledge bounds are updated as follows:

$$x^{t+1}_{\min,j} = \begin{cases} x^t_{\min,j} + L\left(x^t_{l,j}\right)\left[x^t_{l,j} - x^t_{\min,j}\right] & \text{if } x^t_{\min,j} \le x^t_{l,j} < 0.9\mathrm{c}^t_j \\ x^t_{\min,j} & \text{if } 0.9\mathrm{c}^t_j \le x^t_{l,j} < 1.10\mathrm{c}^t_j \end{cases} \tag{7.22}$$

$$x^{t+1}_{\max,j} = \begin{cases} x^t_{\max,j} + R\left(x^t_{l,j}\right)\left[x^t_{l,j} - x^t_{\max,j}\right] & \text{if } 1.10\mathrm{c}^t_j < x^t_{l,j} \le x^t_{\max,j} \\ x^t_{\max,j} & \text{if } 0.9\mathrm{c}^t_j \le x^t_{l,j} < 1.10\mathrm{c}^t_j \end{cases} \tag{7.23}$$

in which the $L(x)$ and $R(.)$ are the fuzzy functions defined as:

$$L\left(x^t_{l,j}\right) = R\left(x^t_{l,j}\right) = \frac{1}{\left(1 + \left(x^t_{l,j}\right)^p\right)}, \quad p \ge 0 \tag{7.24}$$

in the above equations, $x^t_{l,j}$ represents the jth variable of the lth accepted individual with $l = 1, 2, \ldots, n_{\text{accept}}$, and c^t_j indicates the centre of normative interval in the jth dimension.

Algorithm 7.4 presents the pseudo-code of the fuzzy update function. As it can be seen from the fuzzy update process, the strategy is to move the normative lower bound $x^t_{\min,j}$ in the jth dimension from the left toward the kernel, while the normative upper bound $x^t_{\max,j}$ will move toward the kernel from the right. The main goal of this strategy is to keep the normative bounds within the closed interval $\left[0.9c^t_j, 1.10c^t_j\right]$.

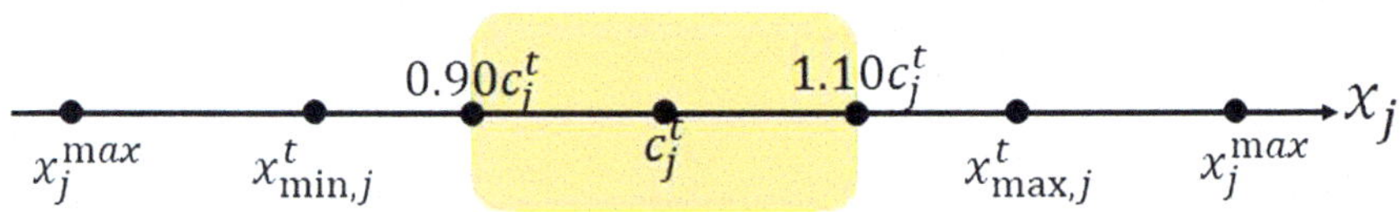

Fig. 7.3 Kernel of the jth normative interval (Ali et al. 2018)

The value of the centre of normative interval in the jth dimension c^t_j should be updated in each generation.

Algorithm 7.4 Pseudo-code of the fuzzy update function proposed by Reynolds and Zhu (2001) and Ali et al. (2018)

Set p;
Calculate $c_j^t = \frac{x^t_{\max,j} - x^t_{\min,j}}{2}$;
for $l = 1{:}\, n_{\text{accept}}$ **do**
  for $j = 1{:}\, n$ **do**
    $L(x^t_{l,j}) = \frac{1}{\left(1+\left(x^t_{l,j}\right)^p\right)}$;
    $R(x^t_{l,j}) = \frac{1}{\left(1+\left(x^t_{l,j}\right)^p\right)}$;
    if $x^t_{l,j} \geq x^t_{\min,j}$ **and** $x^t_{l,j} < 0.90c^t_j$ **then**
      $x^{t+1}_{\min,j} = x^t_{\min,j} + L(x^t_{l,j})[x^t_{l,j} - x^t_{\min,j}]$;
    elseif $x^t_{l,j} \geq 0.90c^t_j$ **and** $x^t_{l,j} < 1.10c^t_j$
      $x^{t+1}_{\min,j} = x^t_{\min,j}$;
    end
    if $x^t_{l,j} \leq x^t_{\max,j}$ **and** $x^t_{l,j} > 1.10c^t_j$ **then**
      $x^{t+1}_{\max,j} = x^t_{\max,j} + R(x^t_{l,j})[x^t_{l,j} - x^t_{\max,j}]$;
    elseif $x^t_{l,j} \geq 0.90c^t_j$ **and** $x^t_{l,j} < 1.10c^t_j$ **then**
      $x^{t+1}_{\max,j} = x^t_{\max,j}$;
    end
    Update the centre of the normative interval as follows: $c_j^{t+1} = \frac{x^{t+1}_{\max,j} - x^{t+1}_{\min,j}}{2}$;
  end
end

7.7.4 *Fuzzy Clustering*

In data analysis, data clustering is referred to the categorisation of a data set into a given number of clusters. Fuzzy clustering is one of the modern data clustering approaches, in which a given data point is categorised into more than one cluster with different membership values based on the fuzzy theory (Kesemen et al. 2016). The fuzzy clustering techniques can be used to create the sub-populations within the population space of CAs. This subsection is focused on the application of the fuzzy c-mean clustering technique in the categorisation of the individuals into different clusters (or sub-populations) in CAs. Hence, the fuzzy c-mean clustering technique is briefly introduced first. Then, the application of this algorithm for clustering the individuals within the CA will be discussed.

7.7.4.1 Fuzzy c-Mean Algorithm

One of the popular fuzzy clustering approaches is the fuzzy c-mean algorithm. Let us assume a set of n-dimensional data represented by $X_1, X_2, \ldots, X_{N_\text{I}}$. The fuzzy c-mean algorithm aims to separate these data into n_c clusters with the cluster centres represented by $V_1, V_2, \ldots, V_{n_\text{c}}$. To this end, the fuzzy c-mean algorithm formalises the problem as an optimisation problem and tries to minimise the following objective function (Höppner et al. 1999; Kesemen et al. 2016):

$$\begin{aligned} & f = \sum_{i=1}^{N_\text{I}} \sum_{j=1}^{n_\text{c}} m_{i,j}^{p} \left\| X_i - V_j \right\|^2, \; 1 < p < \infty \\ & \text{Subject to:} \\ & m_{i,j} \in [0, 1], \quad i = 1, 2, \ldots, N_\text{I}, \; j = 1, 2, \ldots, n_\text{c} \\ & \sum_{j=1}^{n_\text{c}} m_{i,j} = 1, \quad i = 1, 2, \ldots, N_\text{I} \\ & 0 < \sum_{i=1}^{N_\text{I}} m_{i,j} < 1, \quad j = 1, 2, \ldots, n_\text{c} \end{aligned} \tag{7.25}$$

where $m_{i,j}$ represents the membership degree of the ith data to the jth cluster and p is the fuzziness parameter. In the next subsection, it is discussed how the fuzzy c-mean algorithm can be used to cluster the individuals within the population space of CAs.

7.7.4.2 Clustering Individuals Within the CAs

The fuzzy c-mean clustering algorithm explained in the previous subsection can be employed to cluster the individuals in CAs into different sub-populations. In this case, the data set is the N_I individuals represented by vectors $X_1^t, X_2^t, \ldots, X_{N_\text{I}}^t$ at generation t and the clusters are sub-population spaces represented by $p_1^t, p_2^t, \ldots, p_{n_\text{s}}^t$, where n_s is the number of sub-populations and t is the iteration number.

To develop a multi-population version of CA, Alami et al. (2007) employed a fuzzy c-mean clustering approach to divide the population space into several sub-population spaces represented by $p_1^t, p_2^t, \ldots, p_{n_\text{s}}^t$. The procedure proposed by Alami et al. (2007) uses the following fuzzy membership function to evaluate the membership of a given individual to a specific cluster in the search space:

$$m_{i,j}\left(X_i^t, V_j\right) = 1 - \frac{d_{i,j}\left(X_i^t, V_j\right)}{\sqrt{N_\text{I}}}, \; i = 1, 2, \ldots, N_\text{I}, \; j = 1, 2, \ldots, n_\text{s} \tag{7.26}$$

where $m_{i,j}\left(X_i^t, V_j\right)$ is the membership degree of X_i^t to the jth cluster (or sub-population) and $d_{i,j}\left(X_i^t, V_j\right)$ represents the distance between the ith individual and the jth cluster centre in the search space.

To create the sub-populations within a given population space, the fuzzy *c-mean clustering* algorithm iteratively updates the fuzzy membership degrees of the individuals $m_{i,j}$ to find the assigned individuals to each cluster (or sub-population). The algorithmic steps of the fuzzy c-mean clustering approach for creating the n_s sub-populations from individuals X_1^t, X_2^t, ..., $X_{N_I}^t$ in CAs can be summarised as follows:

Step 1: Set the initial c-mean algorithm cycle number to zero ($c = 0$) and initialise random arbitrary membership degrees of the individuals to a given cluster in such a way that $m_{i,j}^c \in [0, 1], i = 1, 2, \ldots, N_I, j = 1, 2, \ldots, n_s$.

Step 2: The initialised membership degrees should satisfy the following constraints:

$$\sum_{j=1}^{n_s} m_{i,j}^c = 1, \;\; i = 1, 2, \ldots, N_I$$

$$0 < \sum_{i=1}^{N_I} m_{i,j}^c < 1 \tag{7.27}$$

Step 3: Calculate the jth cluster centre as follows:

$$V_j^c = \frac{\sum_{i=1}^{N_I} \left(m_{i,j}^c\right)^p X_i^t}{\sum_{i=1}^{N_I} \left(m_{i,j}^c\right)^p}, \;\; j = 1, 2, \ldots, n_s \tag{7.28}$$

where p is the fuzziness parameter and superscript t represents the generation number.

Step 4: Update the membership degrees of the individuals as follows:

$$m_{i,j}^{c+1} = \left(\sum_{k=1}^{n_s} \left(\frac{\left\|X_i^t - V_j^c\right\|}{\left\|X_i^t - V_k^c\right\|}\right)^{\frac{2}{p-1}}\right)^{-1}, \;\; i = 1, 2, \ldots, N_I, \;\; j = 2, 3, \ldots, n_s \tag{7.29}$$

Step 5: If the difference between the membership function values obtained from two subsequent cycles is not significant, stop the algorithm. This termination criterion can be mathematically written as follows:

$$\left\|M^{c+1} - M^c\right\| < \varepsilon \tag{7.30}$$

where M^c and M^{c+1} are fuzzy membership matrixes at cycles c and $c+1$, respectively, and ε is a small number. If the condition in Eq. (7.30) is not satisfied, steps 3–4 must be repeated.

The obtained final fuzzy membership matrixes from the aforementioned algorithm will determine which individuals belong to which sub-population. Interested readers are referred to Kesemen et al. (2016) for more details on the fuzzy c-mean algorithms.

7.8 Chaotic CAs

The chaotic theory in nonlinear dynamics has been a source of inspiration to develop new versions of CAs. Chaos theory studies the irregular and complex behaviours exhibited by nonlinear dynamic systems with a highly sensitive nature to the initial conditions (McBride 2005). In recent decades, chaotic sequences have been used within meta-heuristic algorithms to reduce their dependencies on the initial solutions (Jalili et al. 2014; Das and Saha 2021; Gagnon et al. 2021). The literature review reveals that one-dimensional chaotic map functions have been used to generate chaotic sequences within the search operators of meta-heuristic algorithms. There are a variety of chaotic maps in literature, such as Kent, Logistic, Bernoulli shift, Sine, and Circle maps (Özkaynak 2015).

One of the most popular chaotic maps is the logistic map function originally developed by May (2004) which is defined by the following equation:

$$z^{t+1} = \phi z^t \left(1 - z^t\right) \tag{7.31}$$

where z^t represents the chaotic variable, superscript t indicates the iteration number, and ϕ is a control parameter. The initial solution for Eq. (7.31) is defined as $z^0 \in [0, 1]$. Figure 7.4 presents the ergodic property and probability distribution of the logistic map function calculated for $z^0 = 0.4$, $\phi = 4$ and 5000 iterations.

By subtitling $z^t = \frac{(y^t+1)}{2}$ into Eq. (7.31), the standard logistic map function can be symmetrically mapped into the interval $[-1, 1]$ as follows:

$$z^{t+1} = 1 - 2z^{t2} \tag{7.32}$$

Figure 7.5 shows the ergodic property and probability distribution of the logistic map function within the interval $[-1, 1]$ calculated for $z^0 = 0.4$, $\phi = 4$ and 5000 iterations.

The chaotic maps can also be embedded into CAs to replace the purely random variables with chaotic variables. This section discusses CAs with chaotic operators.

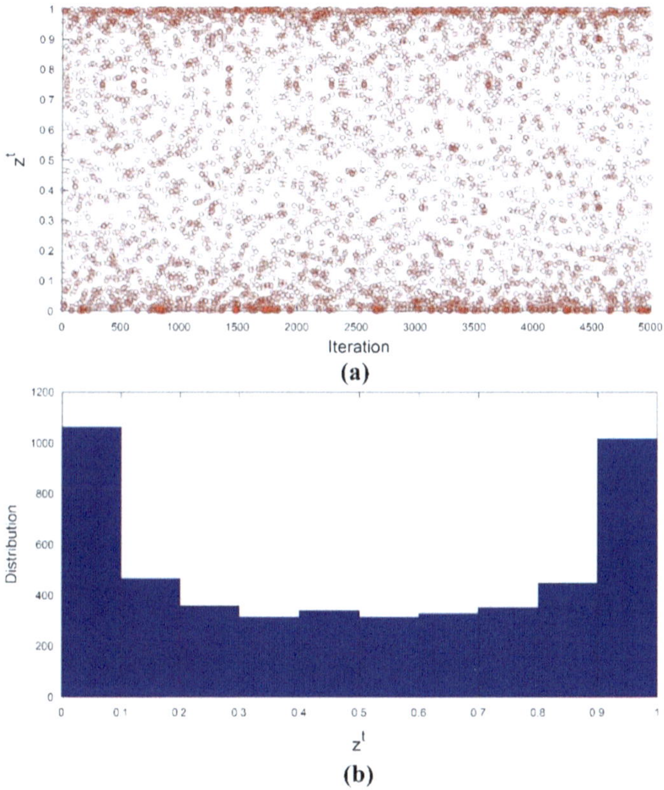

Fig. 7.4 Chaotic logistic map function within the interval [0, 1] with parameter $\phi = 4$ and initial condition of $z^0 = 0.4$: **a** ergodic property and **b** the probability distribution

7.8.1 Belief Space with Chaotic Search

Wang et al. (2009) employed Chaos Culture PSO (CCPSO) for optimal operation of hydropower systems, in which CA and PSO were hybridised based on a sequential model. The algorithm uses the chaotic logistic map to search around the currently accepted elite individuals within the belief space as can be seen from the following

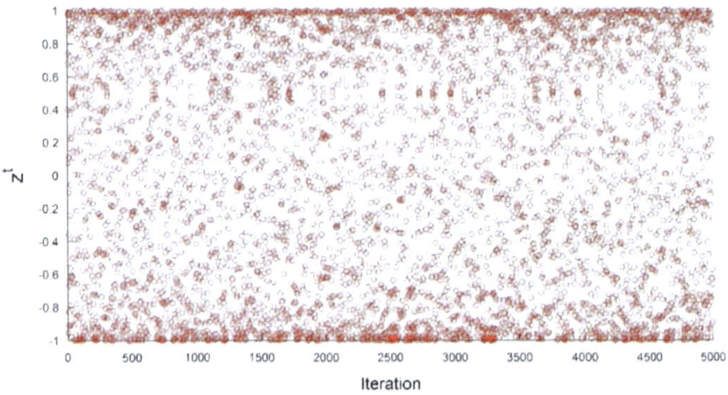

(a)

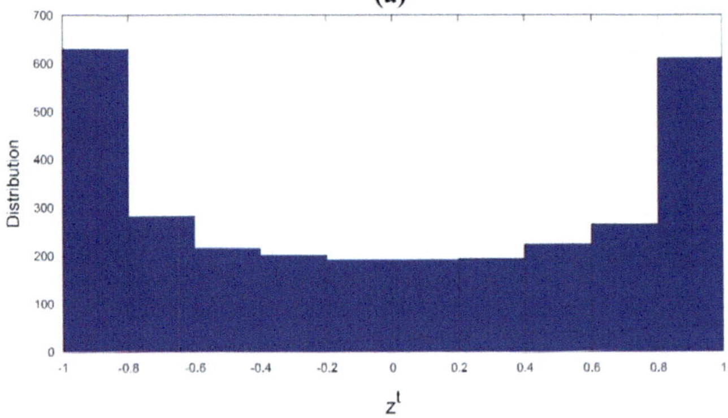

(b)

Fig. 7.5 Chaotic logistic map function within the interval [−1, 1] with parameter $\phi = 4$ and initial condition of $z^0 = 0.4$: **a** ergodic property and **b** the probability distribution

equation:

$$x_{i,j}^{t+1} = \left(1 - \lambda^t\right)x_j^t + \lambda^t\left[x_j^{\min} + z^{t+1}\left(x_j^{\max} - x_j^{\min}\right)\right],$$
$$i = 1, 2, \ldots, n_{\text{accept}}, \quad j = 1, 2, \ldots, n \tag{7.33}$$

in which:

$$\lambda^t = 1 - \left(\frac{t-1}{t}\right)^m \tag{7.34}$$

where λ^t is a parameter that determines the degree of chaotic perturbation at each iteration, z^{t+1} represents a chaotic variable obtained from the chaotic logistic map in

Eq. (7.32) and, $x_j^{\min}$ and $x_j^{\max}$ are the bounds of the search space in the jth dimension. Algorithm 7.5 presents the pseudo-code of the CCPSO algorithm.

Algorithm 7.5 Pseudo-code of the CCPSO algorithm

Initialise the belief space $\mathcal{B}^0$;
Initialise the population space $\mathcal{P}^0$ with random N_{I} individuals as follows: $X_1^0, X_2^0, \ldots, X_{N_{\mathrm{I}}}^0$;
Initialise the velocity vectors $V_1^0, V_2^0, \ldots, V_{N_{\mathrm{I}}}^0$ for individuals;
Evaluate the objective functions of the individuals;
Accept the n_{accept} individuals and update the belief space;
Set $t = 0$;
while *stopping criteria are not met* **do**
    Update the velocities and positions of individuals using PSO operators;
    Evaluate the objective functions of the individuals;
    Perform the chaotic search for the accepted individuals within the search space using equation (7.33);
    Evaluate the objective functions of the solutions obtained from the chaotic search;
    Accept the n_{accept} number of individuals and update the belief space $\mathcal{B}^{t+1}$;
    Set $t = t + 1$;
end

7.8.2 *Chaotic Influence*

He et al. (2011) developed a Chaos Cultural Algorithm (CCA) for function optimisation, in which two new influence functions were proposed based on the chaotic logistic map function. The influence functions are based on the normative and situational knowledge components and use the sequences generated from the chaotic logistic map function. The chaotic influence functions can be written as (He et al. 2011):

$$x_{i,j}^{t+1} = \begin{cases} x_j^{\min} + z^{t+1}\left|l_j - x_{\min,j}^t\right| & \text{if rand} \le 0.33 \\ x_{\min,j}^t + z^{t+1}\left|x_{\min,j}^t - x_{\max,j}^t\right| & \text{if } 0.33 < \text{rand} \le 0.66 \\ x_{\max,j}^t + z^{t+1}\left|x_{\max,j}^t - u_j\right| & \text{if rand} > 0.66 \end{cases} \tag{7.35}$$

$$x_{i,j}^{t+1} = s_{i,j}^t + \left(z^{t+1} - 0.5\right)\left|x_{\max,j}^t - x_{\min,j}^t\right| \tag{7.36}$$

where $0 \le z^{t+1} \le 1$ represents the chaotic variable obtained from the chaotic logistic map function in Eq. (7.32), $x_{\min,j}^t$ and $x_{\max,j}^t$ are the lower and upper normative bounds at iteration t, respectively, $s^t_{i,j}$ is the jth variable of the ith situational vector stored in the situational knowledge component at iteration t and, l_j and u_j are the lower and upper bounds for the jth variable of the problem, respectively.

7.8.3 Chaotic Mutation

Cheng et al. (2009) proposed a chaotic knowledge-based mutation, in which the standard logistic map function defined in Eq. (7.31) was used.

7.9 Summary

Due to fact that the standard operators of CAs may not exhibit satisfactory performance for some classes of problems, researchers have been tried to develop modified versions of CAs. The overall aim of this chapter was to provide a review of recent developments and extensions in different components of the standard CAs. The chapter discussed a set of extended and modified influence and acceptance functions developed in the literature. A real human society consists of different forms of social networks with numerous individuals who are interacting with each other based on complex interrelation behaviours. It was discussed how the social fabric concept can be used to simulate these interactions in the social networks within CAs. It was discussed that the quality of knowledge distribution between the individuals can be significantly enhanced by modelling the population structures with neighbourhood or topological properties. The chapter also discussed the application of fuzzy set theory in the development of new influence, acceptance, and update functions for the CAs. It was also investigated how the c-mean fuzzy clustering algorithm can be used to divide the population space of CAs into several sub-populations. The chapter discussed the application of chaotic sequences within the different components of CAs. The literature survey revealed that a set of new chaotic influence functions have been developed in recent years, in which the chaotic logistic map function is used to generate chaotic sequences.

References

Alami J, el Imrani A, Bouroumi A (2007) A multipopulation cultural algorithm using fuzzy clustering. Appl Soft Comput 7:506–519. https://doi.org/10.1016/j.asoc.2006.10.010

Ali MZ, Salhieh A, Snanieh RTA, Reynolds RG (2012) Boosting cultural algorithms with a heterogeneous layered social fabric influence function. Comput Math Organ Theory 18:193–210

Ali MZ, Reynolds RG (2009) An intelligent social fabric influence component in cultural algorithms for knowledge learning in dynamic environments. In: 2009 IEEE/WIC/ACM international joint conference on web intelligence and intelligent agent technology. IEEE, pp 161–168

Ali MZ, Reynolds RG (2014) Cultural algorithms: a Tabu search approach for the optimization of engineering design problems. Soft Comput 18. https://doi.org/10.1007/s00500-013-1169-5
Ali M, Reynolds R, Ali R, Salhieh A (2011a) Knowledge-based constrained function optimization using cultural algorithms with an enhanced social influence metaphor. In: Computational intelligence. Springer, pp 103–119
Ali MZ, Salhieh A, Snanieh RTA, Reynolds RG (2011b) Boosting cultural algorithms with an incongruous layered social fabric influence function. In: 2011b IEEE congress of evolutionary computation (CEC). IEEE, pp 1225–1232
Ali MZ, Awad NH, Reynolds RG, Suganthan PN (2018) A balanced fuzzy cultural algorithm with a modified Levy flight search for real parameter optimization. Inf Sci 447. https://doi.org/10.1016/j.ins.2018.03.008
Bede B (2013) Mathematics of fuzzy sets and fuzzy logic. Springer, Berlin, Heidelberg
Cheng J, Qian J, Guo Y (2009) Adaptive chaotic cultural algorithm for hyperparameters selection of support vector regression. In: International conference on intelligent computing, pp 286–293
Coello Coello CA, Becerra RL (2002) A cultural algorithm for constrained optimization, pp 98–107
Das S, Saha P (2021) Performance of swarm intelligence based chaotic meta-heuristic algorithms in civil structural health monitoring. Measurement 169:108533. https://doi.org/10.1016/j.measurement.2020.108533
Deb K, Pratap A, Agarwal S, Meyarivan T (2002) A fast and elitist multiobjective genetic algorithm: NSGA-II. IEEE Trans Evol Comput 6:182–197. https://doi.org/10.1109/4235.996017
Dubois DJ (1980) Fuzzy sets and systems: theory and applications. Academic press
Estrada E (2012) The structure of complex networks: theory and applications. Oxford University Press
Gagnon I, April A, Abran A (2021) An investigation of the effects of chaotic maps on the performance of metaheuristics. Eng Rep 3. https://doi.org/10.1002/eng2.12369
He J, Xu F (2011) Chaotic-search-based cultural algorithm for solving unconstrained optimization problem. Model Simul Eng 2011:1–6. https://doi.org/10.1155/2011/239743
Höppner F, Klawonn F, Kruse R, Runkler T (1999) Fuzzy cluster analysis: methods for classification, data analysis and image recognition. Wiley
Jalili S, Hosseinzadeh Y, Kaveh A (2014) Chaotic biogeography algorithm for size and shape optimization of truss structures with frequency constraints. Periodica Polytechnica Civil Eng 58:397–422. https://doi.org/10.3311/PPci.7466
Jalili S, Hosseinzadeh Y, Rabczuk T (2019) Simultaneous size and shape optimization of dome-shaped structures using improved cultural algorithm
Kesemen O, Tezel Ö, Özkul E (2016) Fuzzy c-means clustering algorithm for directional data (FCM4DD). Expert Syst Appl 58:76–82. https://doi.org/10.1016/j.eswa.2016.03.034
Lynn N, Ali MZ, Suganthan PN (2018) Population topologies for particle swarm optimization and differential evolution. Swarm Evol Comput 39:24–35. https://doi.org/10.1016/j.swevo.2017.11.002
May RM (2004) Simple mathematical models with very complicated dynamics. The theory of chaotic attractors. Springer, New York, NY, pp 85–93
McBride N (2005) Chaos theory as a model for interpreting information systems in organizations. Inf Syst J 15:233–254. https://doi.org/10.1111/j.1365-2575.2005.00192.x
Özkaynak F (2015) A novel method to improve the performance of chaos based evolutionary algorithms. Optik (stuttg) 126:5434–5438. https://doi.org/10.1016/j.ijleo.2015.09.098
Reynolds RG, Zhu S (2001) Knowledge-based function optimization using fuzzy cultural algorithms with evolutionary programming. IEEE Transactions on Systems, Man and Cybernetics, Part B (Cybernetics) 31:1–18. https://doi.org/10.1109/3477.907561
Reynolds RG, Kinnaird-Heether L (2003) Survival of the fastest: using cultural algorithms to optimize the design of a controller for a 3D racing game. Comput Math Org Theory 9:227–254
Reynolds RG, Ali M, Jayyousi T (2008) Mining the social fabric of archaic urban centers with cultural algorithms. Computer (long Beach Calif) 41:64–72. https://doi.org/10.1109/MC.2008.25

Reynolds RG, Ali MZ (2008c) The social fabric approach as an approach to knowledge integration in cultural algorithms. In: 2008a IEEE congress on evolutionary computation (IEEE world congress on computational intelligence). IEEE, pp 4200–4207

Reynolds R, Ali M (2008a) Embedding a social fabric component into cultural algorithms toolkit for an enhanced knowledge-driven engineering optimization. Int J Intell Comput Cybern 1. https://doi.org/10.1108/17563780810919131

Reynolds RG, Gawasmeh YA (2012) Evolving heterogeneous social fabrics for the solution of real valued optimization problems using cultural algorithms. In: 2012 IEEE congress on evolutionary computation. IEEE, pp 1–8

Reynolds RG, Kinnaird-Heether L (2017b) Problem solving using social networks in cultural algorithms with auctions. In: 2017a IEEE congress on evolutionary computation (CEC). IEEE, pp 2714–2721

Reynolds RG, Kinnaird-Heether L (2017a) Population mechanics and cultural algorithms in the development of a cultural engine. In: 2017b IEEE symposium series on computational intelligence (SSCI). IEEE, pp 1–8

Wang W, Song Y, Xue Y et al (2015) An optimal vibration control strategy for a vehicle's active suspension based on improved cultural algorithm. Appl Soft Comput 28:167–174. https://doi.org/10.1016/j.asoc.2014.11.047

Wang X, Hao W, Li Q (2017) An adaptive cultural algorithm with improved quantum-behaved particle swarm optimization for sonar image detection. Sci Rep 7:17733. https://doi.org/10.1038/s41598-017-17945-3

Wang Y, Zhou J, Lu Y, et al (2009) Chaos cultural particle swarm optimization and its application, pp 30–40

Xue H (2020) adaptive cultural algorithm-based cuckoo search for time-dependent vehicle routing problem with stochastic customers using adaptive fractional Kalman speed prediction. Math Probl Eng. https://doi.org/10.1155/2020/7258780

Zhang R, Zhou J, Mo L et al (2013) Economic environmental dispatch using an enhanced multi-objective cultural algorithm. Electric Power Syst Res 99:18–29. https://doi.org/10.1016/j.epsr.2013.01.010

Chapter 8
Multi-population Variants of Cultural Algorithms

Abstract The idea of using multiple independent sub-populations in the population space of meta-heuristic algorithms is quite common in the literature. In some cases, the multiple groups of individuals searching different parts of the search space could increase the diversity and keep the algorithm away from the local optimums. In this chapter, the multi-population versions of CAs and their algorithmic frameworks will be discussed.

Keywords Multi-population · Sub-population · Knowledge migration · Shared belief space · Cultural algorithms

8.1 Introduction

In population-based meta-heuristic algorithms, the individuals exchange information about the search space with each other which could provide the advantage of the balance between the exploration and exploitation of the search space. However, due to the random initialisation of individuals in meta-heuristic algorithms, they can be inappropriately spread within the search space which can lead to inefficient exploration and premature convergence. One of the efficient approaches to reducing the risk of premature convergence in meta-heuristic algorithms is the application of multi-population strategies. In this strategy, the individuals in the population space are divided into several sub-populations to enhance the exploration capability of the algorithms. In this approach, the different sub-populations can potentially search different areas of the search space and increase the diversity of individuals.

The multi-population strategy has also been employed to develop new versions of CAs with higher exploration capabilities. Table 8.1 lists the different multi-population versions of CAs in literature alongside brief descriptions of their overall algorithmic frameworks. The main aim of this chapter is to discuss these multi-population variants of CAs developed in the literature.

S. Jalili, *Cultural Algorithms*, Engineering Optimization: Methods and Applications,
https://doi.org/10.1007/978-981-19-4633-2_8

Table 8.1 Multi-population versions of CAs

References	Comment(s)
Guo et al. (2011), Raeesi et al. (2012), Unold et al. (2018)	– Different sub-populations with independent belief spaces. The sub-populations exchange cultural knowledge based on a knowledge migration condition
Raeesi et al. (2014)	– Different sub-populations optimise a given number of variables of the problem based on the up-down and bottom-up approaches. The cultural knowledge migration was not considered between sub-populations
Chen et al. (2008), Chen et al. (2015), Cheng-Jian Lin et al. (2009a)	– Different sub-populations with independent belief spaces. No knowledge migration mechanism was considered between the sub-populations
Guo et al. (2012)	– Multi-population version of CA with dynamically varying sub-populations
Alami et al. (2007)	– Multi-population CA with fuzzy clustering
Stanley et al. (2019)	– Multi-population CA-PSO with one shared global belief space
Oloruntoba et al. (2019)	– Different sub-populations with independent local belief spaces. A global belief space is considered on top of local belief spaces
Zhou et al. (2019)	– A cognitive space is added to the framework of CA
Singh et al. (2018), Zadeh et al. (2015)	– A single global belief space is employed for several sub-populations

The rest of the chapter is organised as follows. Section 8.2 presents the multi-population versions of CAs with the knowledge migration mechanisms. The multi-population CAs with shared global belief space is discussed in Sect. 8.3. In Sect. 8.4, a multi-population CA with different knowledge-based sub-populations is investigated. In Sect. 8.5, the development of multi-population CA based on the dynamic dimension decomposition approach is presented. Finally, Sect. 8.6 provides a summary of the contents discussed in this chapter.

8.2 Multi-population CAs with Knowledge Migration

The population space in CAs can be divided into several sub-populations with independent sub-belief spaces. A set of multi-population versions of CAs have been designed in literature based on the knowledge migration mechanism. These algorithms intend to exchange cultural knowledge between different sub-belief spaces. In the following subsections, the multi-population versions of CAs with knowledge migration mechanisms are discussed.

8.2.1 Multi-population CA Adopting Knowledge Migration

Guo et al. (2011) developed a multi-population version of the Cultural Algorithm Adopting Knowledge Migration (MCAKM) for function optimisation. The population space in the MCAKM consists of multiple sub-populations with local belief spaces. These sub-populations perform the search process independently and occasionally share their experience during the evolution process. This experience-sharing mechanism was called "knowledge migration" by Guo et al. (2011). Figure 8.1 shows the general framework of MCAKM.

MCAKM divides the N_I individuals into n_s sub-populations represented by $\left(\mathscr{p}_1^t, \mathscr{p}_2^t, \ldots, \mathscr{p}_{ns}^t\right)$ at each iteration. Each sub-population has its own belief space represented by $\left(\mathscr{b}_1^t, \mathscr{b}_2^t, \ldots, \mathscr{b}_{ns}^t\right)$. The belief spaces consist of normative and topographical knowledge components. At each iteration, each sub-population independently performs the evolution process and updates its own belief space. To provide a communication mechanism between the different sub-populations, MCAKM performs the knowledge exchange process between the sub-populations once per a certain number of generations, represented by τ. It should be noted that MCAKM performs the knowledge migration process based on ring and mesh topologies defined for sub-populations, which were discussed in Sect. 7.5 of Chap. 7. Algorithm 8.1 presents the pseudo-code of the MCAKM algorithm.

A similar approach, as illustrated in Fig. 8.1, was proposed by Unold et al. (2018) for TSP. In another study, Digalakis et al. (2002) developed a Parallel Co-operating Cultural Algorithm (PARCA) in which a set of isolated populations are separately evolved. The evolution process in each population is performed by different mutation operators. The communication between the different populations is performed by sharing only the best individuals obtained by each population.

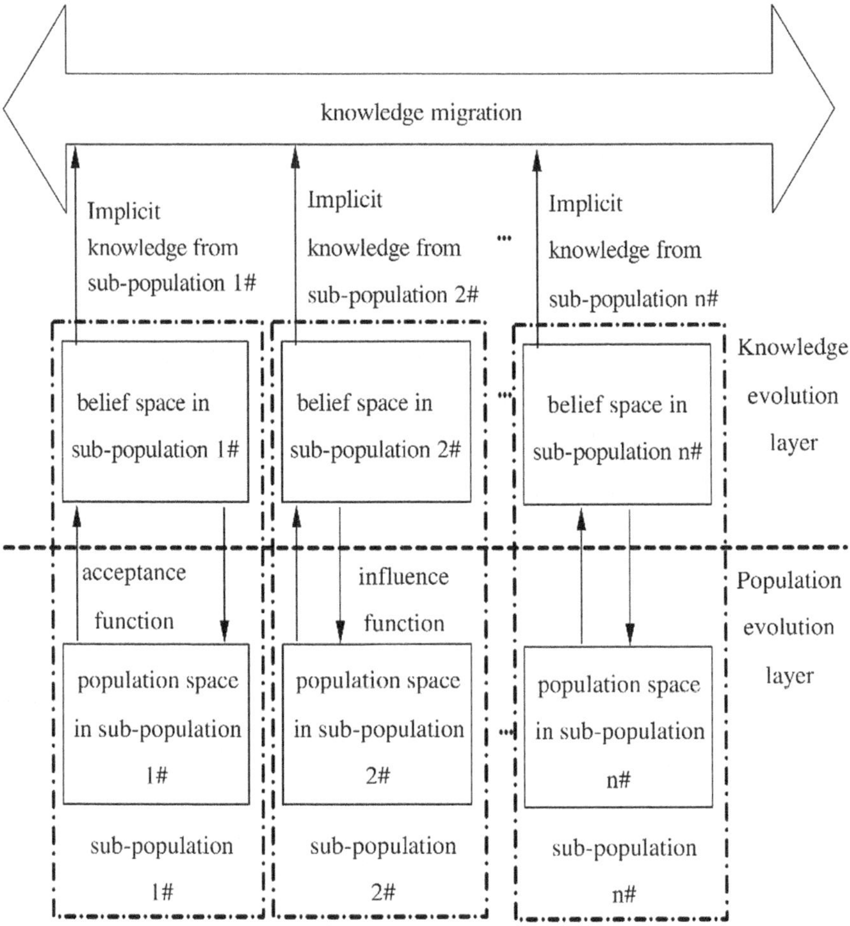

Fig. 8.1 General framework of MCAKM proposed by Guo et al. (2011)

Algorithm 8.1 Pseudo-code of the MCAKM

Initialise the internal parameters of the algorithm;

Initialise the random population space with individuals evenly divided into several sub-populations represented by $\mathcal{P}^0 = (\mathcal{p}_1^0, \mathcal{p}_2^0, \ldots, \mathcal{p}_{n_s}^0)$;

Initialise the local belief space for each sub-population represented by $\mathcal{B}^0 = (\mathcal{b}_1^0, \mathcal{b}_2^0, \ldots, \mathcal{b}_{n_s}^0)$;

Evaluate the individuals and accept the top $p\%$ of individuals in each sub-population and update the belief space $\mathcal{B}^0$;

Set $t = 0$;

while *stopping criteria are not met* **do**

  for $i \leftarrow 1$ to n_s **do**

    Update the individuals in the ith sub-population based on the knowledge components in its own belief space $\mathcal{b}_i^t$;

    Evaluate the objective function of newly generated individuals;

    Form the new generation for the ith sub-population $\mathcal{b}_i^{t+1}$ by performing the greedy selection between the parent and offspring individuals;

  end

  If the knowledge migration condition is satisfied, perform the knowledge migration process between the sub-populations and update the belief space of all sub-populations $\mathcal{B}^{t+1} = (\mathcal{b}_1^{t+1}, \mathcal{b}_2^{t+1}, \ldots, \mathcal{b}_{n_s}^{t+1})$;

  Set $t = t + 1$;

end

8.2.2 *Multi-population Cooperative CA*

Guo et al. (2012) developed a multi-population cooperative CA, in which the population space is divided into several sub-population. Each sub-population is evolved by GA-based operators. The algorithm dynamically changes the size of each sub-population based on a competition concept to improve the diversity of individuals.

8.2.3 *Multi-population CA with Fuzzy Clustering*

Alami et al. (2007) proposed a multi-population version of CA based on the fuzzy clustering approach. The algorithm divides the individuals into a set of n_s sub-populations represented by $\mathcal{p}_1^0, \mathcal{p}_2^0, \ldots, \mathcal{p}_{n_s}^0$ and creates a sub-belief space for each of them indicated by $\mathcal{b}_1^0, \mathcal{b}_2^0, \ldots, \mathcal{b}_{n_s}^0$. The fuzzy clustering approach considers the similarities between the individuals as a criterion to divide them into several sub-populations. Each sub-population independently evolves by the operators of CA and updates its corresponding sub-belief space based on its accepted elite individuals. In each generation, the sub-belief spaces exchange their normative and situational knowledge to share their own experience with other sub-populations. To explain the procedure algorithmically, Algorithm 8.2 presents the pseudo-code for the multi-population CA with a fuzzy clustering approach. The details of the fuzzy clustering approach were explained in Sect. 7.7.4 of Chap. 7.

Algorithm 8.2 Pseudo-code of the CA with Fuzzy Clustering Proposed by Alami et al. (2007)

Initialise the population space $\mathcal{P}^0$ with random individuals;

Evaluate the objective functions of individuals;

Perform the fuzzy clustering to create and divide the individuals into a set of subpopulations represented by $\mathcal{p}_1^0, \mathcal{p}_2^0, \dots, \mathcal{p}_{n_s}^0$;

Initialise sub-belief spaces for each sub-population represented by $\mathcal{b}_1^0, \mathcal{b}_2^0, \dots, \mathcal{b}_{n_s}^0$;

Set $t = 0$;

while *stopping criteria are not met* **do**

  Evolve the sub-populations $\mathcal{p}_1^t, \mathcal{p}_2^t, \dots, \mathcal{p}_{n_s}^t$ independently based on cultural information inside the sub-belief spaces $\mathcal{b}_1^t, \mathcal{b}_2^t, \dots, \mathcal{b}_{n_s}^t$, respectively;

  for $i \leftarrow 1$ to n_s **do**

    Influence the individuals in the ith sub-population based on the ith sub-belief space $\mathcal{b}_i^t$;

    Evaluate the objective function of individuals in the ith sub-population;

    Accept elite individuals from the ith sub-population to update the sub-belief space $\mathcal{b}_i^t$;

    Update the sub-belief space and obtain $\mathcal{b}_i^{t+1}$;

  end

  Exchange the situational and normative knowledge components between the sub-belief spaces $\mathcal{b}_1^t, \mathcal{b}_2^t, \dots, \mathcal{b}_{n_s}^t$;

  Set $t = t + 1$;

end

8.3 Multi-population CAs with Shared Global Belief Space

In the multi-population versions of CAs, an upper level shared belief space can be designed to accumulate the cultural information stored within the different sub-belief spaces. Some studies used this idea within the multi-population versions of CAs.

Sun et al. (2010, 2012) developed a Co-Evolutionary Cultural-Based PSO (CECBPSO) algorithm, which can be categorised as a multi-population algorithm. Figure 8.2 shows the overall framework of CECBPSO. The algorithm consists of five different spaces, including two independent population spaces $\mathcal{P}^t = (\mathcal{p}_1^t, \mathcal{p}_2^t)$ with

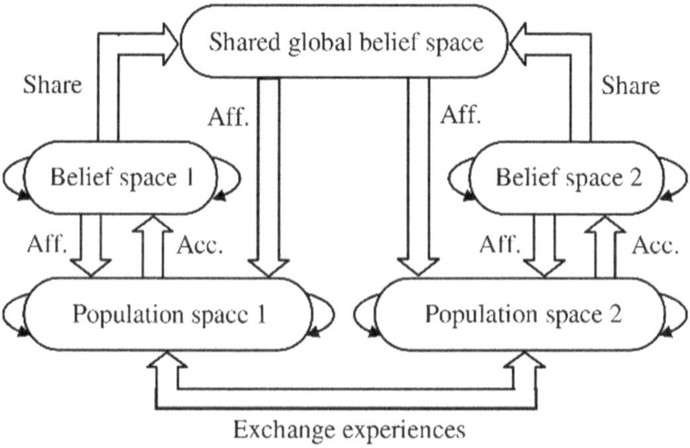

Fig. 8.2 Overall framework of the CECBPSO algorithm (Sun et al. 2010, 2012)

two belief spaces (i.e., $\mathscr{b}_1^t$ and $\mathscr{b}_2^t$) and one shared belief space $\mathscr{b}_s^t$. The CECBPSO tries to take the advantage of multiple populations with different sources of cultural information that can enhance the diversity of individuals.

Let us consider $2N_I$ individuals that are divided into two population spaces with independent belief spaces. In all the population and belief spaces, the individuals are evolved based on the PSO formulations. Then, the elite particles in each population space are selected by the acceptance function to update the corresponding belief spaces. Then, the influence function of CA is performed for each population space based on the cultural information stored in the corresponding belief space. The algorithm uses a "shared global belief space" that stores the global cultural information acquired by all $2N_I$ particles. Thereafter, the two population spaces are influenced by the shared global belief space. The last stage is information exchange between the population spaces, in which the worst particles in each population space are replaced by the best particles selected from the other population space. These steps are repeated until the global optimum solution is achieved or the termination criterion is satisfied. Algorithm 8.3 presents the pseudo-code of the CECBPSO algorithm.

8.3.1 *Clan-Based CAs*

Oloruntoba et al. (2019) proposed a Clan-based Cultural Algorithm (CCA) for the feature selection problem of large data sets. In the CCA, the population space consists of several clans (or sub-populations) with independent local belief spaces. Each clan evolves based on its own belief space. At each iteration, the cultural information of each clan is transmitted to a global belief space. Then, the whole population space is evolved based on global beliefs. The overall framework of the CCA is similar to the framework of the CECBPSO algorithm explained in the previous section.

8.4 Multi-population CA with Different Knowledge-Based Sub-populations

Within the multi-population framework of CAs, each sub-population can be evolved based on a given cultural knowledge component. Xu et al. (2010, 2012) developed a Multi-population Cultural DE (MCDE) in which each sub-population is evolved based on a given knowledge component. In MCDE, new influence functions are proposed based on cultural knowledge and DE schemes. The algorithm uses the following situational and normative knowledge-based influence functions:

Algorithm 8.3 Pseudo-code of the CECBPSO Proposed by Sun et al. (2010, 2012)

Initialise two population spaces $\mathcal{P}^0 = (\mathscr{p}_1^0, \mathscr{p}_2^0)$ and three belief spaces $\mathcal{B}^0 = (\mathscr{b}_1^0, \mathscr{b}_2^0, \mathscr{b}_s^0)$;

Initialise and evaluate the random $2N_I$ individuals in the search space and divide them into two population spaces $\mathscr{p}_1^0$ and $\mathscr{p}_2^0$ and two belief spaces $\mathscr{b}_1^0$ and $\mathscr{b}_2^0$;

Set $t = 0$;

while *stopping criteria are not met* **do**

  Evolve the population spaces $\mathscr{p}_1^t$ and $\mathscr{p}_2^t$ and the belief spaces $\mathscr{b}_1^t$, and $\mathscr{b}_2^t$, based on the PSO formulation;

  Accept the elite particles in $\mathscr{p}_1^t$ and $\mathscr{p}_2^t$, and update their corresponding belief spaces, i.e., $\mathscr{b}_1^t$ and $\mathscr{b}_2^t$;

  Perform the influence function for each population space based on the cultural information in their corresponding belief space;

  The elite particles from all populations and belief spaces are selected by the acceptance function to form the global beliefs in $\mathscr{b}_s^t$;

  Influence the two population spaces based on the cultural information in the shared global belief space $\mathscr{b}_s^t$;

  Perform the information exchanges between the population spaces $\mathscr{p}_1^t$ and $\mathscr{p}_2^t$ by replacing the worst particles in each population space with the best particles in the other population space;

  Set $t = t + 1$;

end

$$x_{i,k}^{t+1} = \mathscr{s}_{i,k} + F \times \left(x_{r_1,k} - x_{r_2,k}\right) \tag{8.1}$$

$$x_{i,k}^{t+1} = \begin{cases} x_{i,k}^t + F\mathrm{rand}\left(x_{\mathrm{max,k}}^{\mathrm{t}} - x_{\mathrm{min,k}}^{\mathrm{t}}\right) & \text{if } x_{i,k}^t < x_{\mathrm{min,k}}^{\mathrm{t}} \\ x_{i,k}^t - F\mathrm{rand}\left(x_{\mathrm{max,k}}^{\mathrm{t}} - x_{\mathrm{min,k}}^{\mathrm{t}}\right) & \text{if } x_{i,k}^t > x_{\mathrm{max,k}}^{\mathrm{t}} \\ x_{i,k}^t + FN(0,1)\left(x_{\mathrm{max,k}}^{\mathrm{t}} - x_{\mathrm{min,k}}^{\mathrm{t}}\right) & \text{otherwise} \end{cases} \tag{8.2}$$

where F is the scaling factor, rand is a random number between 0 and 1, and $N(0,1)$ is a random number obtained from the normal distribution with 0 mean and 1 standard deviation.

By using the influence functions in Eqs. (8.1) and (8.2), MCDE creates two different population spaces with N_{I} individuals in each, including one population space obtained based on the situational knowledge from Eq. (8.1) represented by $\mathcal{P}_{\mathrm{S}}^t$ and one population space created based on normative knowledge from Eq. (8.2) indicated by $\mathcal{P}_{\mathrm{N}}^t$. By combining the population spaces $\mathcal{P}_{\mathrm{S}}^t$ and $\mathcal{P}_{\mathrm{N}}^t$, the algorithm creates a new trial population space $\mathcal{P}_{\mathrm{tr}}^t$ with $2 \times N_{\mathrm{I}}$ number of individuals. Then, MCDE scores the individuals based on their fitness values and selects the top N_{I} individuals to form the new population for the next iteration. Let $U_i^t, i = 1, 2, \ldots, 2 \times N_{\mathrm{I}}$ be the individuals in the trial population space $\mathcal{P}_{\mathrm{tr}}^t$. The MCDE scores the performance of each individual by comparing it to other individuals in the trial population space using the following formulas:

$$sc_{i,j} = \begin{cases} 1 \text{ if } f\left(U_i^t\right) < f\left(U_j'^t\right) \\ 0 \text{ otherwise} \end{cases}, \quad j = 1, 2, \ldots, N_{\mathrm{I}} \tag{8.3}$$

$$SC_i = \sum_{j=1}^{N_{\mathrm{I}}} sc_{i,j} \tag{8.4}$$

where $sc_{i,j}$ represents the score obtained for the i th individual in the trial population $\mathcal{P}_{\mathrm{tr}}^t$ in comparison with the j th individual in the same population, U_i^t is the i th individual in the trial population space $\mathcal{P}_{\mathrm{tr}}^t$, $U_j'^t$ indicates the j th randomly selected individual in normative-based population space $\mathcal{P}_{\mathrm{N}}^t$, and SC_i is the total score of the i th individual in the trial population space $\mathcal{P}_{\mathrm{tr}}^t$. After calculation of scores, the top N_{I} number of individuals is selected to form the new population space for the next iteration $\mathcal{P}^{t+1}$.

Algorithm 8.4 Pseudo-code of New Population Generation in MCDE Algorithm

---

A population space of individuals $\mathcal{P}^t$ with N_{I} individuals are given;

Generate a new population of individuals $\mathcal{P}_{\mathrm{S}}^t$ based on situational knowledge-based influence function in equation (8.1);

Generate another population of individuals $\mathcal{P}_{\mathrm{N}}^t$ based on normative knowledge-based influence function in equation (8.2);

Combine the populations $\mathcal{P}_{\mathrm{S}}^t$ and $\mathcal{P}_{\mathrm{N}}^t$ and form a new trial population $\mathcal{P}_{\mathrm{tr}}^t$ with $2 \times N_{\mathrm{I}}$ individuals;

Calculate the score of each individual SC_i in $\mathcal{P}_{\mathrm{tr}}^t$ using equations (8.3) and (8.4);

Select N_{I} individuals with better score values and form the population space $\mathcal{P}^{t+1}$ for the next iteration;

8.5 Multi-population CA with Dynamic Dimension Decomposition

Raeesi et al. (2014) proposed a heterogeneous multi-population CA with dynamic dimension decomposition strategies. The algorithm divides the individuals evenly into several sub-populations. Each sub-population tries to find the optimum values for a given number of dimensions in the problem. The study proposed two dynamic dimension decomposition approaches, the top-down and bottom-down strategies.

In the top-down strategy, the algorithm starts the search process in all dimensions with a single population space. After a given number of generations, the algorithm divides the population space into several sub-populations, called local CAs. In each sub-population, only some dimensions of the problem are being optimised. On the other hand, the solution-finding process in the bottom-up strategy is reversed. It starts with several sub-populations that are intended to search for only one dimension of the problem. After a certain number of generations, the algorithm merges these sub-populations and continues the search process in all dimensions using a single population. A similar approach was also proposed in Raeesi and Kobti 2013.

8.6 Summary

The chapter investigated the different multi-population variants of CAs in which the population space is divided into several sub-populations with independent sub-belief spaces. The majority of these algorithms try to provide an efficient exploration of search space through modelling a knowledge exchange mechanism between the different sub-populations. Some of the multi-population CAs employ a knowledge migration mechanism, in which the cultural information is directly exchanged between the sub-populations. While other algorithms form an upper level shared global belief space in which the cultural information acquired by different sub-populations is accumulated and used to evolve the whole population space. The chapter discussed that the sub-population spaces can be evolved based on different knowledge components. Moreover, the dynamic dimension decomposition technique was reviewed for multi-population CAs. This class of multi-population CAs uses each sub-population to optimise a given number of variables of the problem at a given time. Although the multi-population strategies could potentially enhance the exploration capabilities of CAs and improve their convergence properties, sometimes their appetite for searching unproductive regions of search space may also lead to unnecessary computational effort. Hence, it seems to be necessary to have a conservative approach in their applications to computationally expensive problems.

References

Alami J, el Imrani A, Bouroumi A (2007) A multipopulation cultural algorithm using fuzzy clustering. Appl Soft Comput 7:506–519. https://doi.org/10.1016/j.asoc.2006.10.010

Chen C-H, Liu Y-C, Lin C-J, Lin C-T (2008) A hybrid of cooperative particle swarm optimization and cultural algorithm for neural fuzzy networks. In: 2008 IEEE international conference on fuzzy systems (IEEE World congress on computational intelligence). IEEE, pp 238–245

Chen C-H, Yang S-Y (2015) Efficient DE-based symbiotic cultural algorithm for neuro-fuzzy system design. Appl Soft Comput 34:18–25. https://doi.org/10.1016/j.asoc.2015.04.026

Digalakis JG, Margaritis KG (2002) A multipopulation cultural algorithm for the electrical generator scheduling problem. Math Comput Simul 60:293–301. https://doi.org/10.1016/S0378-4754(02)00021-6

Guo Y, Cheng J, Cao Y, Lin Y (2011) A novel multi-population cultural algorithm adopting knowledge migration. Soft Comput 15:897–905. https://doi.org/10.1007/s00500-010-0556-4

Guo Y, Liu D, Cheng J (2012) Multi-population cooperative cultural algorithms, pp 199–206

Lin C-J, Chen C-H, Lin C-T (2009) A hybrid of cooperative particle swarm optimization and cultural algorithm for neural fuzzy networks and its prediction applications. In: IEEE transactions on systems, man, and cybernetics. Part C (applications and reviews), vol 39, pp 55–68. https://doi.org/10.1109/TSMCC.2008.2002333

Oloruntoba O, Cosma G, Liotta A (2019) Clan-based cultural algorithm for feature selection. In: 2019 International conference on data mining workshops (ICDMW). IEEE, pp 465–472

Raeesi N. MR, Kobti Z (2012) A multiagent system to solve JSSP using a multi-population cultural algorithm, pp 362–367

Raeesi N. MR, Kobti Z (2013) Heterogeneous multi-population cultural algorithm. In: 2013 IEEE congress on evolutionary computation. IEEE, pp 292–299

Raeesi N. MR, Kobti Z (2014) Heterogeneous multi-population cultural algorithm with a dynamic dimension decomposition strategy, pp 345–350

Singh D, Zadeh PM, Kobti Z (2018) A multilevel cooperative multi-population cultural algorithm. In: 2018 Innovations in intelligent systems and applications (INISTA). IEEE, pp 1–5

Stanley SD, Kattan K, Reynolds R (2019) CAPSO: a parallelized multiobjective cultural algorithm particle swarm optimizer. In: 2019 IEEE congress on evolutionary computation (CEC). IEEE, pp 3060–3069

Sun Y, Zhang L, Gu X (2012) A hybrid co-evolutionary cultural algorithm based on particle swarm optimization for solving global optimization problems. Neurocomputing 98:76–89. https://doi.org/10.1016/j.neucom.2011.08.043

Sun Y, Zhang L, Gu X (2010) Co-evolutionary cultural based particle swarm optimization algorithm, pp 1–7

Unold O, Tarnawski R (2018) A parallel framework for multi-population cultural algorithm and its applications in TSP, pp 470–482

Xu W, Wang R, Zhang L, Gu X (2012) A multi-population cultural algorithm with adaptive diversity preservation and its application in ammonia synthesis process. Neural Comput Appl 21:1129–1140. https://doi.org/10.1007/s00521-011-0749-5

Xu W, Zhang L, Gu X (2010) A novel cultural algorithm and its application to the constrained optimization in ammonia synthesis. In: Life system modeling and intelligent computing. Springer, pp 52–58

Zadeh PM, Kobti Z (2015) A multi-population cultural algorithm for community detection in social networks. Procedia Comput Sci 52:342–349. https://doi.org/10.1016/j.procs.2015.05.105

Zhou Y, Zhang S, Luo Q, Abdel-Baset M (2019) CCEO: cultural cognitive evolution optimization algorithm. Soft Comput 23:12561–12583. https://doi.org/10.1007/s00500-019-03806-w

Chapter 9
Application of Cultural Algorithms to Structural Optimisation

Abstract This chapter investigates the performance of CAs with different influence functions in solving the structural optimisation problems. The statistical performances of different versions of CAs are investigated on two large-scale optimum structural design problems under static and dynamic constraints. The statistical results obtained from the design examples suggest that the types of influence functions can significantly affect the statistical behaviours of CAs.

Keywords Structural optimisation · Cultural algorithms · Size and shape optimisation · Barrel vault structure · Dome structure

9.1 Introduction

Nowadays, complex structures have important applications in different engineering industries, such as building, aerospace, and telecommunication. Recent advances in computational machines made it possible to design and analyse complex structural systems under combinations of load cases for a given application. Cost and structural stability are two conflicting factors in structural design evaluations. The engineers not only have to properly design safe structures to resist the different loading conditions during their lifetime but also should minimise the costs and environmental impacts of construction projects.

Due to the complexity of modern structures and the existence of a variety of design parameters, the achievement of optimal designs for large-scale structures is not an easy task. A large number of design variables as well as the extent of available options for each design parameter make the simple trial-and-error design approach inefficient or even impossible. Hence, efficient computational approaches are required to effectively find the optimum or at least near-optimum designs for structural systems. Meta-heuristics are attractive tools for structural optimisation. In the past two decades, researchers have employed a variety of meta-heuristics to find optimum designs for different structural systems, such as towers, buildings, domes and space structures (Saka 2003; Saka et al. 2016; Kaveh 2017; Jalili and Hosseinzadeh 2018; Talatahari et al. 2021).

S. Jalili, *Cultural Algorithms*, Engineering Optimization: Methods and Applications,
https://doi.org/10.1007/978-981-19-4633-2_9

The main aim of this chapter is to investigate the performance of CAs with different influence functions in finding the optimum designs for large-scale structures. Different influence functions are assumed to be used within the CAs, including four standard influence functions as well as modified, Gaussian, and truncated geometric distribution-based influence functions. In order to assess the performance of CAs, two large-scale structural optimisation problems, double-layer grid and barrel vault structures, under static and dynamic constraints are investigated.

The rest of the chapter is organised as follows. Section 9.2 briefly presents the algorithms employed in this chapter. The optimum design problem of barrel vault structure under static constraints is presented in Sect. 9.3. Section 9.4 investigates the size and shape optimisation problem of the lamella dome structure. In the end, Sect. 9.5 provides the summary and concluding remarks.

9.2 Algorithms

As was discussed in Chap. 7, a variety of influence functions have been developed for CAs. In this chapter, the performance of CAs with different influence functions is investigated. To this end, seven different influence functions are selected to be employed within CAs as listed in Table 9.1, including four standard influence functions as well as modified, Gaussian, and truncated geometric distribution-based influence functions. The related formulations of each influence function are available in Chap. 7. Some of the influence functions include constant parameters, which are assumed with suitable values as listed in the last column of Table 9.1.

The investigated algorithms follow the same algorithmic steps of CAs explained in Sect. 3.7 of Chap. 3. The only difference between the different versions of CAs in this chapter is in the employed influence functions. In the investigated design examples, the number of individuals is taken as 30 for all algorithms, i.e., $N_I = 30$. It is also assumed that 20% of individuals will be accepted to update the belief space in

Table 9.1 Different variants of CAs with different influence functions

Notation	Description	Parameters
CA-I	Standard CA with the influence function in Eq. (3.8)	–
CA-II	Standard CA with the influence function in Eq. (3.9)	$\sigma_{i,j}^{t} = 0.2(l_j - u_j)$
CA-III	Standard CA with the influence function in Eq. (3.10)	–
CA-IV	Standard CA with the influence function in Eq. (3.11)	$\beta = 0.2$
CA-M	CA with the modified influence function developed by Ali and Reynolds (2014) in Eq. (7.1)	–
CA-G	CA with the gravity-based influence function proposed by Wang et al. (2017) in Eq. (7.2)	–
CA-TGD	CA with influence the truncated geometric distribution-based influence introduced by Jalili et al. (2019) in Eq. (7.4)	$q_0 = 1$

each generation. In the next sections, the performance of CAs with different influence functions are investigated on two large-scale structural optimisation problems.

9.3 Optimum Design of Barrel Vault Structure Under Static Constraints

The first problem is the optimum design of the barrel vault structure under static constraints. Figure 9.1 shows the barrel vault structure with 1607 elements under the axial tension and compressive loads. The top and bottom layers of the structure are arcs with internal and exterl radiuses of 45 and 50 m as illustrated in Fig. 9.1. All free nodes of the structure are subjected to a concentrated vertical load of −3.00 kips in the z-direction. Young's modulus and yield stress of the structural elements are assumed as 29,000 ksi and 36 ksi, respectively. The design constraints of the problem limit the displacements of nodes and internal stresses within the elements of the structure. It is assumed that the displacement of all free nodes in all directions must not exceed ±3.5 in. The allowable tension stresses of the elements are considered as $\sigma_i^{\mathrm{t}} = 0.6F_{\mathrm{y}}$, where F_{y} is the yield stress of the steel. The compressive stresses within the elements must not violate the allowable stress values calculated based on the elements' slenderness ratios, which can be written for each element as follows:

$$\sigma_i^{\mathrm{c}} = \begin{cases} \left[\left(1 - \dfrac{\lambda_i^2}{2C_{\mathrm{c}}^2}\right)F_{\mathrm{y}}\right] \Big/ \left(\dfrac{5}{3} + \dfrac{3\lambda_i}{C_{\mathrm{c}}} - \dfrac{\lambda_i^3}{8C_{\mathrm{c}}^3}\right) & \text{for } \lambda_i < C_{\mathrm{c}} \\ \dfrac{12\pi^2 E}{23\lambda_i^2} & \text{for } \lambda_i \geq C_{\mathrm{c}} \end{cases} \tag{9.1}$$

In the above equation, σ_i^{c} represents the allowable compressive stress of the ith element, E represents Young's modulus, F_{y} indicates the yield stress of steel, C_{C} is the slenderness ratio separating the elastic and inelastic buckling regions which is calculated as $C_{\mathrm{C}} = \sqrt{2\pi^2 E/F_{\mathrm{y}}}$, λ_i is the slenderness ratio of the ith element obtained as $\lambda_i = kL_i/r_i$, k is the effective length factor, L_i represents the length of the ith element, and r_i is the radius of gyration of the ith element.

The structure consists of 44 fixed supports with zero displacements in all directions as illustrated in Fig. 9.1. The elements of the structure are categorised into 25 different groups for design purposes. The element groups 1 up to 20 represent the elements on the top and bottom arcs as illustrated in Fig. 9.1. The diagonal elements between the two arcs are categorised into the 21st design group. The top and bottom layers of the structure consist of diagonal and longitudinal elements as displayed in Fig. 9.2, which are categorised into four different design groups. The 22nd design group includes the diagonal elements of the top layer, the 23rd design group represents the longitudinal elements at the top layer, the 24th design group indicates the diagonal elements at the bottom layer, and the longitudinal elements at the bottom layer are represented by the 25th design group. The cross-sections of structural elements represent the design

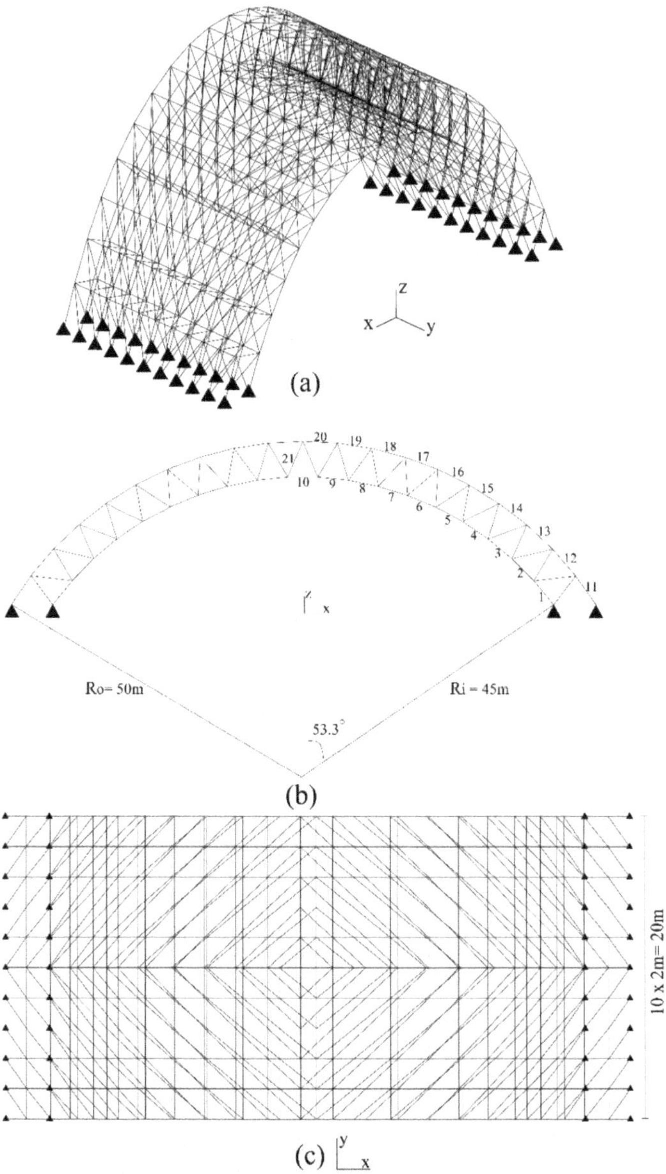

Fig. 9.1 A 1607-bar barrel vault structure: **a** 3D view, **b** side view, and **c** top view

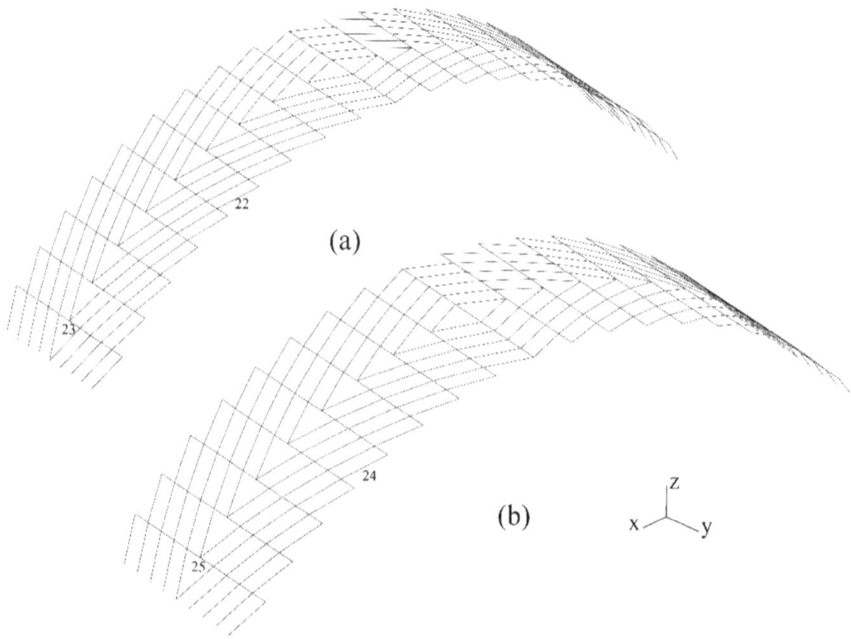

Fig. 9.2 Members' group numbers on the top and bottom layers of the 1607-bar barrel vault structure: **a** top layer and **b** bottom layer

variables for this problem which should be selected from an available discrete set of sections listed in Table 9.2. Thus, the design problem is a discrete optimisation problem with 25 design variables.

The main aim of the optimum design problem of barrel vault structure is to minimise the structural weight under some constraints on stresses and displacements. The optimum design problem can be mathematically expressed by the following equation:

$$\begin{aligned}
&\text{Find: } X = \left[x_1, x_2, \ldots, x_{n_{\text{eg}}}\right] \\
&\text{To minimise: } W^{\text{P}}(X) = W(X)(\psi(X))^{\alpha} \\
&W(X) = \sum_{i=1}^{m} \gamma A_i L_i \\
&x_i \in S,\ S = \{s_1, s_2, \ldots, s_r\},\ i = 1, 2, \ldots, n_{\text{eg}} \\
&\text{Subjected to :} \\
&\begin{cases} C_{\sigma_i}^{\text{t}}(X) = \left|\frac{\sigma_i}{\sigma_i^{\text{t}}}\right| - 1 \le 0 \rightarrow \text{if } \sigma_i > 0 \\ C_{\sigma_i}^{\text{c}}(X) = \left|\frac{\sigma_i}{\sigma_i^{c}}\right| - 1 \le 0 \rightarrow \text{if } \sigma_i < 0 \end{cases} \quad i = 1, 2, \ldots, m
\end{aligned}$$

Table 9.2 An available discrete set of steel pipe sections from the AISC code (Construction 1989)

No	Name	Area (in^2)	Radius of gyration (in)	Unit weight (lb/ft)
1	PIPE1/2SCH40	0.250	0.261	0.852
2	PIPE1/2SCH80	0.320	0.250	1.090
3	PIPE3/4SCH40	0.333	0.334	1.130
4	PIPE3/4SCH80	0.433	0.321	1.480
5	PIPE1SCH40	0.494	0.421	1.680
6	PIPE1SCH80	0.639	0.407	2.170
7	PIPE1-1/4SCH40	0.669	0.540	2.270
8	PIPE1-1/2SCH40	0.799	0.623	2.720
9	PIPE1-1/4SCH80	0.881	0.524	3.000
10	PIPE2SCH40	1.070	0.787	3.660
11	PIPE1-1/2SCH80	1.070	0.605	3.630
12	PIPE2SCH80	1.480	0.766	5.030
13	PIPE2-1/2SCH40	1.700	0.947	5.800
14	PIPE3SCH40	2.230	1.160	7.580
15	PIPE2-1/2SCH80	2.250	0.924	7.670
16	PIPE2XXS	2.660	0.703	9.040
17	PIPE3-1/2SCH40	2.680	1.340	9.120
18	PIPE3SCH80	3.020	1.140	10.300
19	PIPE4SCH40	3.170	1.510	10.800
20	PIPE3-1/2SCH80	3.680	1.310	12.500
21	PIPE2-1/2XXS	4.030	0.844	13.700
22	PIPE5SCH40	4.300	1.880	14.600
23	PIPE4SCH80	4.410	1.480	15.000
24	PIPE3XXS	5.470	1.050	18.600
25	PIPE6SCH40	5.580	2.250	19.000
26	PIPE5SCH80	6.110	1.840	20.800
27	PIPE4XXS	8.100	1.370	27.600
28	PIPE8SCH40	8.400	2.940	28.600
29	PIPE6SCH80	8.400	2.190	28.600
30	PIPE5XXS	11.300	1.720	38.600
31	PIPE10SCH40	11.900	3.670	40.500
32	PIPE8SCH80	12.800	2.880	43.400
33	PIPE12STD	14.600	4.380	49.600
34	PIPE6XXS	15.600	2.060	53.200
35	PIPE10SCH60	16.100	3.630	54.800
36	PIPE12XS	19.200	4.330	65.500
37	PIPE8XXS	21.300	2.760	72.500

$$C_{\delta_j}(X) = \left| \frac{\delta_j}{\delta_j^{\mathrm{all}}} \right| - 1 \leq 0,\ j = 1, 2, \ldots, n_{\mathrm{n}} \tag{9.2}$$

where X represents the vector containing the design variables, x_i is the cross-sectional area for the ith design group, n_{eg} is the number of design groups which is equal to 25 in this problem, $W^{\mathrm{P}}(.)$ indicates the penalised weight of the structure introduced to handle the design constraints, $W(.)$ is the structural weight, $\psi(.)$ is the penalty function, α is the penalty factor, m is the total number of structural elements which is equal to 1607, γ density of steel, A_i indicates the cross-sectional area of the ith element, L_i is the length of the ith element, S is a set of the available discrete set of sections, r is the number of available sections which is equal to 37 in this problem, $C_{\sigma_i}^{\mathrm{t}}(.)$ and $C_{\sigma_i}^{\mathrm{c}}(.)$ are the constraint functions of the tension and comessive stresses for the ith element, respectively, σ_i represents the internal stress of the ith element obtained for the structural analysis under the loading condition, σ_i^{t} is the allowable tension stress of the ith element, σ_i^{c} is the allowable compressive stress of the ith element calculated from Eq. (9.1), $C_{\delta_j}(.)$ is the displacement constraint for the j th ne, δ_j^{all} is the allowable displacement value for the jth node, and n_{n} is the number of nodes.

The penalty function $\psi(.)$ value is calculated based on the values of constraint functions. It is equal to one when the design fully satisfies all the stress and displacement constraints. Otherwise, it would be greater than one. The penalty function can be written as:

$$\psi(X) = 1 + \sum_{i=1}^{m} \left(\begin{cases} \max\left(C_{\sigma_i}^{t}(X), 0\right) \to \text{if } \sigma_i > 0 \\ \max\left(C_{\sigma_i}^{c}(X), 0\right) \to \text{if } \sigma_i < 0 \end{cases} \right) + \sum_{j=1}^{n} \max\left(C_{\delta_j}(X), 0\right) \tag{9.3}$$

The value of penalty factor α in Eq. (9.2) controls the penalisation process which usually starts with a small value and is gradually increased during the solution-finding process to achieve a balance between the exploration and exploitation of the search space. The initial smaller values would provide the algorithm with the opportunity to explore diverse areas of the search space, and the greater values in later stages would facilitate the exploitation process within a specific part of the search space. Similar to the previous studies, it is assumed that the value of the penalty parameter starts from 1.5 and gradually increased to 3.0 (Husseinzadeh Kashan et al. 2019).

The statistical results of different versions of CAs obtained from 30 independent runs for the optimum design problem of barrel vault structure are summarised in Table 9.3. In this design problem, the maximum number of structural analyses set to 5000 is considered as the termination criterion. The performance comparison between the different versions of CAs reveals that the CA-IV performs better than others in different aspects. In terms of the best obtained structural weight, it can be seen from Table 9.3 that the CA-IV yielded a weight of 153,085 lb which is significantly lighter than the corresponding values for other algorithms. The CA-TGD is the most competitive version providing a weight of 181,825.06 lb, which is about 18% heavier than the weight found by CA-IV. The statistical results, including the mean weight, standard deviation, and worst weight, also exhibit remarkable superiority of CA-IV over others. The worst structural weight obtained by CA-IV is about 162,869 lb which is still significantly lighter than the best weights yielded by other algorithms. The standard deviation of the results provided by the CA-IV obtained from 30 independent runs is about 1885 lb, while it is 43,366, 12,261, 39,301, 35,381, 13,474, and 15,919 lb for the CA-I, CA-II, CA-III, CA-M, CA-G, and CA-TGD. The small standard deviation value of CA-IV shows its consistency in finding high-quality final designs for this structure. Moreover, Fig. 9.3 presents statistical intuitive performance comparisons between different versions of CAs. From Fig. 9.3, it can be observed that the ranges and variances of the results obtained by the CA-IV are remarkably smaller than those for other versions of CAs, which shows the statistically stable behaviour of the algorithm during different independent runs.

Figure 9.4 shows the average convergence histories of different versions of CAs obtained from 30 independent runs for this design problem. As can be seen from this figure, the CA-IV exhibits the most promising convergence speed from the early stages of the solution-finding process. The main competitor is the CA-TGD which has a relatively slower convergence speed in the initial stages. However, its convergence rate is gradually accelerated after about 1000 structural analyses which makes it the most attractive version of CA after the CA-IV version.

In order to check whether the optimum design obtained by the CA-IV satisfies the stress constraints of the problem, Fig. 9.5 compares the current stress values in structural elements to the corresponding allowable limits. In some of the elements, the existing stresses are quite close to the allowable values which indicate that the stress constraints control the optimum design process. As can be seen from Fig. 9.5, the design provided by CA-IV fully satisfies the stress constraints which means that the design is a fully feasible solution for the problem.

9.4 Size and Shape Optimisation of Lamella Dome Structure Under Frequency Constraints

In this section, the size and shape optimisation of a lamella dome-shaped structure under frequency constraints is investigated. The structural optimisation under frequency optimisation is a highly nonlinear problem in structural engineering (Hosseinzadeh et al. 2016; Jalili and Hosseinzadeh 2019). The lamella dome shown in Fig. 9.6 is an attractive structural system to cover large spans and areas. The overall geometry of lamella dome structures and their related parameters are illustrated in Fig. 9.7. The coordinates of the nodes in these structures can be expressed as a function of geometrical parameters, including the dome diameter D, the number of rings n_r, the number of nodes on each ring n_{nr}, and the crown height h as displayed in Fig. 9.7. In this study, the distances between the rings are considered constant and it is assumed that the distances between the nodes on each ring are equal. To calculate

Table 9.3 Optimisation results obtained by CAs with different influence functions for the optimum design problem of barrel vault structure

Design group	CA-I	CA-II	CA-III	CA-IV	CA-M	CA-G	CA-TGD
1	PIPE6SCH80	PIPE6SCH40	PIPE8SCH80	PIPE3-1/2SCH80	PIPE8SCH40	PIPE3-1/2SCH80	PIPE3-1/2SCH80
2	PIPE6SCH40	PIPE12STD	PIPE6XXS	PIPE3SCH80	PIPE2-1/2XXS	PIPE3-1/2SCH80	PIPE4SCH80
3	PIPE6SCH80	PIPE4SCH40	PIPE3-1/2SCH40	PIPE3SCH40	PIPE2XXS	PIPE3-1/2SCH80	PIPE6SCH40
4	PIPE3-1/2SCH40	PIPE5SCH40	PIPE4SCH40	PIPE3SCH40	PIPE6SCH80	PIPE3-1/2SCH80	PIPE2-1/2XXS
5	PIPE3-1/2SCH40	PIPE3SCH80	PIPE4SCH40	PIPE2SCH80	PIPE2-1/2SCH40	PIPE2SCH80	PIPE2SCH80
6	PIPE3SCH80	PIPE3SCH40	PIPE3-1/2SCH40	PIPE2SCH80	PIPE3XXS	PIPE3-1/2SCH80	PIPE3-1/2SCH80
7	PIPE10SCH40	PIPE5SCH80	PIPE3-1/2SCH40	PIPE1-1/2SCH80	PIPE3XXS	PIPE3-1/2SCH80	PIPE2SCH40
8	PIPE12STD	PIPE2SCH80	PIPE3SCH40	PIPE1-1/2SCH80	PIPE3-1/2SCH40	PIPE3-1/2SCH80	PIPE2-1/2SCH40
9	PIPE3XXS	PIPE2-1/2XXS	PIPE6SCH40	PIPE1-1/2SCH80	PIPE6SCH40	PIPE3-1/2SCH80	‘PIPE1-1/2SCH80’
10	PIPE3-1/2SCH40	PIPE6SCH40	PIPE5SCH40	PIPE1-1/2SCH80	PIPE6SCH80	PIPE3-1/2SCH80	PIPE1-1/2SCH80
11	PIPE3SCH40	PIPE3-1/2SCH80	PIPE3-1/2SCH40	‘PIPE3SCH40	PIPE2XXS	PIPE3-1/2SCH80	PIPE3SCH40
12	PIPE6SCH80	PIPE2-1/2SCH80	PIPE4SCH80	PIPE2SCH80	PIPE2XXS	PIPE3-1/2SCH80	PIPE3-1/2SCH80
13	PIPE3-1/2SCH40	PIPE4SCH40	PIPE3SCH40	PIPE2SCH80	PIPE3SCH40	PIPE3-1/2SCH80	PIPE5SCH40
14	PIPE3SCH40	PIPE3-1/2SCH80	PIPE3-1/2SCH40	PIPE2SCH80	PIPE3XXS	PIPE3-1/2SCH80	PIPE2SCH80
15	PIPE5SCH40	PIPE3SCH80	PIPE2-1/2SCH80	PIPE2SCH80	PIPE3XXS	PIPE3-1/2SCH80	PIPE2SCH80
16	PIPE3-1/2SCH40	PIPE6SCH40	PIPE3-1/2SCH40	PIPE2SCH80	PIPE6SCH80	PIPE3-1/2SCH80	PIPE4SCH80
17	PIPE3SCH40	PIPE5SCH80	PIPE3SCH40	PIPE2SCH80	PIPE3XXS	PIPE3-1/2SCH80	PIPE2SCH80
18	PIPE2-1/2SCH80	PIPE2XXS	PIPE3XXS	PIPE2SCH80	PIPE2SCH80	PIPE3-1/2SCH80	PIPE2SCH80
19	PIPE3-1/2SCH40	PIPE5SCH80	PIPE3SCH40	PIPE2SCH80	PIPE3XXS	PIPE3-1/2SCH80	PIPE3-1/2SCH40
20	PIPE3SCH40	PIPE2-1/2XXS	PIPE3SCH40	PIPE3SCH40	PIPE3XXS	PIPE3-1/2SCH80	PIPE2XXS
21	PIPE3-1/2SCH40	PIPE3SCH80	PIPE4SCH40	PIPE3-1/2SCH40	PIPE3-1/2SCH40	PIPE3-1/2SCH80	PIPE3-1/2SCH40

(continued)

Table 9.3 (continued)

Design group	CA-I	CA-II	CA-III	CA-IV	CA-M	CA-G	CA-TGD
22	PIPE3SCH40	PIPE2SCH40	PIPE1-1/2SCH40	PIPE1-1/2SCH40	PIPE1-1/2SCH80	PIPE1-1/2SCH40	PIPE1-1/2SCH40
23	PIPE3-1/2SCH40	PIPE3SCH80	PIPE4XXS	PIPE3-1/2SCH40	PIPE3XXS	PIPE3-1/2SCH80	PIPE3-1/2SCH40
24	PIPE3SCH40	PIPE2XXS	PIPE3SCH40	PIPE2-1/2SCH40	PIPE3SCH80	PIPE3-1/2SCH80	PIPE2-1/2SCH40
25	PIPE3SCH40	PIPE2SCH80	PIPE4SCH80	PIPE2SCH80	PIPE3XXS	PIPE3-1/2SCH80	PIPE4SCH40
Weight (lb)	247,905.52	238,035.33	259,532.81	153,085.27	266,043.82	253,263.93	181,825.06
Mean (lb)	340,165.39	258,689.95	321,612.62	153,818.48	351,189.77	283,248.22	206,177.75
Std (lb)	43,365.90	12,260.51	39,301.3	1884.86	35,380.54	13,474.20	15,919.06
Worst (lb)	424,088.91	281,017.14	448,078.78	162,869.03	410,208.47	290,271.00	246,419.95

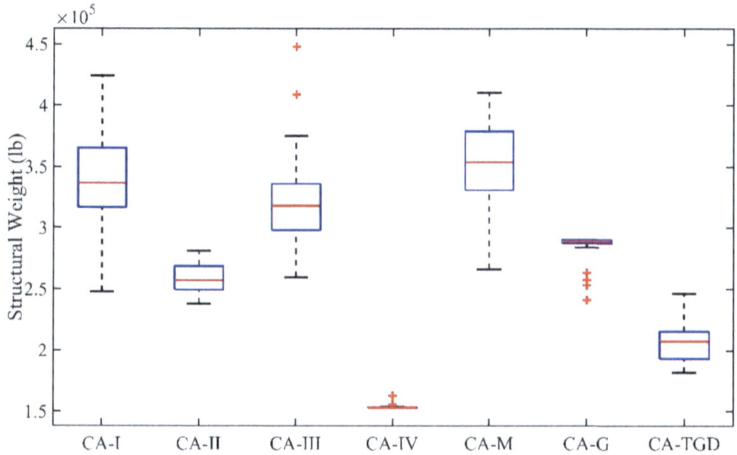

Fig. 9.3 Box plot of the CAs with different influence functions for the optimum design problem of barrel vault structure

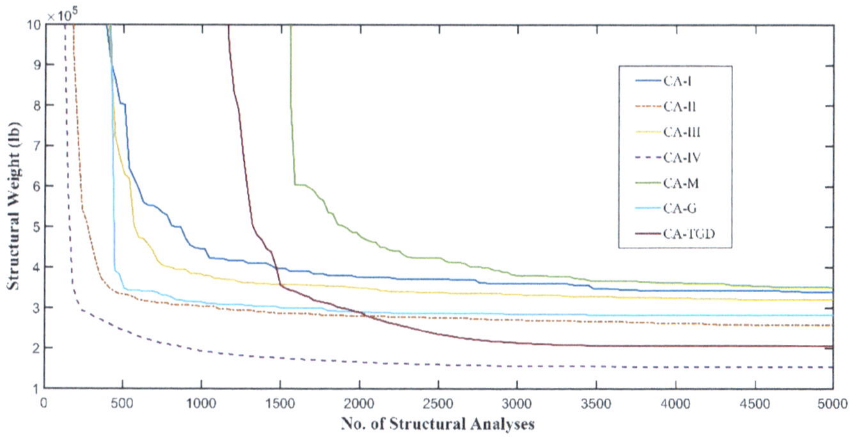

Fig. 9.4 Average convergence histories of CAs with different influence functions for the optimum design problem of barrel vault structure

the nodal coordinates of the lamella dome structure, the angles between each node and the x-axis are required. For the dome structure shown in Fig. 9.7, this angle for the ith nodes is calculated as follows:

- For even-numbered rings:

$$\alpha_i = \frac{360}{n_{\mathrm{nr}}}\left(i - j_{r,1}\right) \tag{9.4}$$

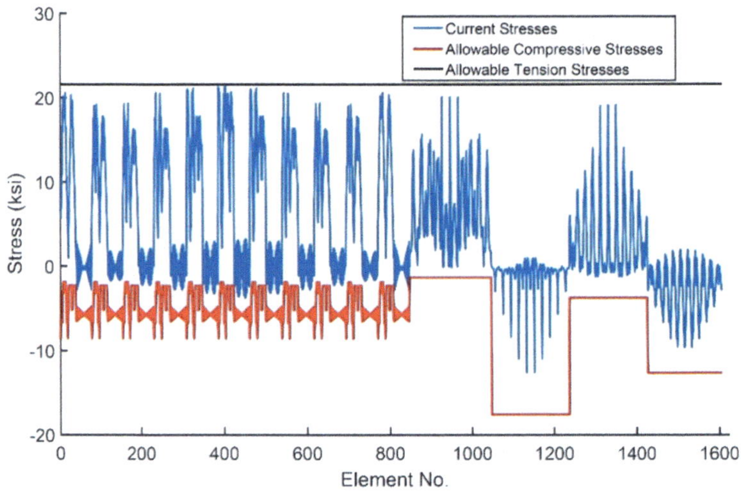

Fig. 9.5 Comparison between the current and allowable stresses in the optimum design obtained from the CA-IV for the barrel vault structure

- For odd-numbered rings:

$$\alpha_i = \frac{180}{n_{\text{nr}}} + \frac{360}{n_{\text{nr}}}\left(i - j_{r,1}\right) \tag{9.5}$$

In the above equations, α_i represents the angle between each node and x-axis, n_{nr} is the number of nodes on each ring, and $j_{r,1}$ denotes the first node number at the rth ring.

The nodal coordinates can be expressed in terms of the angles α_i and other geometrical parameters by the following equations:

$$x_i = ra \ \cos\alpha_i \tag{9.6}$$

$$z_i = -ra \ \sin\alpha_i \tag{9.7}$$

$$y_i = h - R + \sqrt{R^2 - x_i^2 - z_i^2} \tag{9.8}$$

In Eq. (9.8), r is the ring number at which the ith node is located, a represents the distance between the rings in the x-z plane, and R is the radius of semicircle as shown in Fig. 9.7. With the assumption of equal distances between the rings, the parameter a is equal to $\frac{D}{2n_r}$. In addition, the geometry of dome in Fig. 9.7 implies that the radius of the semicircle R can be expressed in terms of the dome diameter D and the crown height h as follows:

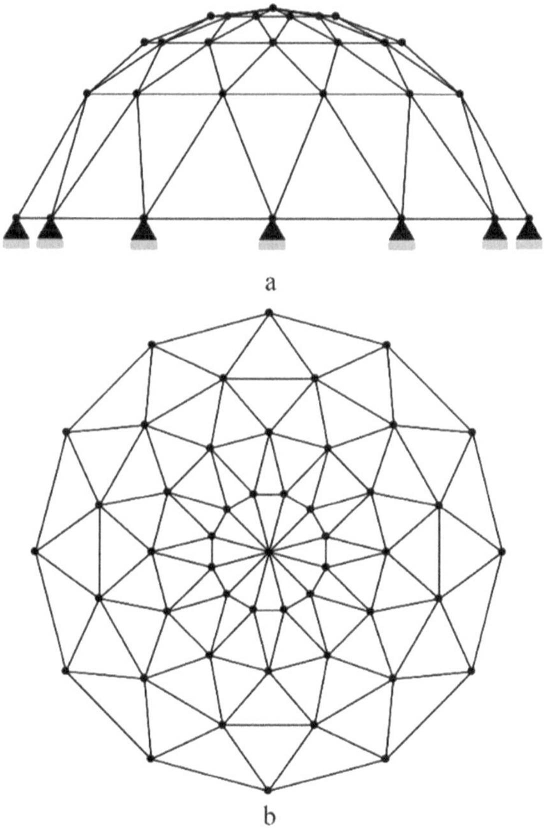

Fig. 9.6 Lamella dome structure: **a** side view and **b** top view (Jalili et al. 2019)

$$R = \frac{\left(D^2 + 4h^2\right)}{8h} \tag{9.9}$$

By using Eqs. (9.6)–(9.8), the nodal coordinates of the lamella dome structure can be calculated.

The design example is a 708-bar lamella dome structure shown in Fig. 9.8. This design example was previously investigated by Jalili et al. (2019). In this section, the performance of CAs with different influence functions is investigated. In the structure shown in Fig. 9.8, the number of rings n_r is equal to 20 and the number of nodes on each ring n_{nr} is assumed to be equal to 12. The diameter of the dome D is taken as 20 m. Due to design considerations and symmetrical geometry of the dome, the structural elements are divided into 40 design groups. The element group numbering pattern is illustrated in Fig. 9.9.

The elements of the structure are made of steel with Young's modulus of 2.1×10^{11} N/m^2 and a density of 7971.810 kg/m^3. It is assumed that a concentrated mass of 100 kg is attached to all free nodes of the structure. The cross-sectional areas are continuous variables, and they can be any value between 1 and 100 cm^2. The

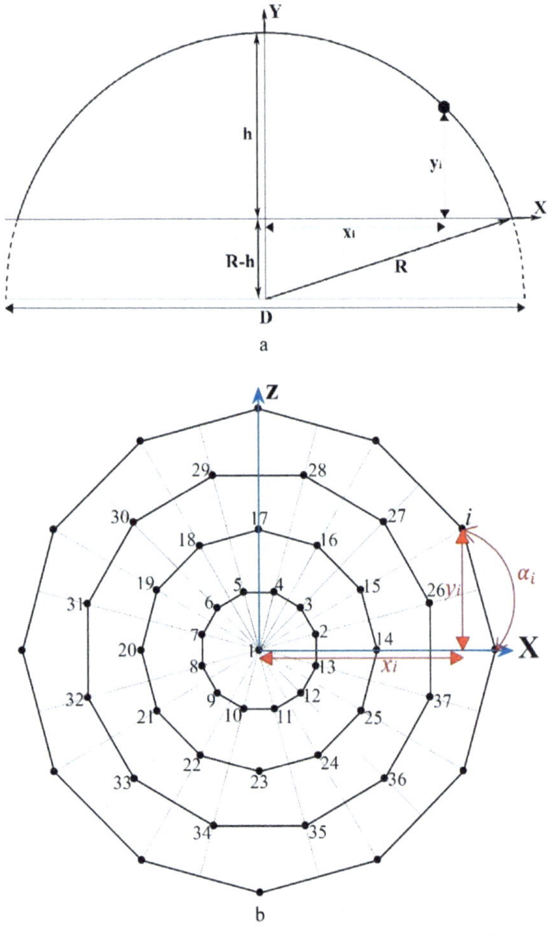

Fig. 9.7 Coordinates of nodes in the lamella dome structure ($n_{\text{nr}} = 12$, $n_{\text{r}} = 4$): **a** side view and **b** top view (Jalili et al. 2019)

crown height h is the shape variable of the structure which should be selected from the range of 1–20 m. Thus, this optimum design problem consists of 41 design variables, including 40 size and 1 shape variables. As the design constraints, the structure is subjected to lower-bound constraints on the first four free vibrational frequencies as follows: $f_1 \geq 3$, $f_2 \geq 6$, $f_3 \geq 6$, and $f_4 \geq 8$.

The size and shape optimisation problem of lamella dome structure looks for the appropriate cross-sectional areas as well as dome shape to minimise the structural weight under several vibrational frequency constraints. The problem can be mathematically written as (Jalili et al. 19):

$$\text{Find: } X = \left[x_1, x_2, \ldots, x_{n_{eg}}, h\right]$$
$$\text{To minimize: } W^P(X) = W(X).(\psi(X))^{\alpha}$$

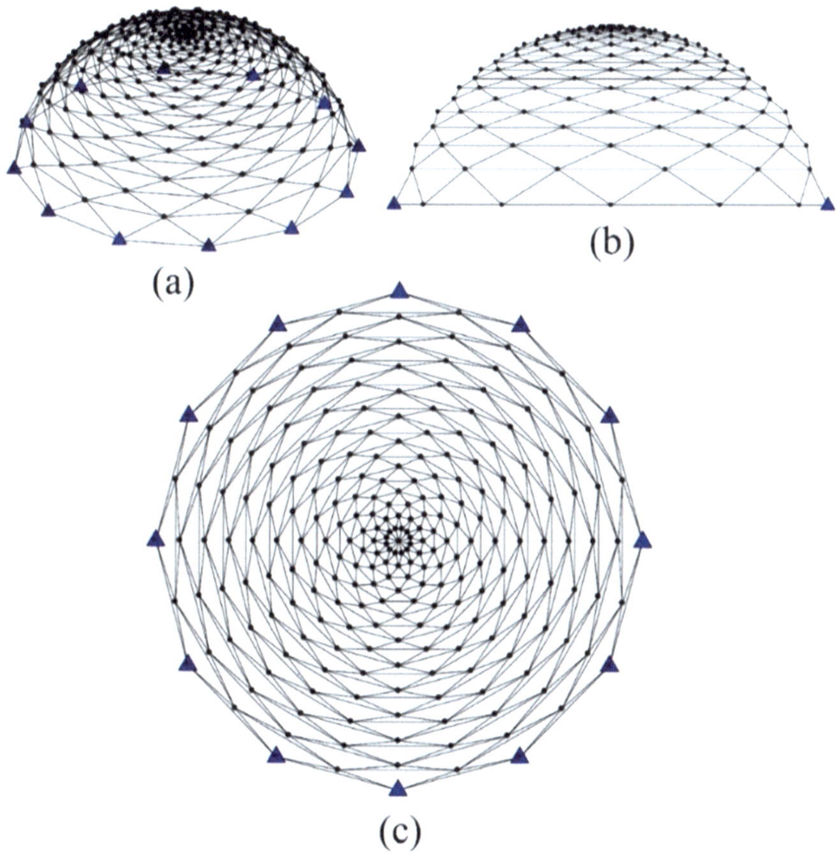

Fig. 9.8 A 708-bar lamella dome structure: **a** 3D view, **b** side view, **c** top view (Jalili et al. 2019)

$$W(X) = \sum_{i=1}^{n_{\text{eg}}} \gamma x_i \sum_{j=1}^{n_i} L_j$$
$$l_i \leq x_i \leq u_i, i = 1, 2, \ldots, n_{\text{eg}}$$
$$h_{\min} \leq h \leq h_{\max}$$
$$\text{Subject to: } C^{\text{L}}_{f_{\text{s}}}(X) = 1 - \frac{f_{\text{s}}}{f^{\text{l}}_{\text{s}}} \leq 0 \tag{9.10}$$

where X represents the vector of design variables, x_i indicates the cross-sectional area of the ith element group, h is the height of the dome which is the shape variable in this problem, $W^{\text{P}}(.)$ is the penalised structural weight defined to consider the effects of the design constraints, $W(.)$ represents the structural weight, $\psi(.)$ indicates the penalty function which is a function of design variables, α is a constant parameter that adjusts the effects of the penalty function on the final objective function value, γ

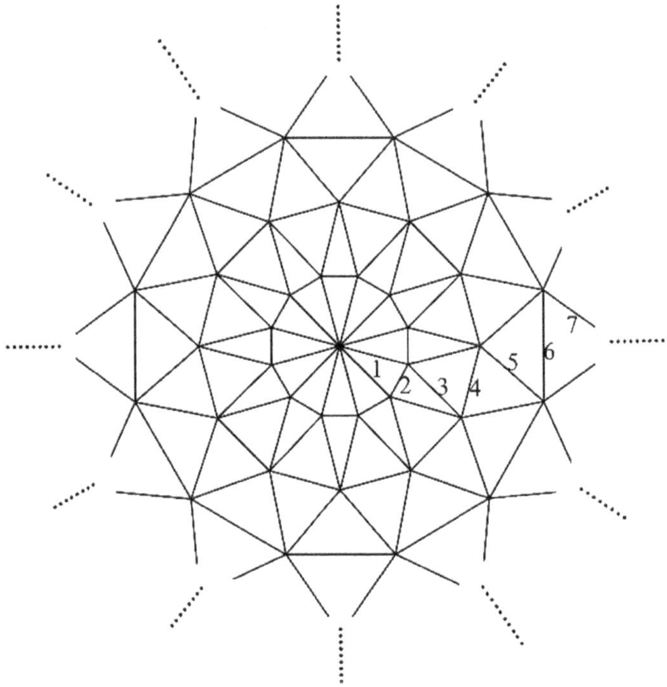

Fig. 9.9 Element group numbering pattern in the 708-bar lamella dome structure

denotes the material density of structural elements, L_j indicates the length of the jth element in the ith design group, l_i and u_i are the lower and upper limits for the cross-sectional areas of the ith element group, respectively, the search space bounds for the crown height h are represented by $h_{\min}$ and $h_{\max}$, $C^{\mathrm{L}}_{f_s}(.)$ indicates the lower-bound constraint of the sth frequency, f^{L}_{s} denotes the lower bound of the sth frequency, and f_s represents the sth frequency of the structure. The value of the penalty function $\psi(.)$ is calculated based on the constraint values by the following equation:

$$\psi(X) = 1 + \sum\nolimits_{s} \max\left(C^{\mathrm{L}}_{f_s}(X), 0\right) \tag{9.11}$$

As can be seen from Eq. (9.10), the value of the penalty function is equal to 1 if all of the design constraints are satisfied. Otherwise, it would be greater than 1 for the infeasible designs violating the constraints.

The optimum cross-sectional areas, crown heights, as well as statistical results, obtained from different versions of CAs are summarised in Table 9.4. In this design problem, the maximum number of structural analyses set to 9000 is considered as the termination criterion. The results reported in Table 9.4 suggest that the CA-TGD and CA-III found better best designs compared to other algorithms, with structural weights of about 25.86 tonnes and 27.40 tonnes, respectively. The mentioned algorithms also provide smaller average weights in comparison with others. Figure 9.10 presents statistical intuitive performance comparisons between different versions of CAs in this problem, which shows that the CA-IV provides statistically consistent performance in comparison with all other algorithms. However, CA-TGD and CA-III perform better than CA-IV in terms of the best and average results.

Table 9.5 lists the free vibrational frequencies obtained from the best designs yielded by different algorithms. From Table 9.5, it turns out that the second and third frequency values obtained by the CA-TGD are very close to their corresponding allowable values which shows the optimality of the obtained design. The second and third frequencies obtained by the CA-TGD are equal to 6.0013 Hz, which are very close to their lower-bound value, i.e., 6 Hz. It also suggests that the second and third frequency constraints control the optimum design process for this structure.

Figure 9.11 illustrates the convergence diagrams of different versions of CAs for the optimum design problem of the lamella dome structure. Each diagram shows the average behaviour of each algorithm during the 30 independent runs with purely stochastic initial solutions. From Fig. 9.11, it can be seen that the CA-IV exhibits a significantly faster convergence speed compared to all other versions of CAs. Despite its initial slow convergence speed, the success of CA-TGD in converging to high-quality designs is remarkable. The CA-TGD starts to exploit the current designs after 5000 structural analyses in a relatively efficient way.

9.5 Summary

The chapter investigated the performance of CAs with different influence functions proposed in the literature. The influence functions employed within the CAs include the standard influence functions as well as modified, Gaussian, and truncated geometric distribution-based influence functions. The performance of different versions of CAs, represented by CA-I, CA-II, CA-III, CA-IV, CA-M, CA-G, and CA-TGD were investigated on two different large-scale structural optimisation problems with relatively large numbers of design variables. The first problem was the optimum discrete design of a barrel vault structure with 1607 elements under stress and displacement constraints. The second problem was the size and shape optimisation of lamella dome structures with 708 elements under multiple frequency constraints, which is a highly nonlinear and complex problem in structural engineering. The statistical results obtained from the two design problems demonstrated that the performance of CAs is highly sensitive to the type of the influence functions. The comparisons showed that some of the influence functions provide a clear advantage over others in finding high-quality results. For example, the CA-IV and CA-TGD exhibited superior performance in terms of statistical and best results in comparison with others for the barrel vault design problem. On the other hand, the success of CA-TGD in finding lighter weights for the lamella dome structure was remarkable. However, the convergence speeds of CA-III and CA-IV were also promising for the

Table 9.4 Results obtained by CAs with different influence functions for the size and shape optimisation of the 708-bar lamella dome structure under frequency constraints

Cross-sections (cm^2) and height (m)	CA-I	CA-II	CA-III	CA-IV	CA-M	CA-G	CA-TGD
A_1	25.5877	46.1763	14.4811	27.5599	42.5439	21.8554	26.2206
A_2	96.0334	63.5214	16.2087	64.8730	71.4578	21.8554	27.1354
A_3	12.0910	50.5904	16.2087	31.0854	34.4666	22.0779	3.9985
A_4	36.7579	47.9026	16.2087	7.5218	93.5806	31.6981	11.8218
A_5	79.8513	32.1139	16.8574	7.5782	22.5428	22.0779	36.7435
A_6	25.5877	30.6967	36.8474	19.2932	49.9320	38.4493	58.9497
A_7	25.5877	9.8040	16.2087	14.7729	36.6028	62.5204	17.9050
A_8	25.5877	68.2036	16.2087	66.7232	28.7863	21.8554	31.0254
A_9	25.5877	29.2758	5.0894	54.1702	43.8686	7.2715	20.1247
A_{10}	15.6370	13.8699	16.2087	29.2314	32.6113	5.0018	21.1720
A_{11}	20.7388	38.3778	16.2087	27.6310	27.3095	26.9633	29.9374
A_{12}	25.5877	17.5089	25.8908	29.4511	22.5428	21.8554	28.7883
A_{13}	20.7388	40.1094	30.0384	4.3710	54.1529	70.6269	40.6521
A_{14}	25.5877	24.4955	16.2087	25.9626	72.8217	29.3367	10.5126
A_{15}	25.5877	20.6253	16.2087	29.7494	67.9128	21.8554	9.3981
A_{16}	22.2317	13.8699	25.8908	27.2628	30.7057	21.8554	11.6230
A_{17}	20.7388	32.6558	16.2087	18.7958	35.8988	18.8706	10.8680
A_{18}	25.5877	5.3678	20.8091	17.9995	78.9708	8.6297	10.2473
A_{19}	20.7388	13.8699	16.2087	19.2010	17.5807	8.6802	18.1915
A_{20}	25.5877	29.6099	16.4985	17.9995	18.8950	22.6878	14.6720
A_{21}	25.5877	21.4777	18.8947	10.8387	27.8954	37.5283	13.4136
A_{22}	20.7388	58.9430	19.7685	31.4327	48.3263	73.2228	45.7521
A_{23}	90.8386	32.9486	16.2087	17.8059	22.5428	15.4110	17.5563
A_{24}	20.7388	9.8305	39.3797	29.3173	45.4527	32.6566	42.6749
A_{25}	20.7388	29.8484	16.2087	48.2098	35.3400	16.8193	24.0933
A_{26}	20.7388	50.5904	17.4255	17.9194	15.2383	58.6394	25.4744
A_{27}	20.7388	19.3827	25.8908	15.4598	20.7719	72.1406	23.6420
A_{28}	20.7388	24.5808	16.2087	17.9995	88.4409	21.8554	23.0955
A_{29}	25.5877	17.4956	25.0874	26.2867	40.1859	39.9725	22.8525
A_{30}	31.1579	50.5904	24.4936	53.5789	29.0739	18.1486	31.1456
A_{31}	20.7388	16.7675	16.2087	21.5574	46.3933	33.5341	28.5652
A_{32}	25.5877	31.8874	29.6943	32.2176	38.0491	50.8927	20.6214
A_{33}	20.7388	43.0035	16.2087	62.5976	34.8988	70.3454	25.7668
A_{34}	42.4989	40.1193	62.2514	43.8958	38.1001	19.9016	31.6765

(continued)

Table 9.4 (continued)

Cross-sections (cm^2) and height (m)	CA-I	CA-II	CA-III	CA-IV	CA-M	CA-G	CA-TGD
A_{35}	45.6295	60.3627	41.4403	29.0124	63.2742	30.4088	20.3856
A_{36}	20.7388	31.8801	75.7572	28.5147	35.3877	87.9286	31.2130
A_{37}	55.2706	50.5904	23.7391	24.9677	75.9013	19.2056	40.2178
A_{38}	37.9797	26.2280	15.9693	19.0390	64.4291	34.0042	16.6359
A_{39}	25.5877	44.4871	16.2087	29.7300	22.8443	28.1915	20.2089
A_{40}	20.7388	20.2585	20.4066	15.7661	53.8197	21.8554	1.0003
Crown height h	9.8246	11.5922	10.9027	9.9342	13.613	10.3897	9.3490
Weight (kg)	32,777.47	35,701.22	27,399.20	31,018.99	46,494.51	38,285.02	25,857.27
Mean (kg)	35,569.24	39,514.49	33,288.53	33,409.52	51,852.85	48,462.51	31,376.55
Std (kg)	1624.06	2386.88	2060.01	997.28	2269.69	6960.27	2420.24
Worst (kg)	37,704.16	45,444.56	36,656.16	34,611.18	55,384.17	48,462.51	36,928.05
No. of analyses	9000	9000	9000	9000	9000	9000	9000

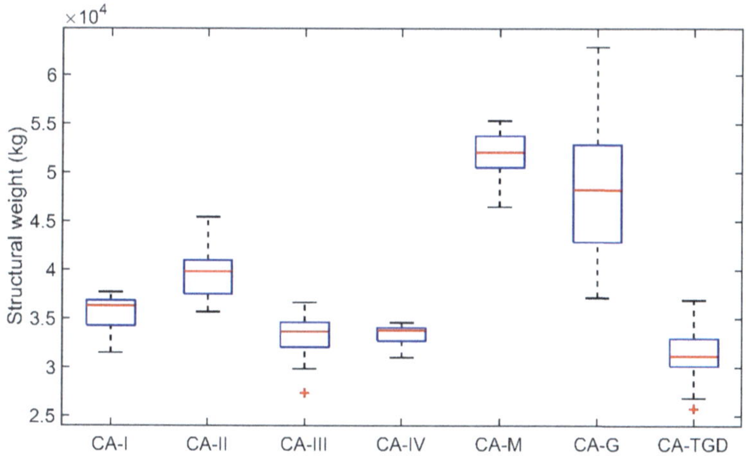

Fig. 9.10 Box plot of the CAs with different influence functions for the size and shape optimisation problem of the 708-bar lamella dome structure

later design problem. The overall conclusion of this chapter is that the influence functions have significant impacts on the statistical performance of CAs. They play the role of information diffusers between the individuals in the search space which can potentially affect the statistical behaviours of CAs. It should be noted that the conclusions about the suitability of influence functions made in this chapter may

Table 9.5 Comparison of the frequencies (Hz) obtained by different versions of CAs for the 708-bar lamella dome structure

Frequency	CA-I	CA-II	CA-III	CA-IV	CA-M	CA-G	CA-TGD
f_1	4.4165	4.1343	4.2826	4.2486	4.2293	4.3727	4.1991
f_2	6.0974	6.0865	6.0142	6.0877	6.0153	6.1909	6.0013
f_3	6.0974	6.0865	6.0142	6.0877	6.0153	6.1909	6.0013
f_4	9.0519	8.6533	8.4656	8.3177	8.9421	8.2682	8.2562
f_5	9.0519	8.6533	8.4656	8.3177	8.9421	8.2682	8.2562
f_6	9.4483	8.7029	9.0747	8.9947	9.2545	8.8349	8.4987
f_7	9.4483	8.7029	9.0747	8.9947	9.2545	8.8349	8.4987
f_8	11.1521	10.1256	10.4075	10.5831	11.6680	9.6693	8.8728

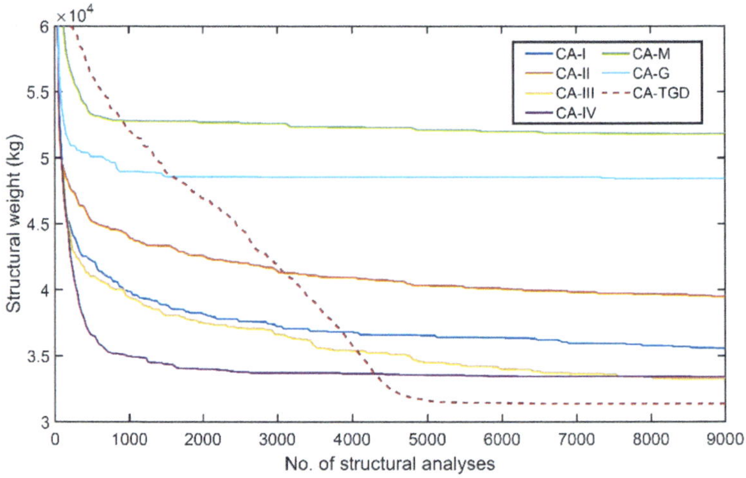

Fig. 9.11 Average convergence histories of CAs with different influence functions for the size and shape optimisation problem of the 708-bar lamella dome structure

only be true for the structural optimisation problems and generalising these conclusions to other types of problems would not be an appropriate approach. Hence, the performance of CAs with different influence functions in solving a specific problem should be carefully investigated to achieve more promising results.

References

Ali MZ, Reynolds RG (2014) Cultural algorithms: a Tabu search approach for the optimization of engineering design problems. Soft Computing 18:. https://doi.org/10.1007/s00500-013-1169-5

Construction A (1989) Manual of steel construction: allowable stress design. American Institute of Steel Construction, Chicago, IL, USA

Hosseinzadeh Y, Taghizadieh N, Jalili S (2016) Hybridizing electromagnetism-like mechanism algorithm with migration strategy for layout and size optimization of truss structures with frequency constraints. Neural Comput Appl 27:953–971. https://doi.org/10.1007/s00521-015-1912-1

Husseinzadeh Kashan A, Jalili S, Karimiyan S (2019) Premier league championship algorithm: a multi-population-based algorithm and its application on structural design optimization, pp 215–240

Jalili S, Hosseinzadeh Y, Rabczuk T (2019) Simultaneous size and shape optimization of dome-shaped structures using improved cultural algorithm

Jalili S, Hosseinzadeh Y (2018) Design optimization of truss structures with continuous and discrete variables by hybrid of biogeography-based optimization and differential evolution methods. Struct Design Tall Spec Build 27:e1495. https://doi.org/10.1002/tal.1495

Jalili S, Hosseinzadeh Y (2019) Combining migration and differential evolution strategies for optimum design of truss structures with dynamic constraints. Iran J Sci Technol Trans Civil Eng 43:289–312. https://doi.org/10.1007/s40996-018-0165-5

Kaveh A (2017) Advances in metaheuristic algorithms for optimal design of structures. Springer, Cham

Saka MP (2003) Optimum design of pitched roof steel frames with haunched rafters by genetic algorithm. Comput Struct 81:1967–1978. https://doi.org/10.1016/S0045-7949(03)00216-5

Saka MP, Hasançebi O, Geem ZW (2016) Metaheuristics in structural optimization and discussions on harmony search algorithm. Swarm Evol Comput 28:88–97. https://doi.org/10.1016/j.swevo.2016.01.005

Talatahari S, Jalili S, Azizi M (2021) Optimum design of steel building structures using migration-based vibrating particles system. Structures 33:1394–1413. https://doi.org/10.1016/j.istruc.2021.05.028

Wang X, Hao W, Li Q (2017) An adaptive cultural algorithm with improved quantum-behaved particle swarm optimization for sonar image detection. Sci Rep 7:17733. https://doi.org/10.1038/s41598-017-17945-3

Appendix
MATLAB Code of CAs

```
%%%% These codes are related to the Cultural Algorithms (CAs) for
the following minimisation problem
%%%% Find X = [x1, x_2, ..., x_n]
%%%% To minimise: f(X)
%%%% Subject to: LB<=X<=UB
% % % % The Update and Influence functions are implemented in two
separate
%%%% functions (i.e., Update() and Influence()) at the end of this
code.
%%%% To change the objective function, the user needs to modify the
f() function at the end of this code.
%%%% Copyright by Shahin Jalili
clc;
clear;
close all;
format shortEng
%% Problem description
n= 2; % Number of decision variables
LB = -6*ones(1,n); % Lower-bounds of search space in
                        the problem
UB = 6*ones(1,n); % Upper-bounds of search space in
                        the problem
%% Parameters of CAs
N_I = 30; % Number of individuals
N_accept = fix(0.2*N_I); % Number of accepted individuals to
                        update the belief space (constant
                        value is considered)
Beta = 0.2; % Constant parameter of influence
                        function
t_max = 3000; % Maximum number of iterations
Sigma = 0.2 *(UB-LB); % Strategy Parameter (assumed as a
                        fraction of search space size)
InfluCode = 3; % The code for applied influence
                        function (1, 2, 3, and 4)
```


S. Jalili, *Cultural Algorithms*, Engineering Optimization: Methods and Applications,
https://doi.org/10.1007/978-981-19-4633-2

```
%% Initialisation %%%%%%%%%%%%%%%%%%%%%%%%%%%%%%%%%%%%%%%%%%%
t = 0;
                        % Iteration number
% Initialisation of Belief Space
S = realmax*ones(1,n+1); % Situational knowledge vector
                        initialised as a vector consisting
                        of n+1 variables, in which the
                     last variable indicates its
                     objective function value
X_min = -realmax*ones(1,n); % Vector of lower normative bound
X_max = realmax*ones(1,n);% Vector of upper normative bound
L = inf(1,n); % Lower normative fitness bound
U = inf(1,n); % Upper normative fitness bound
% Random generation of individuals within the search space
X = []; % Matrix containing the positions
      and objective function values of
      individuals in the population
           space
 for i = 1:n
   X(:,i) = unifrnd(LB(i),UB(i),N_I,1);
 end
% Objective functions:
% Last element of each row of matrix X represents the
 objective function value of the corresponding individual
 for i = 1:N_I
    X(i,n+1) = f(X(i,1:n));
 end
%% Main Cycle of CAs %%%%%%%%%%%%%%%%%%%%%%%%%%%%%%%%%%%%%%%%%
 while t < t_max
   t = t + 1; % New iteration is started
     %% Updating the Belief Space
   [S,X_min,X_max,L,U] =
           Update(X,S,X_min,X_max,L,U,N_accept,n);
     %% Influencing the Population Space
   for i = 1:N_I
     X(i,1:n) = Influence(X(i,1:n),X_min,X_max,S,n,...
                      Sigma,Beta,InfluCode);
     % Modification of individuals if they are outside
     the search space
     X(i,1:n) = OutofRange(X(i,1:n),LB,UB);
     % Objective Function Evaluation
     X(i,n+1) = f(X(i,1:n));
   end
   %% Reporting the progress of the algorithm
   fprintf('Evaluations: %d Objective Function:
       %d\n',(N_I+N_I*t),S(end));
 end
%% Objective Function Definition %%%%%%%%%%%%%%%%%%%%%%%%%%%%
 function f = f(X)
 f = sum(X.^2);
 end
%% Update Function %%%%%%%%%%%%%%%%%%%%%%%%%%%%%%%%%%%%%%%%%%%
function [S,X_min,X_max,L,U]= Update(X,S,X_min,X_max,...
                             L,U,N_accept,n)
```

```
    [~,K] = sort(X(:,end)); % K includes the indexes of the individuals
  sorted based on their objective function values
    % Adjusting the Situational Knowledge
    if X(K(1),end) < S(end) % K(1) represents the index of
                            the best individual in the
                            population space at current
                            iteration
      S = X(K(1),:);
    end
    % Adjusting the Normative Knowledge:
    for l = 1:N_accept
      for j = 1:n
        if X(K(l),j)<= X_min(j) || X(K(l),end)< L(j)
          X_min(j) = X(K(l),j);
          L(j) = X(K(l),end);
        elseif X(K(l),j)>= X_max(j) || X(K(l),end)< U(j)
          X_max(j) = X(K(l),j);
          U(j) = X(K(l),end);
        end
      end
    end
  end
  %% Influence Function %%%%%%%%%%%%%%%%%%%%%%%%%%%%%%%%%%%%%%%%%
  function X = Influence(X,X_min,X_max,S,n,Sigma,...
  Beta,InfluCode)
    for j = 1:n
       if InfluCode == 1
          % Influence Function I
          X(j) = X(j) + (X_max(j)-X_min(j))*randn;
       elseif InfluCode == 2
          % Influence Function II
          if X(j) < S(j)
             X(j) = X(j) + abs(Sigma(j)*randn);
          elseif X(j) > S(j)
             X(j) = X(j) - abs(Sigma(j)*randn);
          else
             X(j) = X(j) + Sigma(j)*randn;
          end
       elseif InfluCode == 3
          % Influence Function III
          if X(j) < S(j)
             X(j) = X(j) + abs((X_max(j)-...
                   X_min(j))*randn);
       elseif X(j) > S(j)
             X(j) = X(j) - abs((X_max(j)- ...
                   X_min(j))*randn);
       else
             X(j) = X(j) + (X_max(j)-X_min(j))*randn;
       end
    else
      % Influence Function IV
      if X(j) < X_min(j)
         X(j) = X(j) + abs((X_max(j)-...
                X_min(j))*randn);
```

```
        elseif X(j) > X_max(j)
            X(j) = X(j) - abs((X_max(j)-...
                    X_min(j))*randn);
        else
            X(j) = X(j) + Beta*(X_max(j)-
                    X_min(j))*randn;
        end
    end
 end
end
%% Modification of the variables that are out of the search
space
function X = OutofRange(X,LB,UB)
   OutMin = find(X<LB); % OutMin indicates the variables
                    that are smaller than the lower
                    bounds of the problem
   OutMax = find(X>UB); % OutMax indicates the variables
                    that are greater than the upper
                    bounds of the problem
    % Random values for the variables less than the lower
    bounds of the problem
    X(OutMin) = LB(OutMin) + rand* (UB(OutMin)-LB(OutMin));
    X(OutMax) = LB(OutMax) + rand* (UB(OutMax)-LB(OutMax));
end
```

Glossary

A XXX
Active Power Filters (APFs) 105
Adaptive Quantum PSO (AQPSO) 99
Angle of Attack (AoA) 84
Ant Colony Optimisation (ACO) 16
Artificial Bee Colony (ABC) 94
Artificial Neural Network (ANN) 91
B XXX
balanced hybrid CA with DE (b-hCA-DE) 148
Biogeography-based Optimisation (BBO) 16
C XXX
CA based on PSO (CAPSO) 84
CA-based Cuckoo Search (CACS) 195
Chaos Cultural Algorithm (CCA) 214
Chaos Culture PSO (CCPSO) 105, 212
Charged System Search (CSS) 19
Clan-based CA (CCA) 125
Clan-based Cultural Algorithm (CCA) 222
Collaborative Filtering (CF) 131
Colliding Bodies Optimisation (CBO) 19
Complementary Metal–Oxide–Semiconductor (CMOS) 136
Cuckoo Search (CS) 64
Cultural Algorithm Adopting Knowledge Migration (MCAKM) 218
Cultural Cooperative PSO (CCPSO) 146
Cultural Coyote Optimisation Algorithm (CCOA) 85
Cultural DE with a Measure of Diversity (CDEMD) 151
Cultural Differential Evolution (MCDE) 222
Cultural Firework (CF) 99, 168, 171
Cultural GA (CGA) 117
Cultural Harmony Search (CHS) 164

S. Jalili, *Cultural Algorithms*, Engineering Optimization: Methods and Applications,
https://doi.org/10.1007/978-981-19-4633-2

Cultural Invasive Weed Algorithm (CIWA) 101
Cultural Invasive Weed Algorithm (CIWA) 174
Cultural PSO (CPSO) 117, 142
Cultural Quantum Shuffled Frog Leaping (CQSFL) 178, 182
Cultural-Based Particle Swarm Optimisation (CECBPSO) 221
D XXX
Decision Tree Induction (DTI) 126
Differential CA (DCA) 99, 153
Differential Evolution (DE) 16
E XXX
Elephant Herding Optimisation (EHO) 167
Enhanced Multi-Objective CA (EMOCA) 94, 201
Evolutionary Algorithms (EAs) 16
Evolutionary Programming (EP) 189
F XXX
Finite Impulse Response Filter (FIR) 98
Firefly Algorithm (FA) 94
Firefly Algorithm (FA) 167
Five-Element Cycle Optimisation (FECO) 167
Flexible Job-shop Scheduling Problem (FJSP) 117
G XXX
Gaussian Particle Swarm Optimisaiton (GPSO-CA) 74
Gaussian PSO (GPSO) 145
Gaussian PSO and CA (GPSO-CA) 145
Genetic Algorithms (GAs) 16
Genetic Programming (GP) 131
Gravitational Search Algorithm (GSA) 19
Greedy Randomised Adaptive Search Procedure (GRASP) 117
H XXX
Habitat Suitability Index (HSI) 17
Harmony Memory (HM) 163
Harmony Memory Consideration Rate (HMCR) 163
Harmony Search (HS) 16
Hidden Markov Model (HMM) 20
High-Level Hyper-heuristic (HLH) 20
Hybrid CA and TS (CA-TS) 166
Hybrid Elephant Herding Optimisation and CA (HEHO-CA) 98
Hybrid Flow-shop Scheduling Problem (HFSP) 117
Hybrid Multi-Objective CA (HMOCA) 154
Hybrid renewable energy systems (HRES) 85
I XXX
Imperialist Competitive Algorithm (ICA) 19
Infinite Impulse Response (IIR) 98
Internet of Things (IoT) 130
Invasive Weed Optimisation (IWO) 131

Invasive Wees Optimisation (IWO) 167
J XXX
Job-shop Scheduling Problem (JSP) 117
L XXX
League Championship Algorithm (LCA) 19
Light-Emitting Diode (LED) 105
Loss of Power Supply Probability (LPSP) 86
Low-Level Heuristics (LLHs) 20
M XXX
Mean Square Error (MSE) 123
Modified CA (MCA) 95
Monte Carlo Tree Search (MCTS) 20
Multi-Armed Bandit (MAB) 20
Multi-Objective CA (MCA) 105
Multi-Objective CA and FECO (MOCAFECO) 183, 186
Multi-Objective Quantum CA (MOQCA) 96
N XXX
Niche CA (NCA) 166
Non-dominated Sorting GA (NSGA-II) 94
O XXX
Optics Inspired Optimisation (OIO) 19
P XXX
Parallel Co-operating Cultural Algorithm (PARCA) 219
Parallel GA (PGA) 117
Particle Swarm Optimisation (PSO) 16
Passive Power Filters (PPFs) 105
PhotoVoltaics (PVs) 85
Q XXX
Quadspace Cultural Genetic Tabu Algorithm (QSCGTA) 117
Quality of Service (QoS) 98, 137
Quantum Inspired Particle Swarm Optimisation (IQPSO) 193
Quantum PSO (QPSO) 99
R XXX
Remotely Operated Vehicles (ROVs) 134
S XXX
Shuffled Frog Leaping Algorithm (SFLA) 167
Simulated Annealing (SA) 16
Suitability Index Variables (SIVs) 17
Supply Chain Management (SCM) 110
Swarm Intelligence (SI) 18
T XXX
Tabu Search (TS) 16
Teaching–Learning-Based Optimisation (TLBO) 16
The Open-source Racing Car Simulator (TORCS) 132
Travelling Salesman Problem (TSP) 14

V XXX
Virtual Machine Placement (VMP) 127
Virtual Machines (VMs) 127
W XXX
***Wax Appearance Temperature* (WAT)** 91
Wax Disappearance Temperature (WDT) 91
Wireless Sensor Network (WSN) 96

Bibliography

Abitha R, Vennila SM (2020) Discovering positive association of ASD attributes with class using multi objective cultural algorithm. ICTACT J Soft Comput 11:2222–2226

Alami J, el Imrani A (2008) Using cultural algorithm for the fixed-spectrum frequency assignment problem. J Mob Commun 2:1–9

Bartholomew-Biggs M (2008) Nonlinear optimization with engineering applications. Springer Science & Business Media

Boyd R, Richerson PJ (1988) Culture and the evolutionary process. University of Chicago Press

Chung C-J, Reynolds RG (1998) CAEP: an evolution-based tool for real-valued function optimization using cultural algorithms. Int J Artif Intell Tools 07:239–291. https://doi.org/10.1142/S0218213098000135

Coello CAC, Becerra RL (2002) Adding knowledge and efficient data structures to evolutionary programming: a cultural algorithm for constrained optimization. In: Proceedings of the 4th annual conference on genetic and evolutionary computation, pp 201–209

Coello CAC, Becerra RL (2003) Evolutionary multiobjective optimization using a cultural algorithm. In: Proceedings of the 2003 IEEE swarm intelligence symposium. SIS'03 (Cat. No.03EX706). IEEE, pp 6–13

Geem ZW, Kim JH, Loganathan GV (2001) A new heuristic optimization algorithm: harmony search. SIMULATION 76:60–68. https://doi.org/10.1177/003754970107600201

Geertz C et al (1973) The interpretation of cultures. Basic Books

Jin X, Reynolds RG (1999a) Using knowledge-based system with hierarchical architecture to guide the search of evolutionary computation. In: Proceedings 11th international conference on tools with artificial intelligence. IEEE Computer Society, pp 29–36

Jin X, Reynolds RG (1999b) Using knowledge-based evolutionary computation to solve nonlinear constraint optimization problems: a cultural algorithm approach. In: Proceedings of the 1999b congress on evolutionary computation-CEC99 (Cat. No. 99TH8406). IEEE, pp 1672–1678

Kennedy J (1995) Particle swarm optimization. In: Kennedy J, Eberhart RC (eds) Proceeding of the IEEE international conference on neural networks. IEEE Service Center, Perth, pp 12–13

Kulkarni VR, Desai V (2020) Sensor localization in wireless sensor networks using cultural algorithm. Int J Swarm Intell Res 11:106–122. https://doi.org/10.4018/IJSIR.2020100105

Lin C-J, Weng C-C, Lee C-L, Lee C-Y (2009) Using an efficient hybrid of cooperative particle swarm optimization and cultural algorithm for neural fuzzy network design. In: 2009 International conference on machine learning and cybernetics. IEEE, pp 3076–3082

Lyman RL (2008) Culture, concept and definitions. In: Encyclopedia of archaeology. Elsevier, pp 1070–1075

Maheri A, Jalili S, Hosseinzadeh Y et al (2021) A comprehensive survey on cultural algorithms. Swarm Evol Comput 62:100846. https://doi.org/10.1016/j.swevo.2021.100846

S. Jalili, *Cultural Algorithms*, Engineering Optimization: Methods and Applications, https://doi.org/10.1007/978-981-19-4633-2

Nabavi SR, Najafi M (2021) An optimal routing protocol using multi-objective cultural algorithm for wireless sensor networks (ORPMCA). Majlesi J Telecommun Devices 10:79–86. https://doi.org/10.52547/mjtd.10.2.79

Nguyen TT, Yao X (2008) An experimental study of hybridizing cultural algorithms and local search. Int J Neural Syst 18:1–17. https://doi.org/10.1142/S0129065708001397

Price K, Storn RM, Lampinen JA (2006) Differential evolution: a practical approach to global optimization. Springer Science & Business Media

Reynolds R, Nazzal A (1997) Using cultural algorithm with evolutionary computing to extract site location decisions from spatio-temporal databases, pp 443–456

Yuan X, Yuan Y (2006) Application of cultural algorithm to generation scheduling of hydrothermal systems. Energy Convers Manage 47:2192–2201. https://doi.org/10.1016/j.enconman.2005.12.006

Index

S. Jalili, *Cultural Algorithms*, Engineering Optimization: Methods and Applications,
https://doi.org/10.1007/978-981-19-4633-2

The manufacturer's authorised representative in the EU is Springer Nature Customer Service Centre GmbH, Europaplatz 3, 69115 Heidelberg, Germany. If you have any concerns regarding our products, please contact ProductSafety@springernature.com

Printed and bound by CPI Group (UK) Ltd, Croydon, CR0 4YY
30/09/2026
02225315-0006